# PEPTIDE CHEMISTRY AND DRUG DESIGN

# PEPTIDE CHEMISTRY AND DRUG DESIGN

Edited by

**BEN M. DUNN**

WILEY

Published by John Wiley & Sons, Inc., Hoboken, New Jersey
Published simultaneously in Canada



***Library of Congress Cataloging-in-Publication Data:***

Peptide Chemistry and Drug Design / edited by Ben M. Dunn.
p. ; cm.
Includes bibliographical references and index.
Summary: "This book details many of the problems and successes of peptides as potential drugs"– Provided by publisher.
ISBN 978-0-470-31761-7 (hardback)
I. Dunn, Ben M., editor.
[DNLM: 1. Peptides–chemistry. 2. Drug Design. 3. Pharmaceutical Preparations. QU 68]
RM301.25
615.1′9–dc23

2014040280

Cover image courtesy of David Craik and Quentin Kaas, University of Queensland

Typeset in 10/12pt TimesLTStd by Laserwords Private Limited, Chennai, India

Printed in the United States of America

10 9 8 7 6 5 4 3 2 1

1 2015

# CONTENTS

# LIST OF CONTRIBUTORS

**Fernando Albericio**, Institute for Research in Biomedicine, Barcelona, Spain; CIBER-BBN, Networking Centre on Bioengineering, Biomaterials and Nanomedicine, Barcelona Science Park, Barcelona, Spain; Department of Organic Chemistry, University of Barcelona, Barcelona, Spain; School of Chemistry, University of KwaZulu-Natal, Durban, South Africa; School of Chemistry, Yachay Tech, Yachay City of Knowledge, Urcuqui, Ecuador

**Sabrina Amar**, Departments of Chemistry and Biology, Port St. Lucie, FL USA

**David J. Craik**, Institute for Molecular Bioscience, The University of Queensland, Brisbane, Queensland, Australia

**Ayman El-Faham**, Department of Chemistry, Alexandria University, Alexandria, Egypt; Department of Chemistry, King Saud University, Riyadh, Kingdom of Saudi Arabia

**Gregg B. Fields**, Departments of Chemistry and Biology, Port St. Lucie, FL USA; Departments of Biochemistry, University of Texas Health Science Center, San Antonio TX, USA

**Nader Fotouhi**, Global Alliance for TB Drug Development, Research and Development, New York, NY, USA

**Athanassios S. Galanis**, Department of Pharmacy, University of Patras, Patras, Greece

**Carrie Haskell-Luevano**, Department of Medicinal Chemistry, University of Minnesota, Minneapolis, MN, USA

**Sónia T. Henriques**, Institute for Molecular Bioscience, The University of Queensland, Brisbane, Queensland, Australia

**Victor J. Hruby**, Department of Chemistry and Biochemistry, The University of Arizona, Tucson, AZ, USA

**Yoshiaki Kiso**, Department of Molecular Pharmacy, Kobe Gakuin University, Kobe, Japan

**Anna Knapinska**, Departments of Chemistry and Biology, Port St. Lucie, FL USA

**Isuru R. Kumarasinghe**, Department of Chemistry and Biochemistry, The University of Arizona, Tucson, AZ, USA

**Jeffrey-Tri Nguyen**, Department of Medicinal Chemistry, Kyoto Pharmaceutical University, Kyoto, Japan

**Eliandre de Oliveira**, Proteomics Platform Barcelona Science Park, Barcelona, Spain

**Trista K. Robichaud**, Departments of Periodontics, University of Texas Health Science Center, San Antonio TX, USA; Departments of Biochemistry, University of Texas Health Science Center, San Antonio TX, USA

**Anamika Singh**, Department of Medicinal Chemistry, University of Minnesota, Minneapolis, MN, USA

**Judit Tulla-Puche**, Institute for Research in Biomedicine, Barcelona, Spain; CIBER-BBN, Networking Centre on Bioengineering, Biomaterials and Nanomedicine, Barcelona Science Park, Barcelona, Spain

**Aikaterini A. Zompra**, Department of Pharmacy, University of Patras, Patras, Greece

# PREFACE

This book is result of many conversations with peptide scientists at a variety of meetings, including American Peptide Society Symposia, meetings of the European Peptide Society, the Japanese Peptide Society, and the Australian Peptide Society. Some of these conversations were with the authors of the chapters in this book. One additional influence was a meeting in Dubai, where I had an excellent dinner with Waleed Danho, then with Roche Nutley. Waleed had given an excellent talk about the value of peptide chemistry and peptides as elements in the drug-discovery process. Over a delicious dinner of baked fish and many other courses, we discussed the history of drug discovery and the role that peptides have played in the past. Waleed made the strong point that peptides still have great value in the discovery process and, with appropriate methods to deal with delivery and metabolism issues, can provide excellent drugs for the future.

At around this time, I was contacted by Jonathan Rose of John Wiley & Sons who asked if I would be interested in editing a book on peptides and drug discovery. Sometimes life provides a nice juxtaposition of ideas and I immediately accepted the invitation. Over the following years, I spoke with many scientists, emailed some more, and worked on putting together the chapters for this book. I want to thank Jonathan as well as Kari Capone of John Wiley for their patience and advice over the years it took to bring this together.

The book starts with a chapter provided by Nader Fatouhi, discussing the current state of peptides in drug discovery. I heard Nader speak at the 23rd American Peptide Symposium in the Kona region of the Big Island of Hawaii. As I felt that his presentation provided an update on the thoughts first revealed to me by Waleed Danho, I asked Nader to contribute the opening chapter of the book, as this sets the stage for what follows. In his chapter, Nader discusses the rising importance of peptides as molecules for drug development as well as the issues facing scientists in this field, including cell penetration, stability, and targeting. Tools and techniques are available to address each of these limitations at this time.

Chapter 2 was contributed by Fernando Albericio and colleagues. This presents modern methods of peptide synthesis in a very readable format. Included are sections on solid supports for solid-phase peptide synthesis, which dominates most research level approaches, linkers, protecting groups, methods for peptide-bond formation, and a variety of methods to modify peptides to limit metabolism. In all cases the latest reagents and techniques are featured, thus making this chapter a great starting point for scientists starting out in the peptide field. The authors go on to discuss synthesis of peptides in solution, which still has great value in certain applications, including production of peptides in bulk. In addition, the combination of both solution- and solid-phase methods is discussed for cases where fragment condensation is used to prepare ever larger peptides. This discussion includes native chemical ligation, which permits selectively linking N-termini and C-termini of fragments, and which has several variations with more coming each year. The chapter concludes with a very valuable discussion of separation methods and methods for the analysis of the products of peptide synthesis. Again, this chapter is recommended as a great starting place for new scientists.

Anamika Singh and Carrie Haskell-Luevano have provided Chapter 3 that discusses the important topic of membrane receptors as targets for drug discovery. Due to the vital role of membrane receptors in cell signaling and control of metabolic events, a significant percentage of drugs in current use exert their function by interfering or stimulating binding and signaling events at membrane receptors, also known as G-protein coupled receptors (GPRCs). This chapter provides a catalog of systems where peptides are known to be involved and where it has been shown that synthetic peptides can modulate function. The Haskell-Luevano lab has provided outstanding research on the melanocortin receptors, but this chapter takes a broader approach and discusses a wide variety of these systems, including structural information as known and as modeled by other labs. Anyone involved in aspects of membrane signaling will find this chapter a highly valuable resource for methods, approaches, and strategies for attacking this important area of biology.

Gregg Fields and colleagues present Chapter 4 to introduce the use of peptides as inhibitors of enzymes. In the first part, the authors introduce enzymes and their classification and present several classical examples of the use of peptides to come up with compounds that provide the desired change in enzyme function to overcome a metabolic defect. In a second section, the area of HIV-1 infection and progression to AIDS is described, with emphasis on the value of peptides as modulators of growth and infection. As the human immunodeficiency virus goes through a complicated life cycle, the authors point out that there are multiple targets for approaching therapy and a combination strategy, known as HAART (highly active antiretroviral therapy) has provided the optimal approach to treatment of affected individuals. The Fields lab has made major contributions to discoveries in the area of matrix metalloproteinases and this chapter presents a thorough discussion of this system. The enzymes in this family provide a great example of the development of inhibitors through a process of discovery of aspects of structure and function that can guide the process. The chapter continues with nice discussions of several other systems where peptide chemistry has been key in new discoveries that have driven the drug-development process.

Jeffrey-Tri Nguyen and Yoshiaki Kiso have provided Chapter 5, which continues the discussion of enzyme inhibitors from the aspect of peptides. The highly productive Kiso lab has led the way in creating a very large catalog of peptide derivatives for use in drug discovery in several systems. They begin this chapter by discussing the advantages and disadvantages of peptides as potential drugs and come down on the side of the beneficial role that peptides play. In particular, they make the important point that the use of peptides can frequently define the pharmacophore, or structural model, which can then be transformed into a small molecule of non-peptide nature for further development as a potential drug. This chapter further focuses on the process of the design of potential inhibitors and reviews the history of discovery from natural sources as well as through *ab initio* design. They discuss the advantages of learning from the natural substrates of an enzyme and introduce the important concept of the transition state analog; the critical role that structural information on the target protein can provide. This chapter provides an excellent discussion of systems where targeting with peptide molecules may provide opportunities for further drug discovery.

Sónia T. Henriques and David J. Craik describe many peptide inhibitors from natural sources in Chapter 6. The introduction to their chapter discusses the value of finding compounds from nature and describes a number of sources, including the antimicrobial peptides from many bacteria. In both bacterial and plant worlds, there is a continual war between competing systems, and this has led to the development through evolution of many natural peptides that serve as defensive molecules. The authors discuss the cyclotides, peptides that are connected end to end and that have multiple disulfide bonds. This arrangement is very stable and the molecules are found in venoms of several species as well as in plants. After this introduction, the authors turn to a discussion of the drug discovery process from their perspective. The chapter continues with an in depth discussion of a variety of systems where many methods are used to modify molecules isolated from nature and where the activity against many targets is tested. The wide diversity of structures and targets is featured in this chapter and the many discoveries have pushed research and drug discovery forward significantly.

Isuru R. Kumarasinghe and Victor J. Hruby have taken on the task of describing methods to limit the metabolism of peptide molecules in humans. This leads to a very detailed discussion of the chemistry of peptide modification. As Victor Hruby is the world leader in this aspect of peptides, the chapter is thoroughly exciting and interesting. A main concern is the digestion of peptides by proteolytic enzymes present in both the digestive tract and the circulation. The first step is to define the pharmacophore residues of a naturally occurring and effective peptide. This will show the absolutely critical functional groups and their stereochemical relationships that must be maintained. Then replacement of some nonessential amino acids by non-natural amino acids, with the D-amino acid isomer, or with peptide-bond isosteres may be sufficient to block degradation by proteases. In addition, cyclization can sometimes provide more stability and also enhance passage of peptides through the blood–brain-barrier. Other strategies include replacement of specific the amino acids with the *N*-methyl derivatives, with topographically constrained derivatives, or with the halogenated derivatives of aromatic amino acids. Finally, the use of the "multiple-antigenic-peptide" approach where many molecules are attached to a carrier with multiple attachment points can produce molecules that, due to their size, are not recognized by proteases. This chapter emphasizes the role of creative synthetic chemistry is the modification of peptides to achieve stability and bioavailability.

The book concludes with Chapter 8, provided by Jeffrey-Tri Nguyen Yoshiaki Kiso, that discusses the important area of peptide delivery. While progress in the past 50 years has permitted peptide chemists to make almost any sequence of amino acids that is desired in high yield and purity, getting those molecules into humans and into the specific area in the body where they can exert a therapeutic effect is a problem that has not progressed as rapidly. Thus, this chapter is very important for future advances in drug discovery based on peptides. Many of the readers may already be familiar with the Lipinski's Rule of Five that includes recommendations for the size of a molecule, the number of hydrogen bonding atoms, and the lipophilicity. These rules are discussed in this chapter, but much more information is provided regarding solubility, membrane transport, and metabolic stability.

In conclusion, this book provides a primer for anyone in the field of drug discovery and specifically in the area of the use of peptides as molecules for both the discovery phase and, in favorable cases, the final phase of the creation of new molecular entities that can be moved into further studies to evaluate their potential as therapeutic drugs. I want to thank the authors of the chapters for their friendship, for many discussions, and for their excellent writing for this book.

Ben M. Dunn, Ph.D.
September 3, 2014

# 1

# PEPTIDE THERAPEUTICS

Nader Fotouhi

*Global Alliance for TB Drug Development, Research and Development, New York, NY, USA*

## 1.1 HISTORY OF PEPTIDES AS DRUGS

The advent of molecular biology and our understanding of the physiological and pathological functions of peptides, coupled with advances in synthetic methodologies and peptidomimetics, marked the beginning of a new era in peptide and protein therapeutics, with the vision that there should be no limit to what can be produced as therapeutics. During that period a number of great peptide drugs such as Sandostatin, Lupron, Copaxone, and Zoladex were developed with great therapeutic benefit. The number of approved peptide drugs, however, remains low.

It was not until the last decade that we have seen a significant surge in the number of peptide therapeutics on the market (Figure 1.1). While 10 peptides were approved between 2001 and 2010, the current decade has thus far witnessed the approval of six new peptide therapeutics – a remarkable yearly increase [1, 2]. The number of peptides in development is also steadily growing roughly doubling every decade (Figures 1.2 and 1.3), and there are 400–600 peptides in preclinical studies. This is due to the advances made in our understanding of peptide stability, peptide synthesis, and formulation over the last three decades. Although the market share of peptide drugs is still relatively small (about 2% of the global market for all drugs), the approval rate for peptide drugs is twice as fast as the rate for small molecules, and the market is growing similarly at a rate that is twice the global drug market [3, 4].

*Peptide Chemistry and Drug Design*, First Edition. Edited by Ben M. Dunn.

| Trade name | Generic name | Target | Indication | Year |
|---|---|---|---|---|
| Forteo | Teriparatide | PTH1R agonist | Osteoarthritis | 2002 |
| Fuzeon | Enfuvirtide | Protein–protein inh. | HIV | 2003 |
| Prialt | Aiconotide | $Ca^{2+}$ channel inh. | Pain | 2004 |
| Byetta | Exenatide | GLP-1 R agonist | T2 diabetes | 2005 |
| Symlin | Pramlintide | Calcitonin agonist | T1/T2 diabetes | 2005 |
| Somatuline | Lanreotide | SST agonist | Acromegaly | 2007 |
| Nplate | Romiplostim | Thrombopoietin agonist | Haematology | 2008 |
| Egrifta | Tesamorelin | GHRF agonist | Lipodystrophy | 2010 |
| Victoza | Liraglutide | GLP-1 R agonist | T2D | 2010 |
| Bydureon | Exenatide LAR | GLP-1 R agonist | T2 diabetes | 2011 |
| Surfaxin | Lucinactant | | IRDS | 2012 |
| Omontys | Peginesatide | Erithropoeitin analog. | Anemia | 2012 |
| SIgnifor | Pasireotide | Somatostatin analog | Cushing's disease | 2012 |
| Kyprolis | Carfilzomib | Proteasome inhibitor | Multiple myeloma | 2012 |
| Linzess | Linaclotide | Guanidyl cyclase 2C agonist | IBS-C and CIC | 2012 |
| Gattex | Teduglutide | Gluc-like peptide analog | SBS | 2012 |

**Figure 1.1** Peptide therapeutics marketed since 2002. (*See insert for color representation of this figure.*)

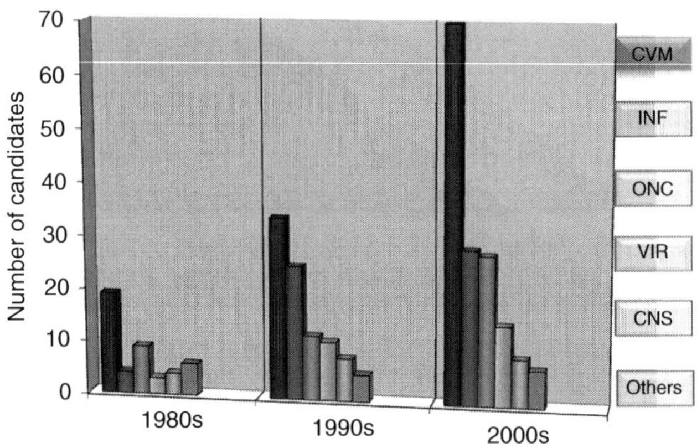


**Figure 1.2** Peptides in development over the last three decades. (*See insert for color representation of this figure.*)

While encouraging, the potential for peptide therapeutics is far greater than what it is today.

## 1.2 FACTORS LIMITING THE USE OF PEPTIDES IN THE CLINIC

A number of factors have thus far limited the explosion that needs to happen in the peptide field. With the exception of a few peptides, the approved drugs so far target the extracellular compartment, and thus have to compete with biologics. Of the

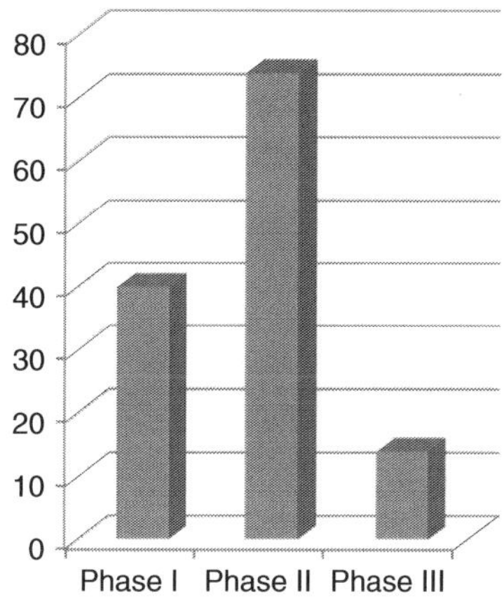


**Figure 1.3** Peptides in clinical trials in 2013.

extracellular targets, GPCRs represent the major class, and in most cases, the peptides are agonist. GLP-1 represents one-third of these GPCR targets. We have seen a great advance in extending the circulating half-life of the peptides through the use of unnatural amino acids and formulation technologies, but have not yet reached the half-life achieved by antibodies. The delivery of peptides is still in the great majority of cases limited to *i.v.* (intravenous), *s.c.* (subcutaneous), or intranasal. Finally, safety is still a concern as better tissue selectivity is required.

To dramatically heighten their impact, peptides need to access the intracellular space to target protein–protein interactions. These interactions represent a vast source of potential targets with significant biological impact (there are estimated 300,000 such interactions in the cell), and will not in the majority of cases be modulated by small molecules. Peptides and biologics, given their relative size and ability to bind to extended surface areas, are the perfect candidates to inhibit protein–protein interactions. The duration of action of peptides needs to be extended, and while peptides are inherently selective against their targets, they need to more selectively distribute to the desired tissue. Finally, the route of administration needs to be expanded to include oral delivery.

## 1.3 ADVANCES THAT HAVE STIMULATED THE USE OF PEPTIDES AS DRUGS

The many great technological advances that started over a decade ago in drug delivery, peptide design, and synthesis are now maturing, and will undoubtedly address these key challenges and revolutionize the field over the next decades. Many of the technological advances are already proving that it is possible to make peptides permeable to cells, target tissues, have longer half-lives, and be orally bioavailable.

The discovery that certain peptides can penetrate cells and can, therefore, be an effective therapeutic on their own or alternatively bring other drugs into cells allowed for the first time to imagine targeting the intracellular compartment (Figures 1.4

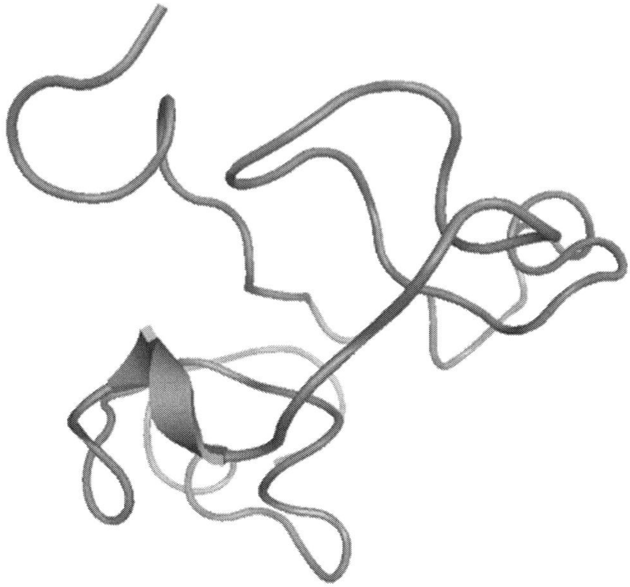

**Figure 1.4** HIV Tat.

and 1.5) [5]. HIV-enveloped protein tat was one of the first to be recognized for its cell-penetrating ability and, therefore, its potential use to carry bioactive cargo into the cell [6]. Since 2004, more than 200 peptides carried into cells by tat or other naturally occurring cell-penetrating peptides (CPPs) have been in various phases of development [7]. However, the more recent advances in the understanding of how these peptides cross the cell membrane through endocytosis and/or macropinocytosis [8] has allowed the generation of CPPs with intrinsic biological activity [9–12]. It is now possible to take a CPP sequence and synthetically modify it to introduce the key amino acids of an effector peptide into its sequence and create potent peptide antagonists of an intracellular protein–protein interaction with good pharmacokinetic properties [13].

## 1.4 DEVELOPMENT OF PEPTIDE LIBRARIES

By looking at the list of CPPs in development, one realizes that they are single cases and have to be synthetically prepared and modified to impart some of the desired stability to be a useful therapeutic. It is hard to compete with the screening of the millions of small molecule compounds in various pharmaceutical companies and more recently in many academic centers.

Until now, the available technologies to screen large libraries of peptides of significant length (possessing secondary structure) would only allow us to generate large libraries of natural amino acid sequences through phage display, and if unnatural

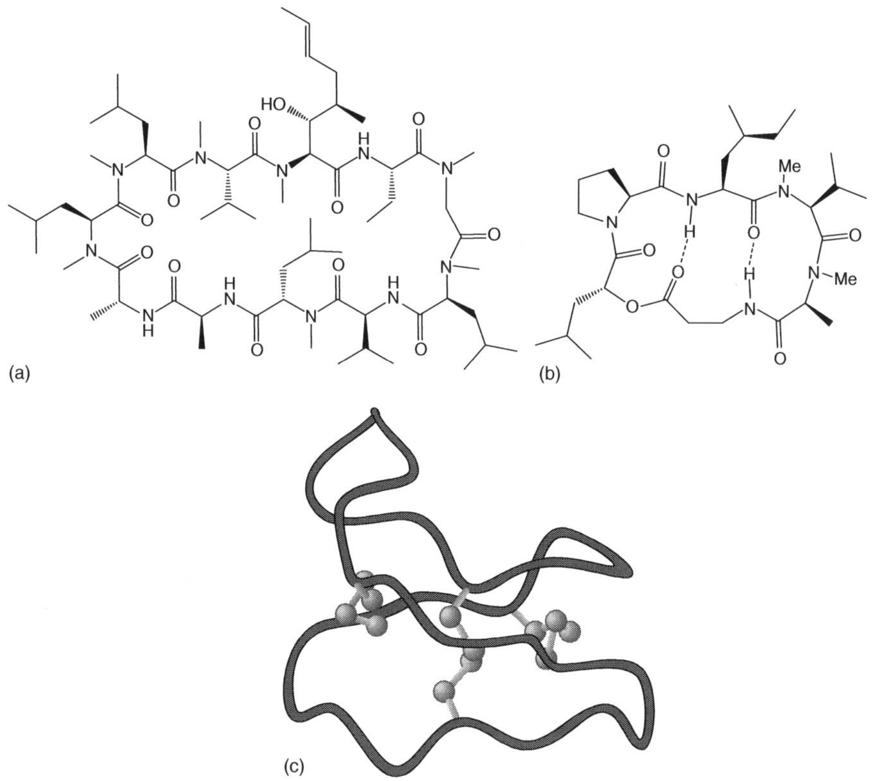


**Figure 1.5** Orally stable and bioavailable peptides (a) Cyclosporin. (b) Destruxin. (c) Kalata B.

amino acids were to be introduced, it had to be done with conventional synthetic methodology, and thus be limited to very low numbers of peptides that can be prepared and screened.

Indeed, over the last decade, there has been an explosion of very elegant technologies that now allow the generation of large to extremely large libraries of linear and macrocyclic peptides with unnatural amino acids and unnatural linkers. For the first time, it is possible to engineer stability, cell permeability, and possibly oral bioavailability at once and screen for the desired properties very rapidly. These major advancements have resulted in the generation of a number of companies that are pushing the limits of these technologies to rapidly screen and identify novel peptide therapeutics against protein–protein interaction targets (Figure 1.5).

Ensemble therapeutics utilizing their DNA-programmed chemistry can generate million-member libraries of small macrocycles with MW of 500–1500. On screening these libraries, they have identified potent and orally bioavailable small molecule inhibitors of IL17 [14]. Through medicinal chemistry optimization, they have now identified picomolar inhibitors with good properties [15]. PeptiDream utilizing Professor Suga's mRNA display technology [16] are generating up to trillion-member libraries of larger macrocycles mimicking cyclosporin. These peptides contain a combination of natural, unnatural, and *N*-methyl amino acids and exhibit good physicochemical properties and membrane permeability [17]. Ra Pharmaceuticals also uses a mRNA display technology developed by Jack Shoztac to generate very large libraries of macrocycles containing unnatural amino acids. They recently presented on their discovery of potent antagonists of mcl-1 and Ras with good cell permeability [18].

## 1.5 MODIFICATION OF PEPTIDES TO PROMOTE STABILITY AND CELL ENTRY

The recent focus on another class of macrocycles, containing multiple disulfides, has generated a lot of excitement in maintaining the stability and membrane permeability of the cyclotide kalata B1, or the knottins (the uncyclized version of cyclotides), in order to create potent peptide drugs. David Craik and colleagues at Cyclotide are systematically exchanging the various loops present on cyclotides with sequences that have important biological function [19]. Recently, the introduction of a myelin oligodendrocyte glycoprotein sequence into a cyclotide resulted in a potent peptide in preventing disease progression in a mouse model of MS [20]. Protagonist is taking advantage of the oral stability of the disulfide-rich peptides for local gut delivery of IL6R antagonists for the treatment of irritable bowel disease (IBD). Moreover, novel technologies developed for the rapid generation and screening of extremely large libraries of knottins and cyclotides will undoubtedly have a major impact on this class of peptide therapeutics. Of note is the Intein-based technology from Julio Camarero capable of introducing unnatural amino acids to facilitate screening [21]. Sutro and MitiBio also have very sophisticated and efficient biosynthetic methods to generate very large libraries.

Finally, Verdine and Wollensky and colleagues [22, 23] as well as the investigators at Aileron Therapeutics have developed a novel stapling technology that imparts stability and membrane permeability to alpha helical structure. Using this technology, Aileron Therapeutics were able to discover very potent dual MDM2/MDMx antagonists with low nanomolar activity in cells and excellent pharmacokinetic properties, resulting in excellent antitumor activity in a mouse xenograft model [24]. Even more interesting is the extended efficacy ATSP-7041 exhibits in cells. While the small molecule MDM2 antagonist showed activity over 24 h, ATSP-7041 was still active beyond 48 hours in the same experiment. This is due to the fact that once the peptide enters the cell, the major elimination pathway is through enzymatic catabolism. Not only can stability be tuned for circulating half-life, it can also be tuned to withstand cellular catabolism to lengthen the desired efficacy. This could offer a significant advantage over (small) molecules that passively diffuse through the cell membrane. Additionally, using the same technology, a GHRH antagonist with much extended half-life was discovered and is currently in Phase I clinical trial [25].

## 1.6 TARGETING PEPTIDES TO SPECIFIC CELLS

One of the greatest challenges in drug discovery is the safety of therapeutics. Main reasons for diminished safety are selectivity against the target and tissue/cell specificity. If one could direct a therapeutic to only the site of pathology, then the therapeutic window of the agent increases and correspondingly decreases the side effects. Peptides, due to their specificity against receptors, are perfect candidates to be able to home into one type of cell/tissue versus another. There has been a tremendous amount of progress in identifying homing peptides (cell-penetrating as well as nonpenetrating) that can then be conjugated to a cargo to deliver it to a specific organ [26].

*In vivo* phage display by Pasqualini and colleagues marked the discovery of the first homing peptide that was able to selectively target the blood vessel of brain and kidney [27]. Since then a number of peptides have been identified that target many other tissues [28]. Arap and colleagues were then the first to perform phage display in humans and discovered a homing peptide to IL11Ra that expresses over 100-fold more on prostate cancer cells versus normal cells [29, 30]. Arrowhead Research is currently in Phase I proof of targeting with a peptide drug conjugate utilizing this homing peptide. Recently, Wen et al., at the Dana Farber, published their first Phase I study result on GRN1005, a peptide drug conjugate that targets the low-density lipoprotein-related protein-1, which mediates blood brain barrier transcytosis. GRN1005 successfully crosses the BBB and delivers its cargo [31].

## 1.7 FORMULATIONS TO IMPROVE PROPERTIES

While the above advances have and will have significant impact, the ability to administer peptides by the oral route will truly allow them to compete with small molecules and biologics as first line therapies. The majority of advances in this area have been the result of very interesting formulation strategies. A number of companies, including ArisGen, Axcess, Chiasma, Emisphere Tech., Enteris Pharmaceuticals, Lipocine, and Merlion Pharmaceuticals, have had successes in enhancing the oral bioavailability of some peptide therapeutics. They employ a combination of stabilizers, absorption enhancers, and carriers to achieve this. The main mode of absorption is through the paracellular space. However, the bioavailability of the peptides formulated remains relatively low.

While significant, cyclosporin remains the only marketed peptide drug that is administered orally and absorbed into the systemic environment. Learning from nature and systematic studies on macrocyclic peptides will have a tremendous impact in discovering peptide drugs with inherent oral bioavailability that could then be enhanced through formulation to achieve bioavailabilities, which would compete with small molecules. As mentioned earlier, PeptiDream and Ra Pharmaceuticals are generating large libraries of macrocyclic peptides mimicking the core structure of cyclosporin. Ensemble therapeutics are generating small macrocylic structures with molecular weights between 500 and 1500 and have already identified an orally bioavailable IL17 R antagonist. Professors Horst Kessler and Locky are doing the first systematic studies on small cyclic peptides to understand the effect of hydrogen bonding and structure on bioavailability [32, 33]. Their work will undoubtedly form the basis of rational designs of orally active peptide drugs.

In conclusion, the great technological advances over the last two decades are well poised to have a major impact on revolutionizing the field of peptide therapeutics. For the first time, tools are available to create stable, cell permeable, long lasting, and orally bioavailable peptides, allowing them to compete with small molecule drugs and biologics, and thus become first line therapies for many diseases with unmet medical needs.

## REFERENCES


1. Reichert J. (2010). Development Trends for Peptide Therapeutics (2010 Report Summary) 2010 *Peptide Therapeutics Foundation*: 1–10.
2. Caspar A, Reichert J. Future directions for peptide therapeutics development. Drug Discov Today 2013;18:807–817.
3. Lax R. The future of peptide development in the pharmaceutical industry. PharManufacturing Int Pept Rev 2010:10–15.
4. Bellmann-Sickert K, Beck-Sickinger AG. Peptide drugs to target G protein-coupled receptors. Trends Pharmacol Sci 2010;31:434–441.
5. Heitz F, Morris MC, Davita G. Twenty years of cell penetrating peptides: from molecular mechanism to therapeutics. Br J Pharmacol 2009;157:195–206.
6. Franklin AD, Pabo CO. Cellular uptake of the tat protein from the human immunodeficiency virus. Cell 1988;55:1189–1193.
7. Langel U et al. Therapeutic potential of cell-penetrating peptides. Ther Deliv 2013;4: 573–591.
8. Duchardt F, Fotin-Mleczek M, Schwarz H, Fischer R, Brock R. A comprehensive model for the cellular uptake of cationic cell-penetrating peptides. Traffic 2007;8:848–866.
9. Verdurmen WPR, Brock R. Biological responses towards cationic peptides as drug carriers. Cell 2011;32:116–127.
10. Low W. Functional cell permeable motifs within medically relevant proteins. J Biotechnol 2007;129:555–564.
11. Jones S, Holm T, Mager I, Langel U, Howl J. Characterization of bioactive cell penetrating peptides from human cytochrome c: protein mimicry and the development of a novel apoptogenic agent. Chem Biol 2010;17:735–744.
12. Milletti F. The Design of Cell-penetrating Peptides with Intrinsic Biological Activity, TIDES: Oligonuleotides and Peptide Therapeutics, from Research through commercialization. Boston, MA, USA 2013.
13. Fotouhi, N. The Evolution of Peptide Therapeutics. The american peptide symposium. 2013.
14. Terret, N. (2013). First in class, orally active small molecule antagonist of IL17. CHI Drug Discovery Chemistry Conference: Constrained Peptides and Macrocylics Drug Discovery Session.

15. Terret N. Producing orally bioavailable biologics. Drug Discov Dev 2013.
16. Passioura T, Suga H. Flexizymes-mediated genetic reprogramming as a tool for noncanonical peptide synthesis and drug discovery. Chem-Eur J 2013;19:6530–6536.
17. Funamoto S, Sasaki T, Ishihara S, Nobuhara M, Nakano M, Watanabe-Takahashi M, Saito T, Kakuda N, Miyasaka T, Nishikawa K, Saido TC, Ihara Y. Substrate ectodomain is critical for substrate preference and inhibition of g-secretase. Nat Commun 2013;4:3529–3541.
18. Treco D et al. Discovery of high affinity cyclic peptidomimetics binding Mcl-1 and Ras. Mol Cancer Ther 2013;12:C212.
19. Craik D, Swedberg JE, Mylne JS, Cemazar M. Cyclotides as a basis for drug design. Expert Opin Drug Discov 2012;7:179–194.
20. Wang CK, Gruber CW, Cemazar M, Siatskas C, Tagore P, Payne N, Sun GZ, Wang SH, Bernard CC, Craik DJ. Molecular grafting onto a stable framework yields novel cyclic peptides for the treatment of multiple sclerosis. ACS Chem Biol 2014;9:156–163.
21. Jagadish K, Borra R, Lace V, Majumder S, Shekhtman A, Wang L, Camarero JAV. Expression of fluorescent cyclotides using protein trans-splicing for easy monitoring of cyclotides-protein interactions. Angew Chem 2013;125:3208–3213.
22. Walensky LD, Kung AL, Escher I, Malia TJ, Barbuto S, Wright RD, Wagner G, Verdine GL, Korsmeyer SJ. Activation of apoptosis in vivo by a hydrocarbon-stapled BH3 helix. Science 2004;305:1466–1470.
23. Bernal F, Tyler AF, Korsmeyer SJ, Walensky LD, Verdine GL. Reactivation of the p53 Tumor Suppressor Pathway by a Stapled p53 Peptide. J Am Chem Soc 2007;129:2456–2457.
24. Chang YS, Graves B, Guerlavais V, Tovar C, Packman K, To KH, Olson KA, Kesavan K, Gangurde P, Mukherjee A, Baker T, Darlak K, Elkin C, Filipovic Z, Qureshi FZ, Cai HL, Berry P, Feyfant E, Shi XGE, Horstick J, Annis DA, Manning AM, Fotouhi N, Nash H, Vassilev LT, Sawyer TK. Stapled $\alpha$-helical peptide drug development: A potent dual inhibitor of MDM2 and MDMX for p53-dependent cancer therapy. Proc Natl Acad Sci U S A 2013;110:1–10.
25. Aileron Therapeutics concludes first-in-human study of ALRN-5281 Stapled Peptide drug. Aileron Therapeutics Company announcement May 2013.
26. Laakonen P. Homing peptides as targeted delivery vehicles. Integr Biol 2010;2:326–337.
27. Pasqualini R, Ruoslahti E. Organ targeting in vivo using phage display peptide libraries. Nature 1996;380:364–366.
28. Arap W, Haedicke W, Bernasconi M, Kain R, Rajotte D, Krajewski S, Ellerby HM, Bredesen DE, Pasqualini R, Ruoslahti E. Targeting the prostate for destruction through a vascular address. Proc Natl Acad Sci U S A 2012;99:1527–1531.
29. Arap W, Kolonin MG, Trepel M, Lahdenranta J, Cardo-Vila M, Giordano RJ, Mintz PJ, Ardelt PU, Yao VJ, Vidal CI, Chen L, Flamm A, Valtanen H, Weavind LM, Hicks ME, Pollock RE, Botz GH, Bucana CD, Koivunen E, Cahill D, Troncoso P, Baggerly KA, Pentz RD, Do KA, Logothetis CJ, Pasqualini R. Steps toward mapping the human vasculature by phage display. Nat Med 2002;8:121–127.
30. Zurita AJ, Troncoso P, Cardo-Vila M, Logothetis CJ, Pasqualini R, Arap W. Combinatorial screenings in patients: the interleukin-11 receptor $\alpha$ as a candidate target in the progression of human prostate cancer. Cancer Res 2004;64:435–439.

31. Drappatz J, Brenner A, Wong ET, Eichler A, Schiff D, Groves MD, Mikkelsen T, Rosenfeld S, Sarantopoulos J, Meyers CA, Fielding RM, Elian K, Wang X, Lawrence B, Shing M, Stephen Kelsey S, Castaigne JP, Yen PY. Phase I study of GRN1005 in recurrent malignant glioma. Clin Cancer Res 2013;19:1567–1576.

32. Beck JG, Chatterjee J, Laufer B, Kiran MU, Frank AO, Neubauer S, Ovadia O, Greenberg S, Gilon C, Hoffman A, Kessler H. Intestinal permeability of cyclic peptides: common key backbone motifs identified. J Am Chem Soc 2012;134:12125–12133.

33. White TR, Renzelman CM, Rand AC, Rezai T, McEwen CM, Gelev VM, Turner RA, Linington RG, Leung SSF, Kalgutkar AS, Bauman JN, Zhang YZ, Liras S, Price DA, Mathiowetz AM, Jacobson MP, Lokey RS. On-resin N-methylation of cyclic peptides for discovery of orally bioavailable scaffolds. Nat Chem Biol 2011;7:810–817.

# 2

# METHODS FOR THE PEPTIDE SYNTHESIS AND ANALYSIS

JUDIT TULLA-PUCHE

*Institute for Research in Biomedicine, Barcelona, Spain; CIBER-BBN, Networking Centre on Bioengineering, Biomaterials and Nanomedicine, Barcelona Science Park, Barcelona, Spain*

AYMAN EL-FAHAM

*Department of Chemistry, Alexandria University, Alexandria, Egypt; Department of Chemistry, King Saud University, Riyadh, Kingdom of Saudi Arabia*

ATHANASSIOS S. GALANIS

*Department of Pharmacy, University of Patras, Patras, Greece*

ELIANDRE DE OLIVEIRA

*Proteomics Platform Barcelona Science Park, Barcelona, Spain*

AIKATERINI A. ZOMPRA

*Department of Pharmacy, University of Patras, Patras, Greece*

FERNANDO ALBERICIO

*Institute for Research in Biomedicine, Barcelona, Spain; CIBER-BBN, Networking Centre on Bioengineering, Biomaterials and Nanomedicine, Barcelona Science Park, Barcelona, Spain; Department of Organic Chemistry, University of Barcelona, Barcelona, Spain; School of Chemistry, University of KwaZulu-Natal, Durban, South Africa; School of Chemistry, Yachay Tech, Yachay City of Knowledge, Urcuqui, Ecuador*

## 2.1 INTRODUCTION

Peptides as drugs show unique characteristics (high biological activity, high specificity, and low toxicity) thereby making them particularly attractive therapeutic agents [1]. However, the role of peptides in drug discovery has suffered *ups* and

*Peptide Chemistry and Drug Design*, First Edition. Edited by Ben M. Dunn.

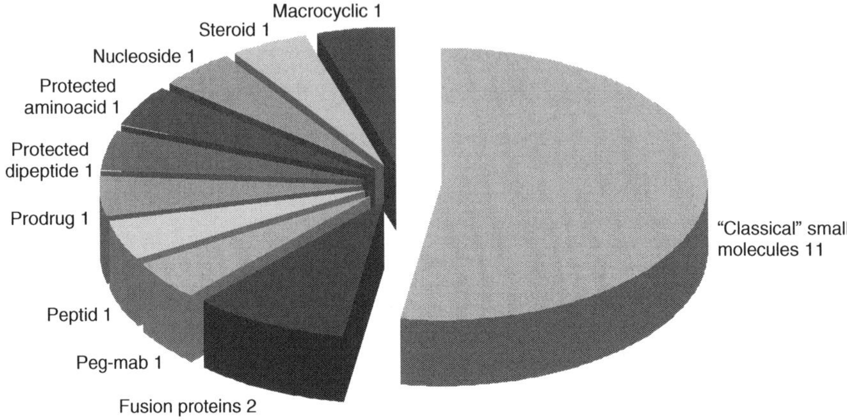


**Figure 2.1** Distribution by chemical structure of the new drugs approved by the FDA in 2008. (*See insert for color representation of this figure.*)

*downs* during the last four decades. A first analysis of the new chemical entities (NCEs) accepted by the Food and Drug Administration (FDA) indicated that while 53 NCEs were introduced as drugs in 1996, only 17 were introduced in 2002. This number increased to 31 in 2004, but decreased again in 2005 with just 18 new drugs, 17 in 2007, and a slight increase to 21 in 2008 (Figure 2.1) [2, 3]. An analysis of these 21 drugs approved in 2008 indicated that almost 50% of the new drugs can be considered *nonclassical*, in the sense that they are *nonclassical small molecules.*

Interestingly, peptides represent approximately 20% of the total number of drugs approved by the FDA in 2008 [3]. Thus, Romiplostim from Amgen, which is a thrombopoietin receptor agonist, is a fusion protein conjugated with a 41 amino acid peptide, containing two disulfide bridges. Degarelix from Ferring, which is a gonadotropin-releasing hormone receptor antagonist, is a 10 amino acid peptide. Alvimopan from Adolor, which is a peripherally acting μ-opioid receptor antagonist, is an N-terminal blocked dipeptide. Lacosamide from Schwarz, which selectively enhances slow inactivation of voltage-gated sodium channels and binds to collapsin response mediator protein 2, is a protected *O*-methylserine [3].

Even more important than the number of peptides accepted by the FDA is the number of peptides that are in clinical phases. In 2008, 39 were in clinical phase I, 77 in phase II, 39 in phase III, and 4 in preregistration [4].

There are several reasons for this renaissance of peptides. The first one is the fact that the number of *classical small molecules* is not increasing enormously. Furthermore, several comparisons with *small molecules* are favorable to peptides. Thus, the well-defined peptide chemistry allows an easier way to prepare analogs. Pharmaceutical companies have also detected a better manpower/milestone ratio. Peptides reach clinical phases more easily. In parallel, advances in the fields of formulation and drug delivery technology, and the fact that these technologies are accepted for the introduction of a peptide into the market for the first time, have fueled this field into the drug market. And last but not least, the great developments in peptide synthetic methods over the past few years have improved accessibility of a wider variety of peptides. This translates into the fact that in 2008 more than 90% of peptide production was by chemical synthesis. Another important supporting fact is that while in the 1980s most pharmaceutical peptides contained less than 10 amino acids, nowadays over 50% of peptides in clinical phase have more than 10 amino acids [4].

The purpose of this chapter is to review the latest advances in peptide chemistry that have boosted the peptide field. Even though, and from a synthetic viewpoint, peptides can be prepared in solid phase or in solution; nowadays, it is possible to say that in almost all peptide syntheses a solid-phase step is involved. Thus, the synthesis of small-to-medium-sized peptides is carried out in the solid phase, and the synthesis of large peptides and/or proteins is performed using a convergent approach. In this case, one of the last steps is carried out in solution, but the fragments either protected for a *classical* strategy or unprotected for a chemical ligation one are prepared in solid phase. Therefore, the solid-phase approach will be covered in detail.

## 2.2 SOLID SUPPORTS

In solid-phase peptide synthesis (SPPS), the most important choice becomes the solid support, which needs to accomplish certain features: (i) stability to mechanic stirring, to a range of temperatures, and to different solvents and reagent conditions; (ii) high swelling, so that reagents can access the active sites; (iii) homogeneity: a narrow range of bead sizes, and (iv) biocompatibility: swelling in aqueous buffers if used in biochemical assays. Solid supports for SPPS can be classified into three types [5]: polystyrene (PS), polyethylene glycol–polystyrene (PEG–PS®), and hydrophilic PEG-based resins (Figure 2.2).*

1. *PS resins.* PS is the most widely used solid support for the synthesis of peptides in solid phase because of its good swelling properties and good level of substitution [6, 7]. PS used nowadays contains 1% cross-linked hydrophobic resins obtained by suspension polymerization from styrene

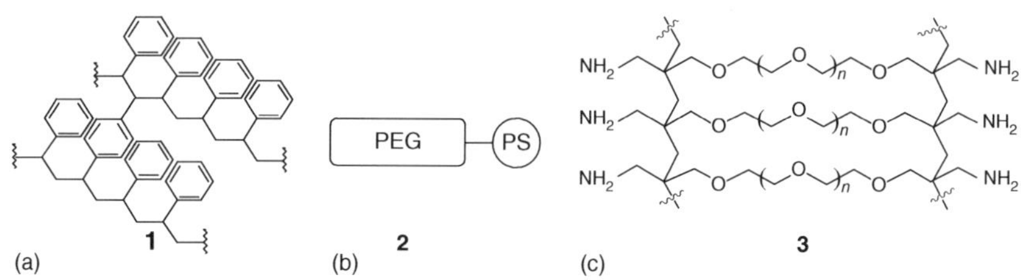


**Figure 2.2** (a) PS supports; (b) PEG-PS supports; and (c) totally PEG based supports.

*Peptide libraries can be prepared on paper membranes as in the SPOT technology [300].

and divinylbenzene. PS swells well in nonpolar solvents such as toluene or $CH_2Cl_2$, but can be used in combination with other more polar solvents such as *N,N*-dimethylformamide (DMF), dioxane, and tetrahydrofuran. Although it is the polymer of choice for the synthesis of small-to-medium-sized peptides, also from an economic viewpoint, it does present certain limitations in some cases, such as in the synthesis of highly hydrophobic or in the aggregation of peptides. In case of *difficult sequences*, more hydrophilic supports and resins show better performance.

2. *PEG-PS resins*. Due to its amphiphilic properties, which allow solvation in both polar and nonpolar solvents, the addition of PEG was investigated. Thus, based on the early work of Mutter [8], PEG-PS supports, which bear both a hydrophobic PS core and hydrophilic PEG chains, were developed independently by Zalipsky, Albericio, and Barany [9] (PEG-PS) and Bayer and Rapp [10] (Tentagel®). The benefits of these resins for the assembly of long peptides prompted the appearance of other supports, such as Champion® I and II (NovaGel®) [11] and ArgoGel® [12]. PEG-PS resins are compatible with both nonpolar and polar solvents.
3. *Hydrophilic PEG-based resins*. Searching to enhance the beneficial swelling properties of PEG's, more hydrophilic PEG resins with a small amount of PS or polyamide (poly(ethylene glycol)-poly(acrylamide) copolymer, PEGA) [13] or acrylate with polymerizable vinyl groups (cross-linked ethoxylate acrylate resin, CLEAR) [14], were developed. While PEGA resin was obtained by inverse suspension radical polymerization of various sizes of linear bis- and branched-tris-2-aminopropyl-PEG samples with acryloyl chloride, CLEAR supports were obtained by copolymerization of branched PEG-containing cross-linkers such as trimethylolpropane ethoxylate triacrylate with amino-functionalized monomers such as allylamine or 2-aminoethylmethacrylate. In the search for more stable and hydrophilic resins, a step forward came with the PEG-based resins developed by Meldal [15], which focused on resins containing only ether bonds, such as polyoxyethylene cross-linked polyoxypropylene (POEPOP), developed from a polymerization of PEG that was partially derivatized with chloromethyloxirane [16]. Although the POEPOP resin is mechanically robust, shows relatively high loading (primary and secondary alcohols), and good performance for organic transformations, the presence of secondary ether bonds implies that this solid support is not totally stable to strong Lewis acids [17]. To overcome this problem, the poly-oxyethylene-poly(3-methylene-3-methyloxethane) copolymer (SPOCC) resin, in which all ether bonds and functional alcohol groups are primary, was developed [18, 19].

At the same time, Côté developed the ChemMatrix (CM) resin [20], a total PEG-based resin comprised of primary ether bonds. Because of its highly cross-linked matrix, CM has surpassed the mechanical stability of other PEG resins. This resin swells well in all of the most common solvents and is, therefore, useful for a broad range of organic chemistries. CM resin performs extremely well compared to PS resins in the solid-phase synthesis of hydrophobic, highly structured peptides such as poly-Arg peptide and β-amyloid (1-42) [21, 22], showing that the presence of PEG chains impairs the aggregation of the growing peptide chain, facilitating the solid-phase synthesis of complex peptides. Furthermore, CM is convenient for the synthesis of oligonucleotides and oligonucleotide peptide conjugates [23].

Compared to earlier PEG-containing resins, these supports are 100% formed by primary ether bonds and thus show improved chemical stability and can reach higher loadings, comparable to those of PS resins. In comparative studies among several resins on the synthesis of human stromal cell-derived factor (SDF)-1α [24] and the acyl carrier protein (ACP 65-74) [25], higher purities were obtained with CM than with PS supports when using similar loadings. Compatibility of all these PEG-based resins with aqueous buffers allows their use for biochemical applications such as on-resin screening of chemical libraries and the development of affinity chromatography [26–29].

## 2.3 LINKERS

A linker is a bifunctional molecule that facilitates the attachment of the growing peptide as well as the final cleavage step. Linkers or handles can be classified into two types: *integral* and *nonintegral* [30]. In the first type, the solid support forms part of, or constitutes, the entire linker/handle, as is the case of, for example, 2-chlorotritylchloride resin (**6**). On the contrary, nonintegral linkers/handles are independent and bifunctional molecules that are attached to the solid support through an ether (e.g., Wang resin, **8**) or more commonly, through an amide bond, and they are more recommended because they provide control and flexibility for the synthetic process [31]. Linkage to the solid support should be totally stable to all synthetic processes, including the final treatment that will detach the target compound from the solid support. Sometimes this bond is not totally stable and the carbocation-containing linker is detached from the solid support, causing further heterogeneity of the crude peptide or causing back-alkylation of the target compound [32, 33]. This is the case of linkers attached to *p*-methylbenzhydrylamine (MBHA) resin when using a *tert*-butyloxycarbonyl (Boc) (**13**)/benzyl (Bzl) (**20**) strategy for preparing peptide amides. To overcome this side reaction, the use of aminobenzyl PS or aminoalkyl resins, which form a more acid-stable bond, is recommended [33]. Similar problems arise with (poly)alkoxybenzyl [34] (Wang (**8**), backbone amide linker (BAL), Rink (**11**))-type resins. Incorporation of the *p*-hydroxybenzyl moiety cleaved from the Wang resin (**8**) into the N of the C-terminal amide of a peptide during trifluoroacetic acid (TFA) cleavage [35], alkylation of the indol ring of Trp-containing peptides by the *p*-hydroxybenzyl moiety [36], and formation of *O*-(4-hydroxy)benzyl derivatives [37], are some of the side reactions encountered. Use of the Wang resin (**8**) for the solid-phase preparation of small molecules has also led to the introduction of impurities due to the undesired cleavage from the resin (no cleavage at the Bzl position) or from a back-alkylation of the *p*-hydroxybenzyl cation in the case of furopyridine and furoquinoline target derivatives [38]. To overcome these problems, two resins have been developed based on the activation of the Bzl position by a MeO group, a noncleavable electron-donating group, in either *ortho* or *para* position. Thus, Gu and Silverman [39] incorporated the precursor of their backbone linker to the resin through a metal-catalyzed coupling reaction and Colombo et al. [34] the precursor of their Wang-type resin through an amide bond. Linkers generally used in SPPS are those labile to acid, (Figure 2.3) although base and photolabile handles are also used in certain applications. High acid-sensitive linkers (Figure 2.3a) can be used to release side-chain protected fragments that are later used to access cyclic peptides or larger peptides by convergent approaches. On the contrary, C-terminal amide peptides can be constructed using Sieber, which is cleavable with low concentration

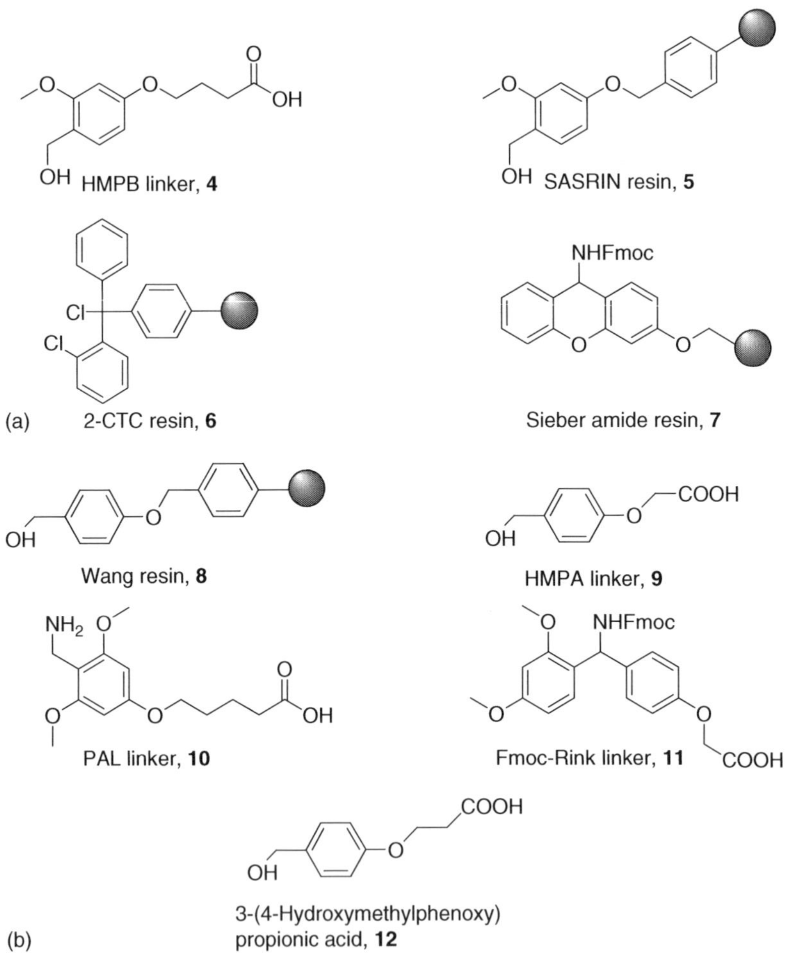


**Figure 2.3** (a) Low and (b) high acid-labile linkers and resins.

of TFA (3–5%), 5-(4-aminomethyl-3,4-dimethoxyphenoxy)valeric acid (PAL) (**10**), and rink linkers (**11**),† which release the amide function on TFA treatment.

## 2.4 PROTECTING GROUPS

*Temporary $N^{\alpha}$ protection.* Since in SPPS the peptides are built on the C to N direction, the temporary α-amino protecting group plays a very important role in the overall strategy. The $N^{\alpha}$-protected amino acid should be a solid that is easy to handle, soluble in the solvents used in SPPS, to prevent or minimize epimerization during coupling, and the protecting group should allow a fast and clean removal [40]. Two main strategies dominate the synthesis of peptides in solid-phase. The first one relies on using Boc (**13**) [41–43] as a temporary protecting group for the $N^{\alpha}$-amino function and Bzl-type protecting groups as permanent protecting groups for side chains. The main drawback of the Boc (**13**)/Bzl (**20**) strategy is the use of HF (hydrogen fluoride) for the final cleavage step, which hampers the application of this methodology to large-scale synthesis. The second strategy and the most employed nowadays uses the 9-fluorenylmethyloxcarbonyl (Fmoc) (**14**) [44] group as a temporary protecting group and *t*-butyl (*t*Bu) (**19**)-type groups for side-chain protection. Fmoc (**14**)/*t*Bu (**19**) strategy allows the use of the milder TFA for the final detachment of the peptide from the resin. Several other $N^{\alpha}$-amino protecting groups have since appeared, such as the trityl group (Trt) (**15**), which is removed by very mild acidic treatment (1% TFA), *p*-nitrobenzyloxycarbonyl (pNZ) (**16**) [45], also removed by acid (6 M $SnCl_2$, 1 mM HCl), the allyloxycarbonyl (Alloc) (**17**) group [46, 47], which can be removed under neutral conditions ($PhSiH_3$ (10 equiv), $Pd(PPh_3)_4$ (0.1 equiv)), and the photolabile 6-nitroveratryloxycarbonyl (Nvoc) (**18**) [48], all of which are orthogonal to the Boc (**13**) and Fmoc (**14**) groups, and allow for the synthesis of cyclic and branched peptides (Figure 2.4). Trt (**15**), pNZ (**16**), and Alloc (**17**) groups have also found an important application in minimizing diketopiperazine (DKP) in sequences prone to the formation of this side product [49, 45, 50].

*Permanent side-chain protection.* These protecting groups need to be stable during the entire elongation of the peptide and are usually removed concomitantly with the cleavage of the peptide from the resin (Figure 2.5). As mentioned earlier, the Fmoc strategy uses mainly *t*Bu (**19**) and Boc-type protecting groups. For Asp/Glu/Ser/Thr/Tyr, *t*Bu (**19**) is usually used, whereas the Boc (**13**) group is applied to Lys. For His/Asn/Gln the Trt (**15**) group is employed, and for Arg, the bulky pentamethyl-2,3-dihydrobenzofuran-5-sulfonyl (Pbf) (**24**) group is used. In the Boc (**13**) strategy, the Bzl (**20**) group is usually used for Asp/Glu/Ser/Thr/Tyr, although lately the cyclohexyl (cHx) (**21**) group is replacing the Bzl (**20**) group in Asp/Glu, as it better prevents aspartimide formation. Asn and Gln are usually being used without protection. The Lys side-chain is usually protected with the benzyloxycarbonyl (Cbz, Z) (**23**) or the 2-chlorobenzyloxycarbonyl (2-Cl-Z) group and for His/Arg, the *p*-toluenesulfonyl (Tos) (**22**) group is used. As for

†Rink resin is also found in the market.

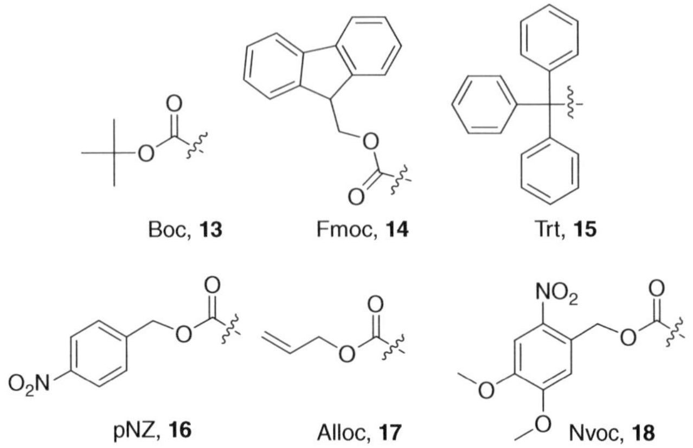


**Figure 2.4** $N^{\alpha}$-amino protecting groups in peptide synthesis.

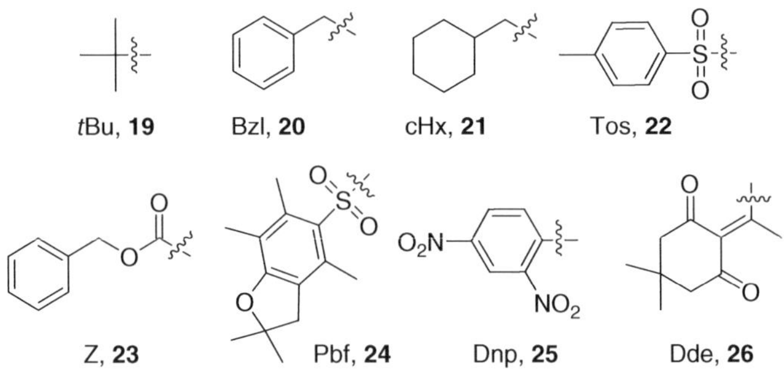


**Figure 2.5** Side-chain protecting groups.

His, another option is to employ the 2,4-dinitrophenyl group (Dnp) (**25**), which is removed by thiolysis prior to the HF cleavage step. The Alloc (**17**) and (1-(4,4-dimethyl-2,6-dioxocylohex-1-ylidene)-3-ethyl) (Dde) (**26**) group, which is removed by hydrazine, introduce an extra degree of orthogonality, and are used as side-chain protecting groups for Lys to access cyclic and branched peptides.

### 2.4.1 The Special Case of Cysteine

An important number of peptides and proteins possess disulfide bridges, which maintain the structure and biological activity of the molecule [51, 52]. Thus, cysteine residues need special side-chain protecting groups that will allow a postelongation transformation to the corresponding disulfide bridges either in solid phase or in solution. Developing groups for the side-chain of Cys and studying the optimal coupling conditions for these derivatives (Cys residues are very prone to racemization)

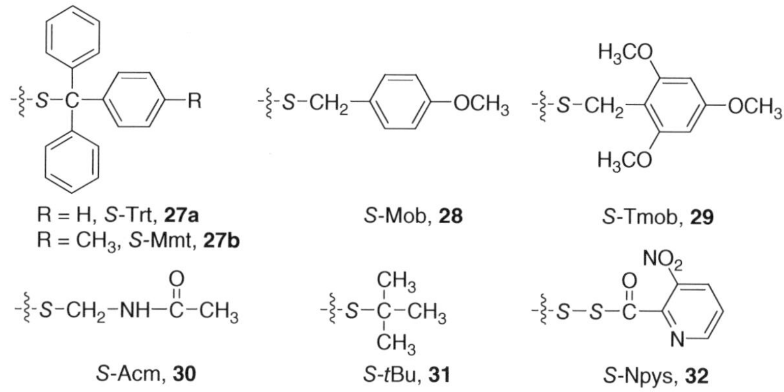


**Figure 2.6** Protecting groups for the thiol function of cysteine.

[53], and deprotection (to prevent back alkylation of the carbocation) has been an active field of research (Figure 2.6) [54, 55]. The most versatile group is maybe *S*-Trt (**27a**), which allows both a mild acidic cleavage (~10% TFA is needed) to later oxidize the thiol moieties, and also the direct formation of disulfide bridges by iodine treatment from the protected Cys. An even milder acid protecting group is the *S*-methyltrityl (*S*-Mmt) (**27b**), which can be removed with less than 1% of a TFA solution. Other groups that form disulfide bridges through iodine-mediated oxidation are *S*-4-methoxybenzyl (*S*-Mob) (**28**), *S*-2,4,6-trimethoxybenzyl (*S*-Tmob) (**29**) [56], and *S*-acetamidomethyl (*S*-Acm) (**30**) [57], for example, the latter having the advantage of being totally stable to strong acidic conditions. This protecting group has been widely used in the formation of disulfide bridges in solid-phase [58]. *S*-*tert*-butyl (*St*Bu) (**31**) [59] and *S*-3-nitro-2-pyridine-sulfenyl (*S*-Npys) (**32**) [60] groups can be removed by thiolysis. *S*-Npys group is not stable toward piperidine treatment and thus needs to be used in Boc chemistry strategies or introduced as Boc-Cys(Npys)-OH residue at the N-terminal position when using Fmoc approaches. Combination of an acidic-labile *S*-protecting group (e.g., *S*-Trt (**27a**), *S*-Mmt (**27b**)), which is removed upon cleavage, and *S*-Npys allows the construction of totally regioselective disulfide bridges and has been applied to the cyclization of peptide in solid phase using microwave-assisted heating synthesis [61] and to the synthesis of peptide-protein conjugates [62].

When acidic-labile protecting groups are used, the final cleavage renders the thiol-free peptide, which is usually oxidized in solution using dimethyl sulfoxide (DMSO) conditions [63] or other mild oxidizing conditions such as air or oxidized/reduced glutathione mixtures (Figure 2.6) [55].

In small peptides bearing more than one disulfide bridge, regioselective strategies are usually sought [64, 65], although usually various strategies needs to be tested to know the most favorable arrangement of Cys protecting groups. In bigger peptides and proteins, the kinetics of folding dominate the process and usually the concomitant oxidation of all Cys gives the correct folded molecule [66].

## 2.5 METHODS FOR PEPTIDE BOND FORMATION

Procedures used to combine two amino acid residues to form a peptide are referred as coupling methods. This reaction requires a previous activation of the carboxyl component to allow the efficient attack of the amino component (Scheme 2.1).

Even though a large *letter soup* corresponding to the different reagents and additives used for constructing the peptide bond is described in the literature, this chapter covers the most commonly used ones to perform the coupling between *N*-alkoxycarbonylamino acid (Fmoc, Boc, Alloc) and a peptide-resin in a stepwise synthesis mode.

### 2.5.1 Peptide-Bond Formation from Carbodiimide-Mediated Reactions

The most popular method of forming peptide bonds is the carbodiimide method, using dicyclohexylcarbodiimide (DCC, **33**). Carbodiimides contain two nitrogen atoms that are slightly basic; this is sufficient to trigger a reaction between the carbodiimide and an acid generating the *O*-acylisourea (**34**) (Scheme 2.2) [67–74].

The *O*-acylisourea from an *N*-alkoxycarbonylamino acid or peptide undergoes aminolysis to give the peptide (path A, Scheme 2.2). However, under certain conditions, some of the *O*-acylisourea undergoes an attack by a second molecule of the acid to give the symmetrical anhydride (**35**) (path B, Scheme 2.2). The latter is then aminolyzed to give the peptide. A third option is that some *O*-acylisourea cyclizes to the oxazolone (**36**) [75, 76] (path C, Scheme 2.2) which, although less reactive, also gives peptide by aminolysis, but can lead to a loss of chirality. However, a fourth and undesirable course of action is possible due to the nature of the *O*-acylisourea, which rearranges to produce the *N*-acylurea (**37**) that is a stable inert form of the acid (path D, Scheme 2.2). This reaction is irreversible and consumes the starting acid without generating peptide. A copious precipitate of *N,N′*-dicyclohexylurea separates within a few minutes in any reaction using *N,N′*-dicyclohexylcarbodiimide (**33**), which poses a problem in solid-phase synthesis because it cannot be removed by filtration. This has led to its replacement by *N,N′*-diisopropylcarbodiimide (DIC, **39**), or *N-ethyl-N′*-(3-dimethylaminopropyl)carbodimide (EDC, **40**), which gives an urea that is soluble in organic solvents. EDC, which is also referred to under the general term of *water soluble carbodiimide*, is most commonly used in solution chemistry, due its actual name and the fact that the urea formed is soluble in aqueous solvents.

Additives such as HOXt (**41–44**) increase the efficiency of carbodiimide-mediated reactions. Even though they render an active ester that is less reactive than the *O*-acylisourea, its formation prevents *N*-acylurea formation. This beneficial effect

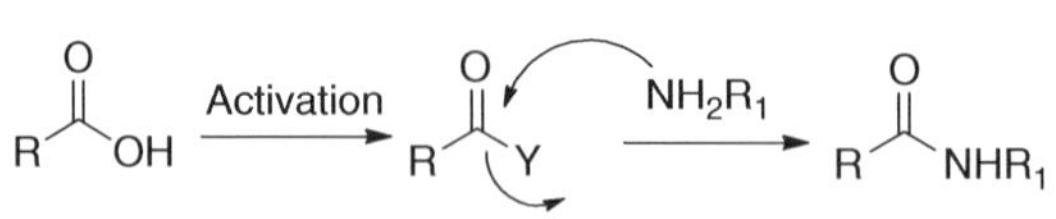


**Scheme 2.1** Peptide-bond formation process.

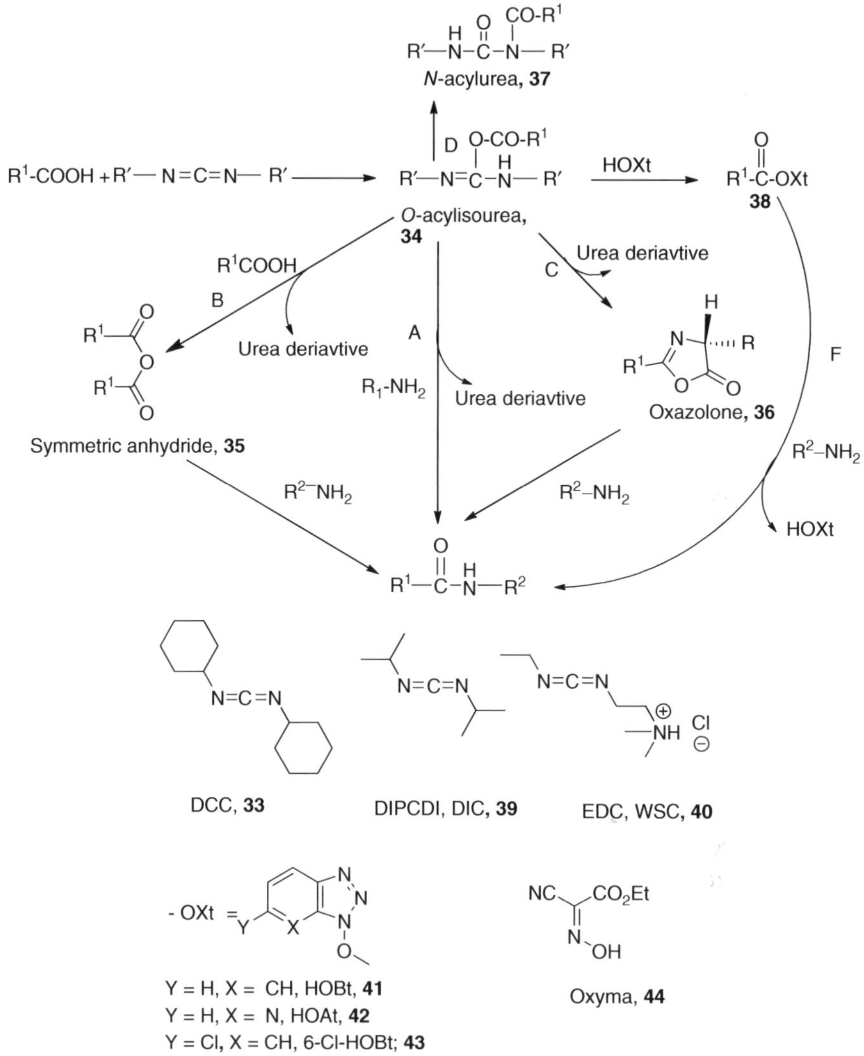


**Scheme 2.2** Mechanism of peptide-bond formation from carbodiimide-mediated reaction. Most commonly used carbodiimides and additives.

is attributed to its role as an acid that protonates the *O*-acylisourea, thus preventing the intramolecular reaction from occurring and shifting the reaction to form the corresponding active esters (**38**) (path F, Scheme 2.2), and decreasing the degree of racemization in numerous cases [76, 77]. Compared to other additives, 1-hydroxy-7-azabenzotriazole (HOAt) (**42**) forms superior active esters in terms of yield and degree (less) of racemization in both solution and solid-phase synthesis [78]. The key behind the outstanding behavior of HOAt is the nitrogen atom

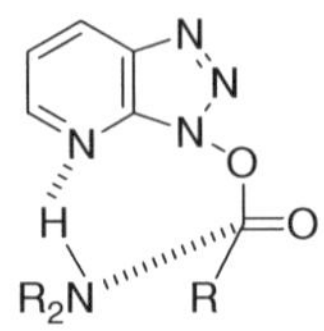

**Figure 2.7** Neighboring group effect for HOAt.

located at position 7 of the benzotriazole, which provides a double effect. First, the electron-withdrawing influence of a nitrogen atom (regardless of its position) improves the quality of the leaving group, thereby leading to greater reactivity. Second, the placement of this nitrogen atom specifically at position 7 makes it feasible to achieve a classic neighboring group effect (Figure 2.7), which can both increase reactivity and reduce the loss of configurational integrity [77].

A decade ago, 6-chloro-1-hydroxybenzotriazole (6-Cl-HOBt) (**43**) had been introduced into solid-phase synthesis. This additive is a good compromise between HOAt (**42**) and HOBt (1-hydroxybenzotriazole) (**41**) in terms of reactivity and price [79].

Very recently, El-Faham and Albericio [80] reported a safe and highly efficient additive, ethyl 2-cyano-2-(hydroxyimino)acetate (Oxyma, **44**) that is used mainly in the carbodiimide approach for forming the peptide bond. Oxyma **44** displays a remarkable capacity to suppress racemization and impressive coupling efficiency in both automated and manual synthesis. These effects are superior to those shown by HOBt (**41**) and comparable to HOAt (**42**). Stability assays show that there is no risk of capping the resin in standard coupling conditions. Finally, calorimetry assays Differential scanning calorimetry and Accelerated rate calorimetry (DSC and ARC) confirm the explosive potential of the benzotriazole-based additives and demonstrate the lower risk of explosion induced by Oxyma [80]. This point is highly relevant because all benzotriazole derivatives, such as HOBt (**41**) and HOAt (**42**), exhibit explosive properties [81].

### 2.5.2 Peptide-Bond Formation from Preformed Symmetric Anhydrides

An alternative to the classical method of synthesis using carbodiimides is symmetric anhydride mediated reaction, in which the carbodiimide and acid are first allowed to react together in the absence of *N*-nucleophile. One-half of an equivalent of carbodiimide is employed. This generates half an equivalent of symmetrical anhydride (**35**) (Scheme 2.2, path B), the formation of which can be rationalized in the same way that the reaction of acid with carbodiimide is rationalized, namely, protonation at the basic nitrogen of the *O*-acylisourea by the acid, followed by attack at the activated carbonyl of the acyl group by the carboxylate anion. Aminolysis at either carbonyl of the anhydride gives the peptide bond. The symmetrical anhydride is less reactive and consequently more selective in its reactions than the *O*-acylisourea, but more reactive that the OBt ester. When the reagent is DCC (**33**), the reaction is carried out in $CH_2Cl_2$, the *N,N'*-dicyclohexylurea is removed by filtration after 15–30 min, the solvent is sometimes replaced by DMF, and the solution is then added to the second amino acid. The anhydrides are particularly effective for acylating secondary amines [82–84].

### 2.5.3 Peptide-Bond Formation from Acid Halides

The most obvious method for activating the carboxyl group of an amino acid for peptide bond formation at room temperature or below would appear to be through a simple acid chloride [85]. Acid chloride method was first introduced to peptide chemistry by Fisher [86] in 1903. Fmoc-amino-acid chlorides are generated by reaction of the parent acid with thionyl chloride in hot $CH_2Cl_2$. Fmoc-amino-acid chloride acylates the amino group in the presence of a base that is required to neutralize the hydrogen chloride that is liberated. The base is necessary, but its presence complicates the issue, converting the acid chloride to the 2-alkoxy-5(4*H*)-oxazolone (**36**), which is aminolyzed at a slower rate and can lead to a loss of chirality. One deficiency of these systems is that acid-sensitive side chains, such as those derived from *t*-butyl residues, cannot be accommodated.

Gilon has reported the use of BTC (*bis*(trichloromethyl)carbonate, triphosgene) as a chlorinating reagent in SPPS for the acylation of *N*-alkylresidues [87]. Coupling reactions mediated by BTC gave good results for Fmoc-amino acids containing acid-labile side-chains.

Acid fluorides, on the contrary, are known to be more stable to hydrolysis than acid chlorides and in addition are not subject to the limitation mentioned with regard to *t*-butyl-based side-chain protection. Thus, Fmoc-based SPPS can be easily carried out through Fmoc amino acid fluorides [88–94].

### 2.5.4 Peptide-Bond Formation from Phosphonium Salt-Mediated Reactions

Kenner and coworkers [95] were the first to describe the use of acylphosphonium salts as coupling reagents. After HOBt (**41**) was discovered as a racemization suppressant, a new coupling reagent, known as BOP (benzotriazol-1-yloxytris(dimethylamino)-phosphonium hexafluorophosphate) (**47**), was introduced in 1975 [95, 96]. Later, chlorotri(pyrrolidino)phosphonium hexafluorophosphate (PyCloP) (**48**), bromotri (pyrrolidino)phosphonium hexafluorophosphate (PyBroP) (**49**), and benzotriazol-1-yloxytri(pyrrolidino)-phosphonium hexafluorophosphate (PyBOP) (**50**) (Figure 2.8) were introduced. In these compounds, the dimethylamine moiety is replaced by pyrrolidine (Figure 2.8) [95, 97, 98]. Pyrrolidine reagents prevent the generation of poisonous hexamethylphosphoramide (HMPA, **51**) by-product [99]. In a further study, Coste [100, 101] reported that halogenophosphonium reagents often give better results than other phosphonium-HOBt reagents for the coupling of *N*-methylated amino acids.

Practically speaking, the active species is the OBt ester, whose formation is achieved in the presence of 1 equiv of a tertiary base such as diisopropylethylamine (DIEA), *N*-methylmorpholine (NMM) [102–104], or collidine (TMP) [105]. Presence of an extra equivalent of HOBt accelerates the coupling and also reduces the loss of configuration [99].

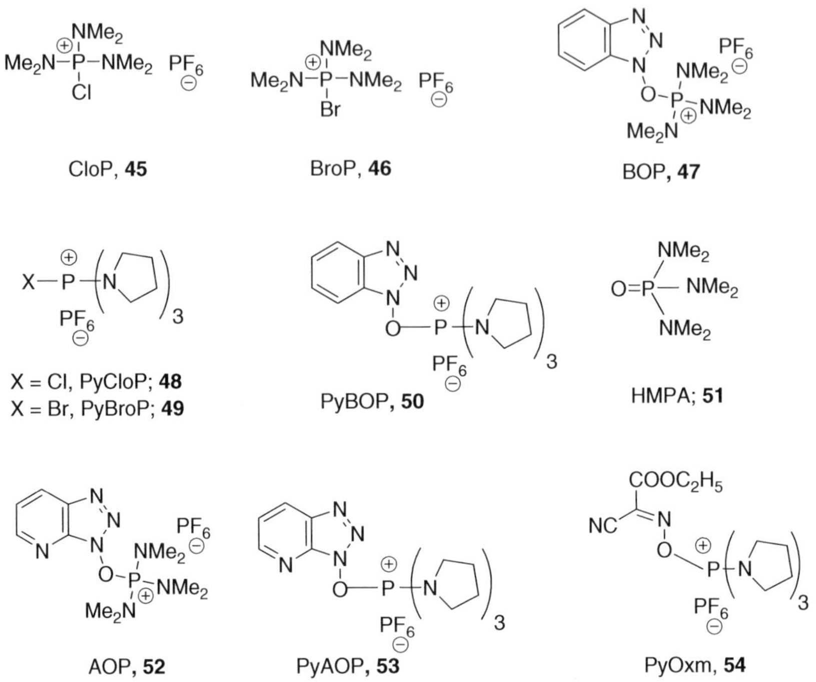


**Figure 2.8** Structure of phosphonium salts.

Phosphonium salts derived from HOAt (**42**), (7-azabenzotriazol-1-yl)oxytris-(dimethylamino) phosphonium hexafluorophosphate (AOP, **52**) and (7-azabenzotria zol-1-yloxy)tris-(pyrrolidino)phosphonium hexafluorophosphate (PyAOP, **53**), have also been prepared and are generally more efficient than BOP (**47**) and PyBOP (**50**) as coupling reagents [78, 106–110].

Recently, El-Faham et al. [111] introduced a new family of phosphonium salt of Oxyma (**44**), *O*-[(cyano-(ethoxycarbonyl)methyliden)-amino]yloxytripyrrolidinopho sphonium hexafluorophosphate (PyOxm; **54**), which has been demonstrated to be an efficient racemization suppressing coupling reagent for the assembly of hindered peptides, performing better than classical benzotriazole derivatives BOP (**47**) and PyBOP (**50**). Cyclization models revealed the advantages on the use of PyOxm (**54**), which rendered a higher percentage of cyclic peptide than other known phosphonium salts [111].

### 2.5.5 Peptide-Bond Formation from Aminium/Uronium Salt-Mediated Reactions

Initially, the product obtained by reaction of HOBt (**41**) with tetramethylchlorouronium salt (TMUCl) was assigned to a uronium-type structure, presumably by analogy

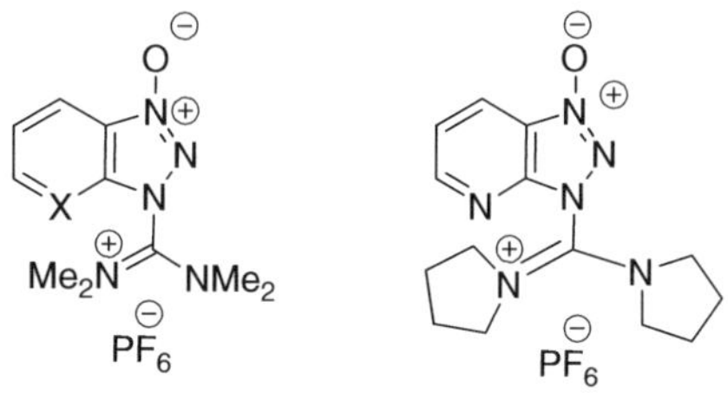

X = CH, *N*-HBTU; **55**
X = N, *N*-HATU; **56**

*N*-HAPyU,**57**

**Figure 2.9** Structures of aminium/uronium salts.

with the corresponding phosphonium salts, which bear a positive carbon instead of the phosphonium residue [112].

Several years ago [113–115], an X-ray analysis showed that salts crystallize as aminium salts (guanidinium *N*-oxides), rather than the corresponding uronium salts. This occurs for *N*-[(1*H*-benzotriazol-1-yl)(dimethylamino)methylene]-*N*-methyl methanaminium hexafluorophosphate *N*-oxide (*N*-HBTU, **55**), *N*-[(dimethylamino)-1 *H*-1,2,3-triazolo[4,5-*b*]pyridin-1-ylmethylene]-*N*-methylmethanaminium hexafluorophosphate *N*-oxide (*N*-HATU, **56**), and 1-(1-pyrrolidinyl-1*H*-1,2,3-triazolo[4,5-*b*] pyridin-1-ylmethylene) pyrrolidinium hexafluorophosphate *N*-oxide (HAPyU, **57**) [116–122]. NMR studies in the case of HAPyU show that the same structure is found in solution [113, 116] (Figure 2.9).

El-Faham and Albericio described a new family of immonium-type coupling reagents based on the differences in the carbocation skeletons of coupling reagents (Figure 2.10), which correlated with differences in stability and reactivity [123–126]. Dihydroimidazole derivatives are highly unstable to air, whereas the salts derived from dimethyl morpholino are the most stable, and the pyrrolidino derivatives are of intermediate stability. Regarding both, coupling yield and retention of configuration, derivatives of Oxyma (1-[(1-(cyano-2-ethoxy-2-oxoethylideneaminooxy)-dimethylamino-morpholinomethylene)] metha naminium hexafluorophosphate, COMU, **61**) have been confirmed to show superior

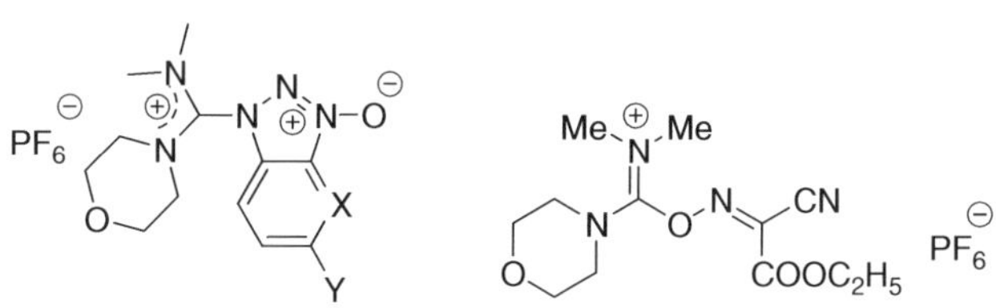

Y = H, X = CH; HDMB, **58**
Y = H, X = N; HDMA, **59**
Y = Cl, X = CH, HDMC, **60**

COMU, **61**

**Figure 2.10** Structure of aminium/uronium salts derived from morpholinium skeleton.

performance to those of HOBt in all cases and the same or sometimes better performance as HOAt [125].

Mechanistically, aminium/uronium salts are thought to function in a manner similar to phosphonium analogs. Formation of carboxyl uronium salts that generate an active ester is achieved in the presence of one equivalent of a tertiary base such as DIEA, NMM [102–104], or TMP [105]. The presence of an extra equivalent of HOXt could accelerate coupling and reduces the loss of configuration [105].

## 2.6 SOLID-PHASE STEPWISE SYNTHESIS

Regardless of the chemistry used (Boc (**13**) or Fmoc (**14**)), the synthetic scheme for the synthesis in solid-phase is identical (Scheme 2.3). If a nonintegral linker is used, the first step will be its attachment to the resin, usually through the formation of an amide bond. Next will follow the coupling of the first amino acid through an ether or an amide bond. Removal of the protecting group is carried out with 25–50% TFA in $CH_2Cl_2$ in Boc chemistry, whereas piperidine–DMF (1:4) is used for cleaving the Fmoc (**14**) group. Collection of the filtrate enables the quantification of the loading of

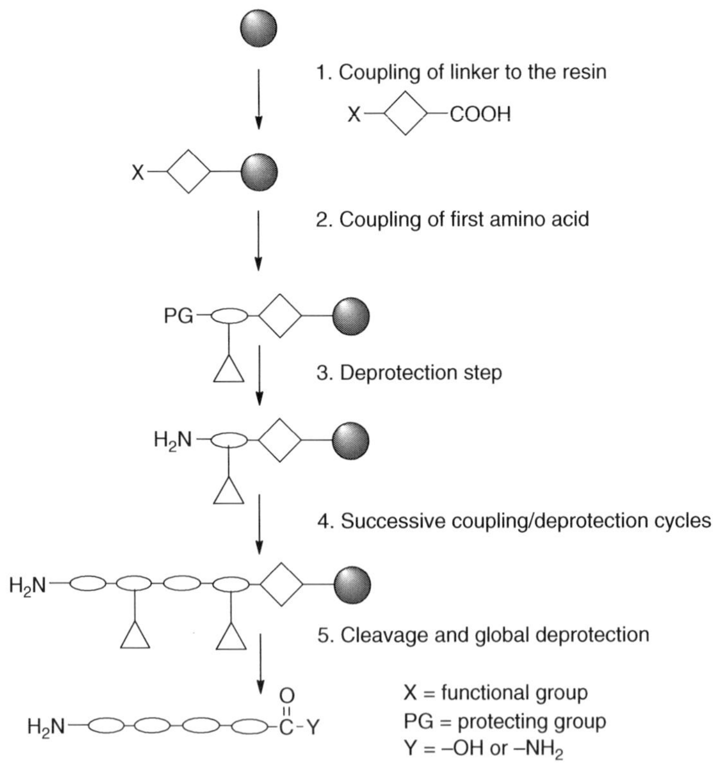


**Scheme 2.3** Schematic representation of solid-phase peptide synthesis.

the resin by ultraviolet (UV) measurement of the Fmoc (**14**) decomposition product dibenzofulvene. Successive coupling and deprotection cycles are carried out until the desired length is reached, and the final release from the resin is performed by HF (Boc (**13**) chemistry) or by a TFA cocktail (Fmoc (**14**) chemistry). If a low acid-containing cocktail is used for the detachment from the resin, then a second TFA treatment is needed to free the protected side-chains. In either case, the crude peptide is treated by cold ether, precipitated, centrifuged, and finally lyophilized.

### 2.6.1 Long Peptides

Synthesis of long peptides faces additional challenges, the most important being the aggregation of the growing chain. Therefore, the use of potent coupling reagents, even if recouplings are performed, becomes insufficient to obtain a crude peptide of good quality when hydrophobic interactions are present, and additional tools are required. To maximize coupling and deprotection yields, the use of DMSO [127], *magic mixtures* [128], the addition of chaotropic salts [129], or the introduction of Pro residues [130], have been applied. However, a more general approach has been the application of PEG (**3**) resins, backbone amide protection, pseudoprolines (ψPros), and the *O*-acyl isopeptide or depsipeptide methodology, all of which are discussed below.

***2.6.1.1 PEG (3) Resins*** As previously explained, hydrophobic resins such as 2-chlorotrityl chloride resin (2-CTC) are not well suited for the synthesis of long peptides. In contrast, the amphiphilic nature of PEG allows the aggregation of the growing chain to be minimized. Thus, totally PEG-based (**3**) resins such as SPOCC [18] and CM [20] have been successfully applied to the synthesis of complex peptides. Striking examples are the synthesis of the highly aggregating β-amyloid (1-42) peptide, where a crude of 91% purity was achieved [21], the synthesis of regulated on activation, normal T cell expressed and secreted (RANTES) [131], a complex aggregated chemokine, where a combination of CM resin and ψPros at key positions rendered the final crude peptide, otherwise unfeasible to assemble, and the 99-residue HIV-protease [132].

***2.6.1.2 Backbone Amide Protection*** Sheppard and coworkers [133] introduced the *N*-(-2-hydroxy-4-methoxybenzyl) (Hmb) group as reversible backbone protection

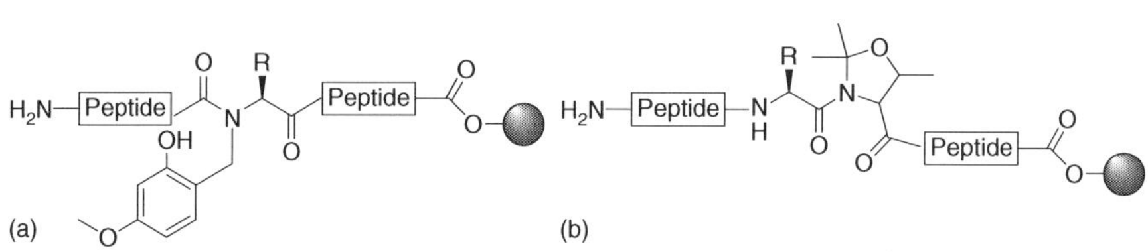


**Figure 2.11** Incorporation of (a) Hmb auxiliary and (b) pseudoproline, into a growing peptide.

(Figure 2.11a). The idea was to break the amide pattern, which promotes hydrogen bonding and ultimately leads to β-sheet structures and aggregation of the growing peptide chain. Final global deprotection also cleaves the Hmb auxiliary. The Hmb auxiliary has a long-range effect and only needs to be introduced at about every sixth residue to inhibit aggregation. This strategy has been successfully applied to the synthesis of the influenza peptide [134] and the human $\alpha_{1E\text{-}3}$ calcium channel subunit fragment 985–1004 [135], among others. The main problem of this approach is the difficulty in the coupling of the residue following the Hmb auxiliary. In this sense, efforts have been directed to synthesizing less sterically hindered auxiliaries, such as the 2,4-dimethoybenzyl (Dmb)Gly [136] and the dicyclopropylmethyl (Dmcp) groups [137].

***2.6.1.3 Pseudoprolines*** Pseudoprolines are also structure-disrupting secondary amino acids that in this case are introduced as ψPro dipeptides (Figure 2.11b) [138–140]. In ψPro dipeptides, the C-terminal amino acid is an oxazolidine-protected Ser, Thr, or Cys. In elongating a peptide, ψPros prevent the aggregation of the growing chain in a similar way as Pro. ψPros are introduced as Fmoc-protected dipeptides to prevent incomplete couplings over the oxazolidine moiety. Once the peptide elongation is completed, TFA treatment results in cleavage of the oxazolidine moiety, thereby recovering the natural amino acids. As with Hmb protection, they also have a long-range effect, the proline motif provides a deliberate change of native structure usually in the region of 6–10 residues after its incorporation. When comparing this strategy with Hmb protection, the introduction of ψPros has been proven to be superior [141]. In our laboratory, the synthesis of the highly complex chemokine RANTES was accomplished using a combination of ψPros and a PEG resin [131]. The RANTES sequence contains nine replaceable residues: $Ser^{68}$, $Ser^{64}$, $Thr^{43}$, $Ser^{35}$, $Thr^{30}$, $Thr^{8}$, $Thr^{7}$, $Ser^{5}$, and $Ser^{4}$. After careful examination of the secondary structure of the peptide, four of these residues were replaced by ψPros: $Ser^{64}$ (to initially alter chain conformation into the C-terminal part); $Thr^{30}$ and $Thr^{43}$ (which are crucial during folding due to hydrogen bonding formation, and located in distinct β-sheets); and $Thr^{7}$ (positioned in the N-terminal region). Using ψPros, other otherwise inaccessible peptides have also been assembled [140, 142–144]. The main limitation of this approach is that a Ser, Thr, or Cys residue needs to be present in the sequence.

***2.6.1.4 O-acyl Isopeptide*** This method, also named the depsipeptide method, has been successfully applied in the synthesis of *difficult sequences* [145–148]. Based on previous work on the synthesis of the more soluble *o*-acyl prodrug analogs [149, 150], this technique involves the assembly of the *O*-acyl isopeptide and its later conversion to its peptide counterpart under physiological conditions (Scheme 2.4). Due to the better solubility of the *O*-acyl isoform, this is obtained in better yields and purities. Presence of the ester is believed to change the secondary structure of the peptide. In fact, in circular dichroism (CD) studies of *O*-acyl isopeptides or *switch* peptides, transitions to their peptide counterparts, controlled induction or reversal of secondary structure, and self-assembly of small peptides have been observed [148, 151, 152].

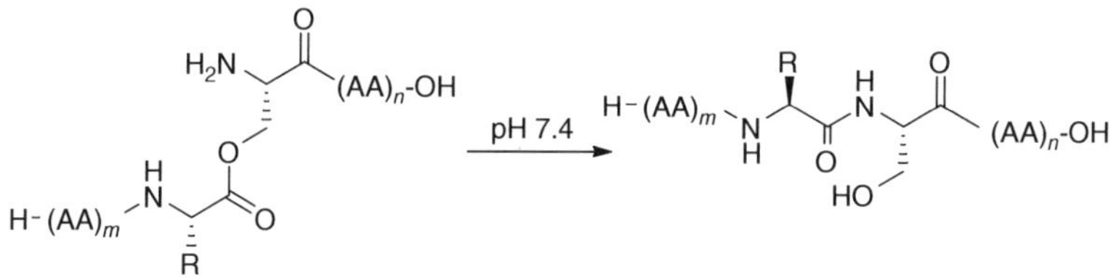


**Scheme 2.4** Conversion of the *O*-acyl isopeptide to the *N*-peptide at pH 7.4 on a Ser residue. *Source*: Adapted from Reference 145.

These studies have provided a tool to disrupt amyloid-derived peptide assemblies [153] and to identify antiamyloid agents [154].

To prove the concept, Sohma et al. synthesized the highly hydrophobic peptide Ac-Val-Val-Pns-Val-Val-$NH_2$ (Pns, phenylnorstatine) on a Rink resin (**11**), obtaining only a 6.9% yield after purification. Analysis of side-products indicated incomplete Fmoc deprotection and incomplete acetylation arising from aggregation. In contrast, when *O*-acyl isopeptide was constructed on the same resin and then converted to the parent peptide in phosphate buffered saline (PBS) at pH 7.4, the yield increased to 54% although a small amount of racemized product (3.2%) was observed.

To further the scope of this methodology, the application to the synthesis of longer peptides, such as Aβ(1-42), on a 2-CTC resin (**6**) was undertaken [155, 156].

To prevent racemization, which was one of the limitations of the method (in some cases levels around 20% were reached), preformed isodipeptide units, such as Boc-Thr(Fmoc-Val)-OH, were introduced [157]. Based on these previous results, a completely convergent approach to suppress racemization was also developed [158]. Thus, for a given peptide, an N-terminal fragment, bearing a C-terminal *O*-acyl isopeptide, was coupled to a C-terminal fragment. Owing to the presence of the urethane-protected Ser/Thr residue, oxazolone formation, and, therefore, racemization, is avoided.

Coin et al. [159] have also used this methodology to assemble the highly β-structured sequence Valine-Threonine $(VT)_{10}$-$NH_2$ by constructing a depsi bond at each Thr, and to synthesize an analog of the rsp5-domain (WW) domain FBP28, $[Asn^{15}]$FBP28-$NH_2$. In this case, extensive DKP formation was found depending on the nature of the two amino acids following the depsi bond. To overcome this side reaction, the Bsmoc group, which can only be removed using 2% piperidine, was used [160, 161]. This methodology has been recently used to synthesize a peptide–polymer conjugate [162], which self-assembles with the formation of microstructures on the recovery of the native peptide backbone by *O–N* acyl migration.

## 2.7 SYNTHESIS IN SOLUTION

As mentioned in the introduction, and although peptide synthesis nowadays is commonly performed in solid phase, classical peptide synthesis in solution remains one of the major chemical approaches especially used by pharmaceutical companies to prepare peptides introduced into the market in the 1970s and 1980s. For these small peptides, the market price of which is moderate, the solution method dramatically reduces production costs, thanks to the starting materials needed and reagents-related expenses, and most importantly due to the low cost of the isolation and purification techniques this method requires.

The main advantage of this approach compared to the SPPS is the isolation and characterization of each intermediate of the synthesis in solution. This is of great importance due to the fact that during the process every step can be relatively easily controlled and due to the less laborious final purification process. Particularly, purification at each step is usually performed by simple crystallization and/or simple chromatography in a silica gel column that provides the intermediates in reasonable purity, avoiding the laborious and expensive isolation methods, such as high performance liquid chromatography (HPLC). Thus, the purification and isolation of the final product is the major benefit of solution peptide synthesis. As a potential drawback, the purification of the intermediates requires extended time compared to SPPS for every cycle of the process. This disadvantage is somehow reduced if several segments of the peptide are synthesized in parallel and a convergent strategy is applied, instead of linear step-by-step synthesis in solution. Another important advantage of solution peptide synthesis is the reduced cost compared to SPPS. In solution peptide synthesis, amino acids derivatives reactants and coupling reagents are commonly used in a 1:1 ratio but not in excess as in SPPS. Furthermore, the use of the relatively expensive solid supports and linkers is avoided. Finally, in solution it is possible and always desirable to keep the use of side protecting groups to a minimum. Unprotected amino acids can be successfully used in solution without side-reactions during the process [163].

In solution synthesis (step-by-step or convergent), except for the reversible masking of the N-amino group of the first amino acid or fragment, orthogonal protection of the carboxyl group of the second amino acid or fragment that participates in the reaction is required (Scheme 2.5). The carboxy terminal protecting group is essentially a replacement for the solid support of SPPS.

### 2.7.1 $N^\alpha$ Protection of the N-Terminal Amino Acid Derivative or Fragment

The preferred groups for $N^\alpha$-amino protection are the Boc (**13**) [41–43] and the Cbz (Z, **23**) [164, 165] because of the volatile by-products formed during the deprotection step. Alternative Fmoc group (**14**) [44] is commonly avoided due to

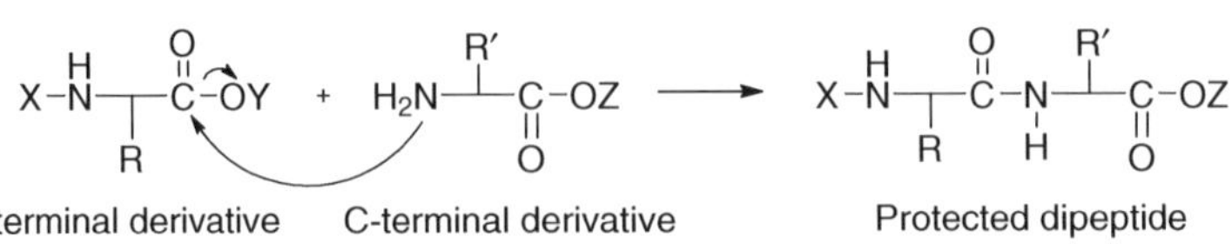


**Scheme 2.5** Basic principles of dipeptide synthesis. The $N^\alpha$-amino function of the N-terminal residue should be protected (X-group) and the carboxyl group activated (OY-group represents an active ester). Instead, the C-terminal derivative has to be protected to the carboxy terminal (Z-group).

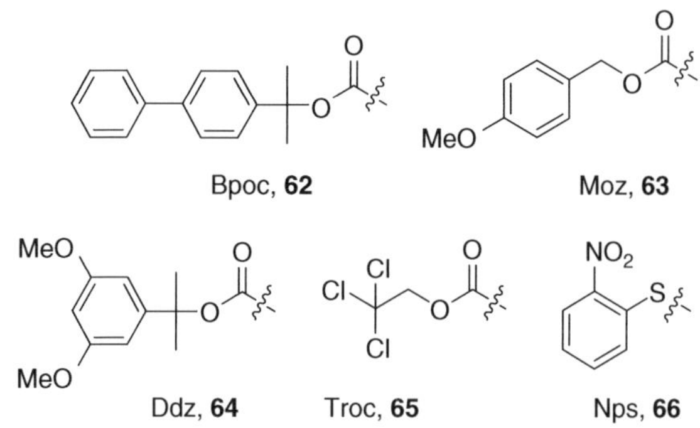


**Figure 2.12** $N^{\alpha}$-protecting groups for peptide synthesis in solution.

the formation of nonvolatile by-products (dibenzofulvene) [166]. Boc group can be commonly removed at each step using TFA acid or HCl in acetic acid (HOAc) [42, 43], whereas the Z (**23**) group is removed under mild conditions by catalytic hydrogenolysis [167]. The Z (**23**) group can also be removed by acidolysis, but strong acids, such as HBr in HOAc, are required, [168]. Substituted derivatives of the Z (**23**) group are also used in solution synthesis as $N^{\alpha}$-protecting groups that are more acid labile such as the 2-(4-biphenylyl)isopropoxycarbonyl (Bpoc, **62**) [169]; the 4-methoxy-benzyloxycarbonyl (Moz, **63**) [170], which can be removed by TFA; and the α,α-dimethyl-3,5-dimethoxybenzyloxy (Ddz, **64**) group [171], which can be removed by photolysis [172]. A less common $N^{\alpha}$-protecting group in solution peptide synthesis is the 2,2,2-trichloroethoxycarbonyl (Troc, **65**)-groups [173], which is removed under treatment with Zn in HOAc [174]. The most known nonurethane type $N^{\alpha}$-protecting group is the *o*-nitrophenylsulfenyl (Nps, **66**) group [175]. The advantage of the Nps-group, except for the cleavage with acids [176], is that it can be selectively removed by nucleophilic reagents [177, 178]. These reagents avoid the problems encountered with protecting groups requiring acids for their cleavage. Thiolytic cleavage of the Nps group with a number of reagents has been described, and has been shown to enable rapid deprotection (Figure 2.12) [179].

### 2.7.2 Carboxy-Group Protection of the C-terminal Amino-acid Derivative or Fragment

The carboxylic group of the C-terminal amino acid or fragment is commonly masked as an alkyl or aryl ester. Alternatively, the C-terminal protecting group can be hydrazides or protected hydrazides. Regarding peptides with amide function at the C-terminal, protection is commonly not essential. Choice of the C-terminal protecting group always depends on the selection of the temporary N-terminal $N^{\alpha}$-protecting group, as it should be stable to the removal conditions of the temporary protection of $N^{\alpha}$-amine, such as with the linker in SPPS (Figure 2.13). The most

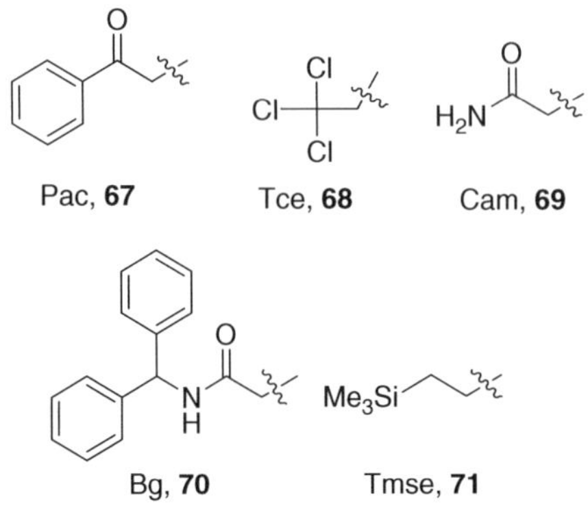


**Figure 2.13** Carboxy C-terminal protecting groups.

frequently used C-terminal protecting groups in solution synthesis are *t*Bu (**19**) and Bzl (**20**) esters. Other alternative esters of the C-terminal are methyl, phenacyl (Pac, **67**), phenyl and their substituted derivatives.

Orthogonal protection of the carboxylic acid with the Bzl (**20**) group is preferred, whereas the $N^{\alpha}$-fragment or amino-acid group is masked with the Boc (**13**) group. Bzl ester is stable to the acidic conditions required for the removal of the Boc (**13**) group. In SPPS, the Bzl (**20**) group is removed by acidolysis with strong acids, while in solution synthesis, the most frequently used milder method is catalytic hydrogenolysis [166] after elongation of the peptide chain. Bzl (**20**) group can also be removed by saponification [180].

Use of the *t*Bu (**19**) ester as a protecting group is particularly effective when the Z (**23**) group is used as $N^{\alpha}$-protecting group. Alternatively, the *t*Bu (**19**) ester provides orthogonal protection with the use of the Bpoc group as $N^{\alpha}$-protecting group. O*t*Bu ester is probably the most useful C-terminal protecting group and can be removed by acidic hydrolysis using moderately strong acids, such as TFA or HCl solution [181]. However, it is a sufficiently stable group to weak acids, allowing the washing steps in standard workup procedures in solution peptide synthesis.

Among other protecting groups used, the Me (methyl ester) group is a reasonable choice when the target compound is a C-terminal amide peptide, since the treatment of the ester with ammonia provides the elongated peptide in good yields. Me ester can also be used for the protection of the C-terminal carboxylic group, but unmasking is problematic, since its removal by saponification can lead to unacceptable amounts of epimerization [182]. However, Me esters, as many alkyl or aryl esters, are quite useful, as they can be easily converted into acyl azides through the formation of hydrazides by hydrazynolysis (Scheme 2.6) [183]. Formation of acyl azides was frequently used in *classical* convergent strategy solution synthesis. Thus, the Me group is commonly preferred for protection of the C-terminal of a peptide in two cases: (i) when the targeted peptide should be permanently protected at the C-terminal and (ii) when, after peptide chain elongation, the modification of the C-terminal is required using the azide method. Alternatively, the use of Et (ethyl ester) ester instead

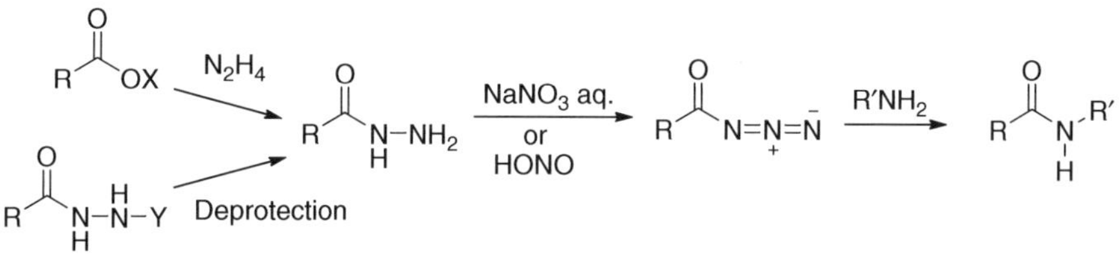


**Scheme 2.6** Coupling of an amino acid to the C-terminal of a peptide by the azide method starting from a hydrazide protected (Y = hydrazide protecting group, e.g., Bzl, *t*Bu, Boc etc.) peptide or from an aryl /alkyl ester (X = H, Me, Et, Bzl, etc.).

of Me is used, but the side-products of saponification are more difficult to manage. Several substituted Me and Et esters have also been used in solution peptide synthesis such as the trichloroethyl (Tce, **68**) ester that is removed by hydrolysis under mild conditions or by elimination reaction induced by Zn in HOAc [184], and is compatible with Boc $N^{\alpha}$-protecting group. Carboxyamidomethyl (Cam, **69**) [185], *N*-benzydrylglycolamide (Bg, **70**) [186] and 2-trimethylsilylethyl (Tmse, **71**) [187] esters have also been used, with the later presenting a labile to fluoride ion properties [188]. Phenyl ester has also been used as C-terminal carboxy terminal protecting group [189], as it is stable to acidolysis and catalytic hydrogenolysis and can be used while the N-terminal is protected with Boc (**13**) and Z (**23**) group. Ph ester can be removed either by saponification or, under mild conditions, by peroxide ion and alkali in the presence of dimethyl sulfide in order to avoid the oxidation of sensitive residues (Met, Cys, etc.) [190].

And last but not least, hydrazide and substituted hydrazide derivatives can be used as carboxy C-terminal protecting groups [191]. Although hydrazide (**72**) is reactive to acylation by the activated amino-acid derivatives, and cannot provide sufficient protection to carboxy C-terminal, it is quite useful for segment coupling into a convergent strategy because it can be easily converted into acyl azide. Thus, only protected hydrazides can be used as protecting groups of C-carboxy terminal for peptide chain elongation at the N-terminal (Figure 2.14). Protecting groups of hydrazide derivatives are then removed, releasing the hydrazine that can be converted into the azide for further coupling of an amino acid derivative or a peptide segment at the C-terminal by

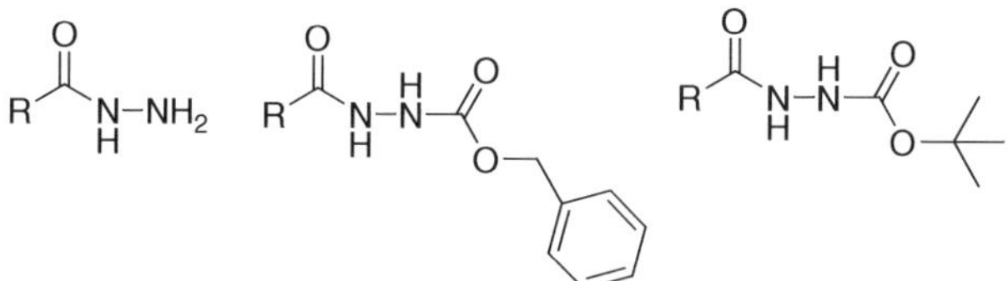


**Figure 2.14** Hydrazide and protected hydrazides that provide protection to the C-terminal carboxy group.

nucleophilic attack (Scheme 2.6) [192, 193]. This method has found extensive use in the synthesis of long peptides using the fragment conversation strategy.

### 2.7.3 Peptide Bond Formation

Basic methods and a few examples of the coupling reagents that are used in peptide synthesis are discussed in depth in Chapter 4. Reagents that are used in SPPS are also frequently used in solution peptide synthesis. Thus, all the coupling methods outlined earlier can be used in solution synthesis, such as symmetric or mixed anhydrides (using carbodiimides) or the formation of active esters by the use of either carbodiimides (DCC (**33**), DIC (**39**), and EDC (**40**)) in the presence of additive reagents (HOBt (**41**), HOAt (**42**), 6-Cl-HOBt (**43**), Oxyma (**44**)), acid halides (acid chlorides or acid fluorides), or phosphonium (BOP (**47**), PyCloP (**48**), PyBroP (**49**), PyBOP (**50**)) or aminium/uronium (HBTU (**55**), HATU (**56**), COMU (**61**)) salts in the presence of a base. The most preferred coupling mixtures among them are the EDC (**40**)/HOBt (**41**), or HOAt (**42**) and the use of phosphonium or uronium salts in the presence of additives (HOBt (**41**), HOAt (**42**), or Oxyma (**44**)). The EDC (**40**) is a water soluble carbodiimide and the corresponding urea that is formed during reaction can be easily removed by washing of the organic phase with acidic water [194].

## 2.8 HYBRID SYNTHESIS–COMBINATION OF SOLID AND SOLUTION SYNTHESIS

Pharmaceutical companies prefer the manufacturing of medium-large-sized peptides (<25 amino acids) and proteins by means of a hybrid strategy. Hybrid synthesis is the combination of the well-established methods of solution and SPPS. Specifically, in this approach, the desired peptide is obtained after condensation in solution of two or more appropriate peptide fragments prepared mainly through solid phase synthesis. Assembling of two fragments can be performed either by the classical condensation method using coupling reagents or by the most modern technique of native chemical ligation. In the first approach, the coupling of protected peptide fragments is assembled in typical organic solvents, followed by global deprotection of side chain protected groups (Figure 2.15).

In the native chemical ligation method, the condensation of the fragments is carried out in water after their full deprotection, where the N-fragment is modified as thioester at the C-terminal and the C-fragment to include a Cys residue at the N-terminal. Both approaches are valuable for the preparation of complex and long peptides and proteins and combine the benefits of both solid- and solution-phase strategies. The number and time of reactions are decreased when compared to the step-by-step solution synthesis. The yields and the purities of the intermediate fragments are often higher as they are commonly prepared by SPPS with the use of reagent excess during coupling reactions. Intermediate protected fragments can be purified if required. Furthermore, benefiting from one of the advantages of solution phase synthesis, the fragments can be fully characterized before condensation, leading to easier control and optimization

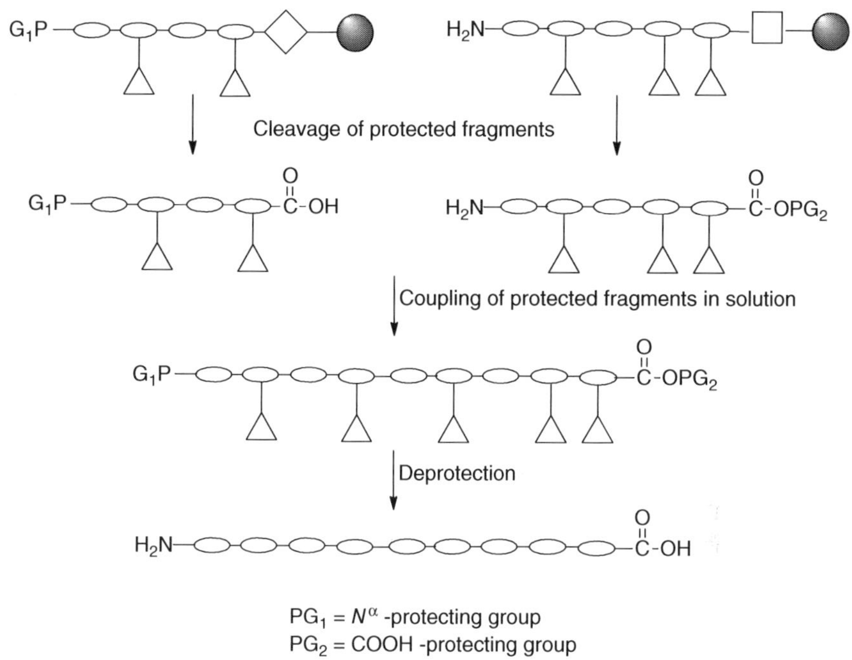


**Figure 2.15** General procedure representation of a hybrid approach using the classical segment condensation method.

of the coupling reaction of the fragments. Even if coupling of peptide fragments is not completed, this kind of impurities is usually easily separated. Therefore, the purification step in the hybrid approach can be more efficient when compared to SPPS for the projected peptide.

### 2.8.1 Classical Segment Condensation

The hybrid method using classical segment condensation became available when new solid supports and resin handles provided peptide chemists with the armory to synthesize fully protected peptide fragments, such as 2-CTC (**6**) and Sieber amide (**7**) resins. An important consideration during the development of the segment condensation approach is the evaluation and design of the appropriate fragments, taking into account their preparation first and second, their condensation. Thus, the solubility of the fragments in condensation reactions has to be taken into account (an estimation or a prediction can be made but it is of no assurance without actual preparation). Furthermore, the potential epimerization during the condensation reaction is a common risk that should be avoided or reduced. Thus, Gly or Pro as C-terminal residues of the fragments is the priority if possible, in order to totally avoid racemization. If this is not possible, then Ala or Leu should be the preferred option as they are less prone to racemization [195]. Evaluation of the linker and the solid support is another crucial point to take into consideration. Side-chain-protecting groups have to be stable into the cleavage conditions of the peptide fragments of the projected peptide. Thus, Fmoc/*t*Bu chemistry is commonly preferred, in combination with super acid-sensitive linkers/resins. Their development and commercial availability has allowed synthetic access to even longer peptide sequences. In the hybrid approach, the condensation of the protected fragments is carried out using the classical coupling reagents, which are described in Chapter 4 and Section 6.3.

### 2.8.2 Native Chemical Ligation

An alternative modern approach that can be applied to the assembly of peptide or protein fragments is native chemical ligation [196]. Two peptide fragments, one containing a C-terminal thioester and the other containing an N-terminal cysteine residue, are assembled by chemoselective coupling. Native chemical ligation is a reversible transthioesterification, followed by amide formation. Intramolecular nucleophilic attack of an α-amino group on the initial thioester product occurs only when the thiol is on the side chain of an N-terminal Cys, thus regenerating the thiol functional group of the Cys side chain and giving a final ligation product containing a native peptide bond at the site of ligation (Scheme 2.7). Initial thiol–thioester exchange step is fully reversible, whereas the second amide-forming step is irreversible under the reaction conditions. Because of this, eventually, only the desired amide-containing product is formed, even in the presence of internal Cys residues in either peptide segment. This reversible-irreversible two step reaction mechanism is the essence of the native chemical ligation method.

The main benefits of the native ligation strategy are the reduced solubility problems that commonly appear in the fragment-condensation strategy, and the absence of reagents that should be purified. In this approach the side chain protecting groups of the fragments are removed before fragment ligation. The reaction takes place in an aqueous environment in neutral pH, in order to give a native peptide bond at the ligation point. Limitation of this hybrid technology is the mandatory use of a Cys residue at the N-terminal of the first fragment, and the synthesis of an appropriate C-terminal thioester in the second fragment, which in some cases provides poor yields [197]. Advances in the field include, for example, the use of conformationally assisted ligation [198], removable auxiliaries [199, 200], Staudinger

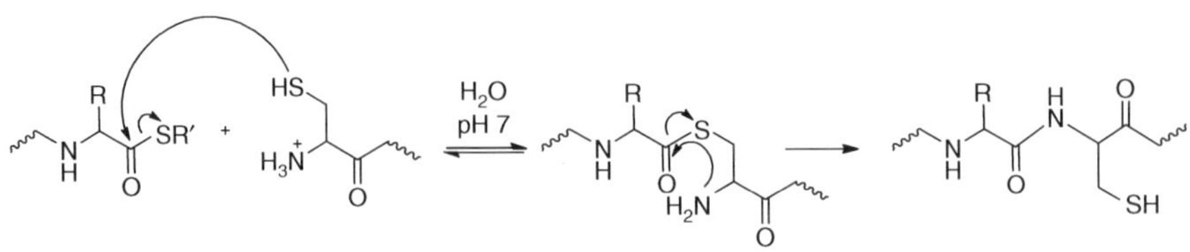


**Scheme 2.7** Principles of chemical native ligation. Side chain protection of the fragments is not essential, but favors the reaction in aqueous media.

ligation [201–203], thiolalkylation [204], desulfurization methods [205–208], and sugar-assisted ligation (SAL) [209, 210]. More interestingly, a side-chain-assisted chemical ligation has been reported lately, with no limits to the assembled amino acids [211]. Continuous improvement of ligation strategies provides an additional tool to peptide chemists to overcome more immediate challenges in view of the significantly increased demand for larger peptides.

## 2.9 CYCLIC PEPTIDES

Synthesis of cyclic peptides has been the subject of numerous studies since the introduction of SPPS due to their potential as therapeutic agents [212–214]. Cyclization offers many advantages, including increased proteolytic resistance and also enhanced biological activity, when compared to their linear counterparts in some cases [215, 216]. Cyclic peptides can be classified depending on their linkage: (a) *head-to-tail* type, when the N- and C-terminus are joined; (b) *side-chain-to-*C*-terminal* or N*-terminal-to-side-chain*, when a side-chain is linked to the C- or N-terminus; and (c) *side-chain-to-side-chain*, when two side-chains are joined. Linkage is usually an amide bond but can also be a disulfide or another type of functionality (Figure 2.16).

Cyclization can be performed either in solution or in solid-phase [214, 217–219]. Cyclizations in solution have the limitation that they must be carried out under high dilution conditions because of the risk of dimerization and oligomerization, and thus, high volumes of solvents are consumed. In contrast, in solid phase, and due to site isolation, intramolecular cyclization is favored. Furthermore, work-ups can be avoided since coupling reagents can be washed away, thereby releasing the final cyclized peptide in good purity.

To obtain the most common *head-to-tail* cyclic peptides in the solid-phase, the *N*- and *C*-terminus must be free and not anchored to the resin. The following two strategies are commonly used: (i) the side-chain anchoring approach and (ii) the BAL.

Side-chain anchoring approach: Here, the amino acid side-chain is linked to the solid support and the C- and N-terminus are orthogonally protected. Once chain elongation is finished, the deprotection of both ends and subsequent cyclization and cleavage delivers the final cyclized product. Numerous amino acids have been used for side-chain anchoring, including Asx/Glx [220–226], Lys/Orn [220, 227], Ser/Thr [220, 228], Tyr [220, 224, 229], His [230, 231], and Cys [232, 233], on the usual supports and linkers for peptide synthesis.

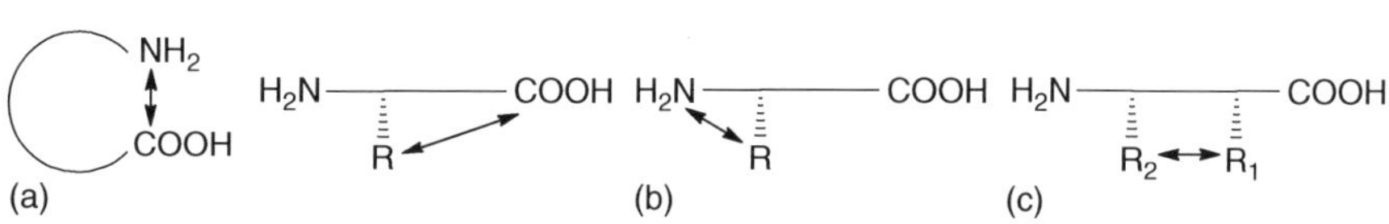


**Figure 2.16** (a) *Head-to-tail* cyclic peptides; (b) *side-chain-to-*C*-terminal* and N*-terminal-to-side-chain* cyclic peptides; and (c) *side-chain-to-side-chain* cyclic peptides.

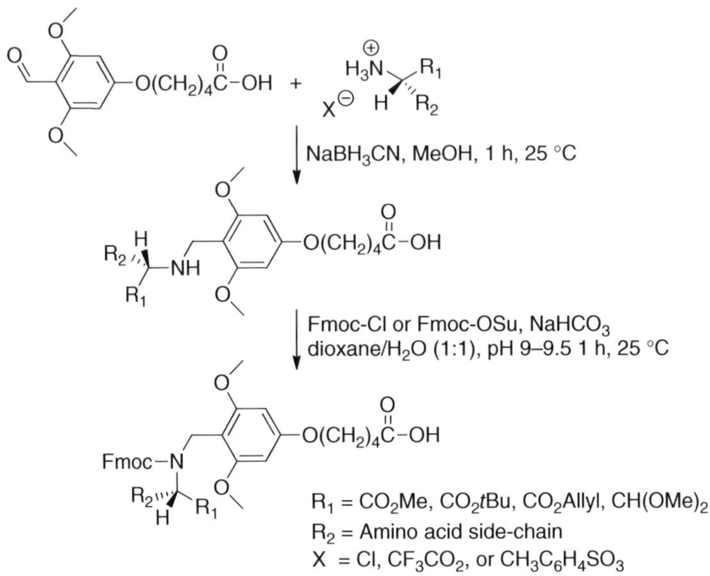


**Scheme 2.8** Synthesis of the preformed BAL linker derivative. *Source*: Adapted from Reference 234.

BAL anchoring: The original concept of the BAL linker [234, 235] involves the attachment of a backbone amide nitrogen to an appropriate handle. Since linkage does not rely on the C-terminal carboxyl, the BAL linker allows the preparation of a large variety of C-terminal modified peptides [236], including cyclic peptides. The starting point is the 5-(4-formyl-3,5-dimethoxyphenoxy)valeric acid (PALdehyde) (Scheme 2.4), an intermediate in the synthesis of the PAL linker. Reductive amination of the PALdehyde with the amine of the C-terminal residue or a salt of it, with sodium cyanoborohydride provided the corresponding secondary amine intermediate, which could be protected as Fmoc derivative and later coupled to the resin (Scheme 2.8).

In a more general approach, the preformed handle can be coupled first to the solid phase, and the C-terminal protected amino acid introduced by reductive amination. For the synthesis of cyclic peptides, the allyl ester of the corresponding amino acid is commonly used. After elongation of the peptide chain by standard Fmoc chemistry, the cyclic peptide is constructed as described above for the side-chain anchoring approach.

Recently, other linkers based on the same concept have been developed [237].

## 2.10 DEPSIPEPTIDES

Natural depsipeptides of marine and terrestrial origin have been frequently found to be potential leads for the pharmaceutical industry [238]. Over the past decade

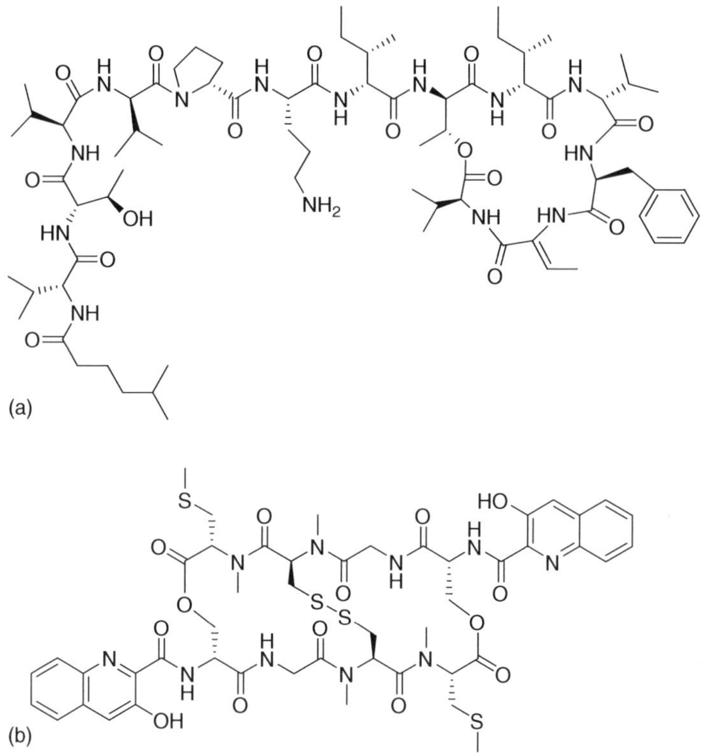

**Figure 2.17** Structures of (a) Kahalalide F and (b) Oxathiocoraline.

solid-phase methodology has been applied to the synthesis of depsipeptides [239]. A major milestone was the synthesis of the antitumoral depsipeptide Kahalalide F [240] (Figure 2.17a), which is active against a wide range of tumor types and currently in clinical trials. Assembly of Kahalalide F was performed on 2-CTC resin (**6**) with Fmoc-D-Val-OH as a first amino acid. After synthesizing the linear tetrapeptide Fmoc-D-aIle-D-aThr(OH)-D-aIle-D-Val-O-CTC, the Thr side-chain was esterified with Alloc-Val-OH. The linear chain was then elongated until the introduction of 5-methylhexanoic acid and then the branch finished by introducing dipeptide Alloc-Phe-Z-didehydroaminobutyric acid (Dhb), previously prepared in solution. After releasing the branched peptide from the resin, the final cyclization was performed in solution, giving the depsipeptide in good yields. Using this strategy a large number of analogs could be obtained [241]. The most important limitation in the synthesis of depsipeptides by the Fmoc/*t*Bu strategy is the potential liability of the ester bond in response to piperidine exposure, and to the formation of DKP after the deprotection of the second amino acid after the ester bond. Oxathiocoraline [242, 243] (Figure 2.17b), for example, presented a severe case of DKP formation due to the sequence *N*Me-Cys(Acm)-*N*Me-Cys(Me), which follows the ester bond. *N*Me increases the presence of cis-configuration in the dipeptide, thereby making it prone to DKP formation. This DKP formed so rapidly that it could not be overcome by regular means. By taking advantage of the symmetry of the molecule, an intermolecular dimer was constructed in solid phase, restricting the mobility of the peptide chain and thereby effectively preventing DKP formation by making the amino function of the *N*Me-Cys unable to reach the carboxyl of the *N*Me-Val. The same strategy was also successfully applied to the synthesis of the natural product Triostin A [244]. In the synthesis of cotransin [245], DKP formation could be reduced by a flash treatment using the stronger 1,8-Diazabicyclo[5.4.0]undec-7-ene (DBU) bases (10% DBU solution in DMF) or tetrabutyl ammonium fluoride (TBAF) (0.15 M in DMF).

For depsipeptides with simpler architectures, a protocol that allows automatic synthesis of the linear chain by Boc chemistry was developed [246]. The following modifications are introduced in the preprogrammed modules for Boc chemistry of a regular automatic peptide synthesizer: (i) The α-hydroxy acid is coupled unprotected by preactivation with HOBt (**41**). (ii) The O-acylation is performed with DCC (**33**) and 4-(*N*,*N*-dimethylamino)pyridine (DMAP), without the presence of HOBt (**41**); when two α-hydroxy acids are introduced consecutively, and the second is tetrahydropyranyl (THP)-protected. (iii) Following O-acylation, a capping step ($Ac_2O$/DIEA) is not performed in order to prevent acyl transfer. (iv) the THP-protecting group is removed under the same conditions used for the Boc group. The remaining couplings are performed using *in situ* neutralization protocols and capping steps.

## 2.11 SEPARATION AND PURIFICATION OF PEPTIDES

The great variability of peptides in terms of size, shape/conformation and charge makes it impossible to use one standard procedure for peptide purification, mostly when looking at an optimized method. Additionally, potential impurities can occur during peptide assembly in SPPS even though great improvements have been achieved in peptide chemistry. The nature of impurities, apart from chemical failures, may be generated to diastereomers, hydrolysis products of labile amide bonds, deletion sequences formed predominantly during SPPS and insertion peptides, and by-products formed during the removal of protection groups in the final step of the synthesis. Polymeric forms of the desired peptide, especially in cyclic peptides, are also known. Much of what applies to the purification of synthetic peptides from synthetic procedure is also relevant to peptide isolation from complex biological material. Consequently, in any case, the purification of a crude peptide to homogeneity can hold great challenges; however it is an absolute prerequisite for successful and meaningful structural and functional analyses. In this context, it should be mentioned that all potential impurities probably could not be removed by a single chromatographic method, but rather by a combination of methods.

This section will deal with separation and purification of peptides. The most commonly used chromatographic methods for separation and purification of peptides are gel-filtration chromatography, ion-exchange chromatography (IEC), and reverse-phase high performance liquid chromatography (RP-HPLC), and will be further discussed.

### 2.11.1 Gel-Filtration Chromatography

Gel-filtration chromatography is a method for separating peptides and proteins based on their size [247]. This method is also known as gel permeation, molecular sieve, gel-exclusion, and size-exclusion chromatography. In this method, the chromatographic matrix consists of porous beads, and their size defines the size of macromolecules that may be fractionated. Proteins and/or peptides that are too large to enter the bead pores have less volume to pass through, and consequently elute first from the column. Sequentially, smaller macromolecules that can insert some, but not all, of the pores are retained slightly longer in the matrix and exit the column next. Finally, small molecules filter through most of the pores, and they emerge from the column with an even larger elution volume. The main advantage of this method is that no binding of the sample is required, no harsh elution conditions are used, and thus, it rarely inactivates enzymes, and often is used as a significant step in peptide or protein purification. In practice, the method is simple and no sophisticated equipment is required. A limitation of the method is the slow separation of the macromolecules, mainly because the column should be narrow and long in order to achieve sufficient component separation. The whole process can be accelerated by the use of pumps or more sophisticated high pressure chromatography equipment and/or matrices allowing faster flow rates, whereas this can hold higher risk for sufficient separation. Another crucial limitation is poor resolution given that in this method the peptide/protein does not bind to the matrix. Thus, it is important to select the proper matrix in terms of pore size. For that reason, gel-filtration chromatography is used mainly as a separation tool when only a small amount of contaminants is present. In many cases, it is used for sample desalting or for changing the buffer of the sample [248].

### 2.11.2 Ion-Exchange Chromatography

Separation of peptides and proteins in IEC is based on the net charge of the macromolecule [249]. In this method, selecting the appropriate ion exchange matrix is significant for sample separation. An anion-exchange matrix is derivatized with positively charged groups, whereas cation exchange contains negatively charged groups. Generally, weekly acidic or weekly basic groups are preferred to achieve binding of peptides with low affinity and thus, interactions can be disrupted without the use of harsh conditions. Most anion-exchange matrices are substituted with a diethylamino ethyl (DEAE) group or a quaternary amine (Mono-Q) while cation-exchange contains a carboxymethyl (CM) group or usually a sulfomethyl group (Mono-S). In an anion exchange procedure, the anion-exchange matrix is initially positively charged in equilibrium with the negatively charged counterion. For DEAE-matrix the counterion is normally $Cl^-$ and for CM-matrix the counterion is usually $Na^+$. The pH of the starting buffer is crucial because it determines the charge on the peptides that are to be separated. The starting buffer pH should be at least one pH unit above or below the pI of the peptide that is to be bound to the matrix to ensure adequate binding. Peptides that are oppositely charged to the matrix at the starting pH will bind to it, so displacing the counterions, whereas peptides with the same charge as the matrix or with no net charge will not bind and consequently elute first. To elute the peptide of interest a higher concentration of counterion is added to the column. The different peptides bound to the matrix have different affinities for the ion exchanger due to differences in their net charge. These affinities can be altered by varying either the pH or the ionic strength of the column buffer and can provide a very sensitive method for peptide/protein separation on the basis of charge. A high level of purification could be achieved with minimal loss of sample if the matrix and column buffer are carefully selected.

### 2.11.3 Reverse-Phase High Performance Liquid Chromatography

The reasons why reverse-phase (RP) HPLC became such a widely used and well-established tool in separation and purification of peptides are resolution, exquisite sensitivity, and speed of process [250]. In terms of resolution, RP-HPLC is able to separate peptides of nearly identical sequences of small and/or even larger proteins. Separation mechanism of small molecules involves continuous partitioning of the molecules between the mobile phase and the hydrophobic stationary phase. However, peptides that are too large to partition to the hydrophobic phase are absorbed into the hydrophobic surface and remain there until the concentration of organic modifier reaches the critical concentration needed to cause desorption and elution from the column [251]. The sensitivity of peptide/protein desorption to specific concentrations of the organic modifier accounts for the high selectivity of RP-HPLC method in separation of peptides.

The most favored solvent used as an organic modifier to elute peptides is acetonitrile (ACN), since it is volatile and easily removed from collected fractions, has low viscosity and hence minimizes column back – pressure, is totally miscible with water and has low UV absorption at low wavelengths (UV cutoff: 188 nm). Other possible solvents are isopropyl alcohol and methanol and mixtures of these solvents with ACN. Besides the requirement for an organic solvent to be used as a surface tension modifier, ion-pair reagents [252] (trifluoroacitc acid, TFA; heptafluorobutyric acid, HFBA) are utilized at low pH (e.g., pH 2.1) to suppress interactions between free silanol groups on the silica surface and basic amino-acid residues. In some cases and in order to achieve sharper peaks, triethylamine is added to suppress those interactions. Presence of ion-pair reagents greatly influences the retention time of peptides [253, 254]. In RP-HPLC, isocratic elution, step elution, or gradient elution modes can be utilized to purify peptides and proteins. The ideal gradient system should be easy to operate, provide consistent retention times, sharp peaks, and a rapid turnaround time to initial eluent conditions for fast throughput from analysis to analysis [255]. The most preferable for separation of peptides is the gradient elution. Increasing the concentration of the organic solvent as peptides elutes results in sharper peaks and better resolution. Typical changes in organic solvent concentration (gradient slope) are on the order of 0.5–2% change/min. However, shallow gradients with slopes of <0.5% have proven to be very effective in separating complex mixtures of peptides [256].

Regarding the chromatographic stationary phase, there are several chemical and physical factors (ligand composition, ligand density, surface heterogeneity, surface area, pore diameter, pore diameter distribution, particle size, particle size distribution, and particle compressibility) that contribute to the variation in the resolution and the recovery of peptides and proteins in HPLC systems, and thus by means of careful selection of stationary phase a high level of purification could be possible [257]. Stationary phases typically used in RP-HPLC for peptide separation are silica-based packing materials of 3–10 μm average particle diameter and ≥300 Å pore size, with *n*-butyl, *n*-octyl, or *n*-octadecyl ligands.

Other factors that can influence separation of polypeptides in RP-HPLC are pH, temperature and flow rate of mobile phase. Eluent pH can be a useful tool in optimizing peptide separations as protonation or deprotonation of acidic or basic side chains of peptides influence their retention times. Changes of temperature strongly affect separation of peptides and for that reason should be optimized in any method for best separation [258]. The temperature of the stationary phase and eluents should be constant (±0.1 °C) using a thermostatically controlled system in order to ensure reproducibility in the resolution [259]. The flow rate of the mobile phase slightly affects peptide separation because, as previously mentioned, peptide desorption is the result of reaching a precise organic modifier concentration. However, it should be noted that when refining a separation process of small peptides where resolution is limited, slight improvements may be gained with minor changes in the flow rate of the mobile phase. Flow rate affects other aspects in separation such as detector sensitivity and column back pressure.

The peptide bond absorbs strongly in the far-UV region of the spectrum ($\lambda = 205-215$ nm). Therefore, UV detection (in this wavelength range) is the most widely used method for detection of peptides and proteins in HPLC. Furthermore, in some cases, it is possible to use longer wavelengths to detect the presence of Phe (257 nm), Trp (280), and Tyr (274 nm) and also to some extent cysteine absorbs light above 250 nm [260].

All these important factors ultimately determine the selection of the optimal separation conditions or the resolution of peptide and protein mixtures.

## 2.12 CHARACTERIZATION OF PEPTIDES THROUGH MASS SPECTROMETRY

Mass spectrometry (MS) is a technology used to measure the molecular mass (MM) ($M$) of a molecule. Actually, MS measures the mass-to-charge ratio of molecules, ($m/z$; $(M + z)/z$, where $M$ is the uncharged mass of analyte). MS is also a tool used to perform structural elucidation in tandem (MS/MS) experiments, in which molecules are fragmented inside of the instrument and the products formed are analyzed. Molecular structure can also be evaluated by ion mobility, which is a quite recent improvement in MS. Ions having the same MM, but different conformational structures, can be separated by collisions with a bath gas in a uniform electric field [261]. MS can also be a useful tool to quantify molecules. However, it is not the main application of this technology and some particular methodology has to be used depending on the type of quantification and the type of molecules one wants to analyze, as we will explain below.

To carry out any type of MS analysis, the first requirement is to transfer the molecules to a gas-phase and ionize them with a positive or negative charge. Once ionized, all ions are submitted to electrical or magnetic fields that guide them to the mass analyser, where separation between the different ions occurs. Thus, ions reach the detector according to their mass-to-charge ratios. In general, a mass spectrometer consists of a sample inlet, an ionization source, one or more mass analysers, a detector (the two or three last ones are under high vacuum, depending on the ion source, which can be under vacuum or at atmospheric pressure) and one data system (Figure 2.18). The ionization source and the analyser are the main parts of the equipment, and they define the characteristics of the mass spectrometer. We discuss below some of the different MS/MS instrument configurations (with different capabilities in terms of ionization method, mass-to-charge range, resolution and sensitivity) that have been developed in the last years.

### 2.12.1 Ionization Source

For peptides and proteins studies, ionization can be done in a matrix-assisted laser desorption/ionization (MALDI) source [262] or in an electrospray ionization (ESI) [263] source. Both are soft ionization techniques allowing observation of intact molecules. Introduction of soft ionization techniques to *volatilize* biomolecules, such as ESI and MALDI was the main improvement that enabled analysis of protein and peptide structure by MS and MS/MS [264]. Ionization in electrospray sources occur by passing a solubilized sample through a high voltage needle at atmospheric pressure [265]. In this step very small charged droplets are produced and are immediately evaporated helped by high temperature in the sample cone. Desolvation and ionization processes occur prior to the entrance into the high vacuum of the mass spectrometer. ESI typically induces a range of charge states and the resulting spectra can have many ions for each analyte. ESI is the most commonly used approach to couple a liquid chromatography system to a mass spectrometer allowing analysis of complex samples.

In MALDI, samples are co-crystallized, typically onto a metallic sample plate, with an organic matrix compound that usually has a conjugated aromatic ring structure. It is thought that laser energy is absorbed by the matrix molecules that

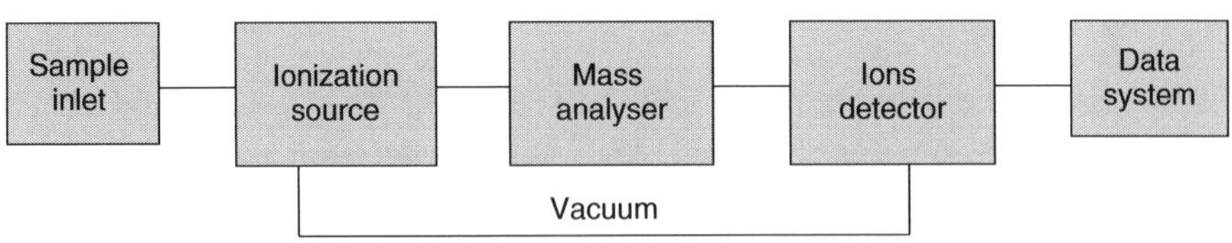


**Figure 2.18** Schematic representation of a mass spectrometer.

transfer the charge to the analyte [266–268]. Although multiple charged ions can be produced, mostly single-charged ions are observed in MALDI spectra. The matrix is chosen according to the analyte. Peptides usually co-crystallize efficiently with α-cyano-4-hydroxycinnamic acid (CHCA) matrix; most of the ions observed are 600–3500 Da. For bigger peptides, the matrix used is sinapinic acid (SA), larger ions give better signals with this matrix, although the efficiency of ionization depends on the laser. 2,5-dihydroxybenzoic acid (DHB) is a matrix of choice when analyzing small molecules, although the background or the signals of matrix are still a problem not solved in the MS analysis of small molecules. Use of more than one matrix is recommended to confirm the results when the MM of the analyte match the matrix signals. Moreover, the improvement of peptide signals has been observed by the combination of two matrixes in the same MS assay [269].

### 2.12.2 Mass Analysers

Analyser is the component that defines some aspects of the mass equipment such as the accuracy to measure the molecular weight of the analyte. Moreover, the analyser defines the mass-to-charge ratio that can be measured and the ability to run tandem mass experiments (MS/MS), that is, fragment peptides to obtain information about their amino-acid sequence. The choice of the best mass instrument to achieve expected results depends on the analyte and the experiment one wants to perform. One point to take into account is the resolution needed to separate neighbor mass. The higher the resolution the better separation between close $m/z$. Measurement of exact MM is an analytical example of the need for a very high resolution mass instrument. High resolution and mass accuracy are closely related concepts because the achievement of an accurate mass depends on the ability of the mass instrument to resolve close neighboring masses. Nevertheless, they should not be confused because a high resolution mass measurement alone does not imply an accurate mass measure. Resolution can be measured by different ways, although peak width definition is one of the most widely used. Resolution ($R$) is defined as the ratio of the mass ($m$) to the difference in mass ($\Delta m$) ($R = m/\Delta\mathrm{m}$), defined by the width of a signal at 50% of the peak height (full width at half maximum, FWHM) [270] (Figure 2.19).

Amino acids are mainly composed of four elements, carbon, hydrogen, nitrogen, and oxygen, which exist naturally as a mixture of isotopes. For instance, carbon is a mixture of $^{12}C$ and $^{13}C$ isotopes in an abundance of 98.9% and 1.1%, respectively. Abundance of the isotopes is reflected in the composition and MM of the molecules. It is reflected in the mass spectrum by the combination of an isotopic mixture of the compound. There are two types of mass measurement for a given compound: average mass and monoisotopic mass. Average mass reflects the contribution of all isotopes in the compound. In the mass spectrum, it is taken at the centroid of the isotope mixture (Figure 2.20a). Monoisotopic mass is the mass of the first peak in the isotope mixture (Figure 2.20b). Monoisotopic mass can only be measured if the $^{12}C$ and $^{13}C$ isotopes of the peptide mixture can be sufficiently resolved, that is, if the mass analyser has enough resolution to separate the isotopes, that is, the 1 Da of difference in mass between them.

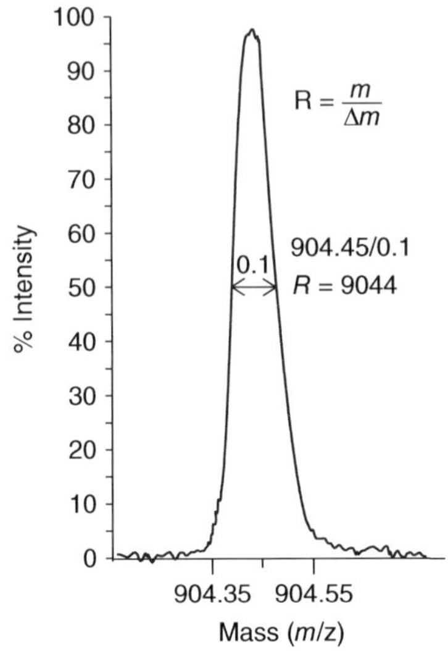


**Figure 2.19** Scheme of resolution mass calculation based on the FWHM.

The most commonly used analysers in biomolecule studies are quadrupoles (Q), time-of-flight (TOF), and quadrupole ion traps (ITs). Fourier transform ion cyclotron resonance (FTICR) is not very common, but deserves mention because of its high resolution (for reviews, see References 271, 272).

*Quadrupole mass analysers* are one of the most common mass analysers. Separation of ions is efficiently done by applying radio frequency (RF) and DC voltages; only a narrow mass-to-charge range is allowed to cross the quadrupole (four parallel metal rods) to reach the detector [273]. On the contrary, quadrupole resolution and mass range are usually limited. Most of the commercially available instruments usually have a range of 0–4000 *m/z*; however, there is already a commercial mass instrument with amplified mass range to 32,000 *m/z* [274]. ESI ionization provides multiple charge ion formation, thus enabling quadrupole mass measurement of intact proteins.

*TOF* is one of the simplest mass analysers. It measures the *m/z* ratio of an ion by measuring the time required for such ion to cross the length of a field free tube. Ions are accelerated to the tube entrance using a short-voltage gradient. Flight time is proportional to the square root of the *m/z*. TOF resolution can be improved by ion manipulations (delayed extraction of ions from the source, two stage sources with complex voltage gradients, and reflector technology). This last one consists of including an ion mirror at the end of the flight tube, which reflects ions back through the flight tube to the detector. The ion mirror increases the length of the flight tube and also corrects for small energy differences among ions [268]. Commercial TOF instruments can achieve a resolution of 10,000 or greater.

*Ion traps* are the mass analysers allowing MS/MS experiments. Ion separation occurs by focusing ions into a small volume with an oscillating electric field; ions

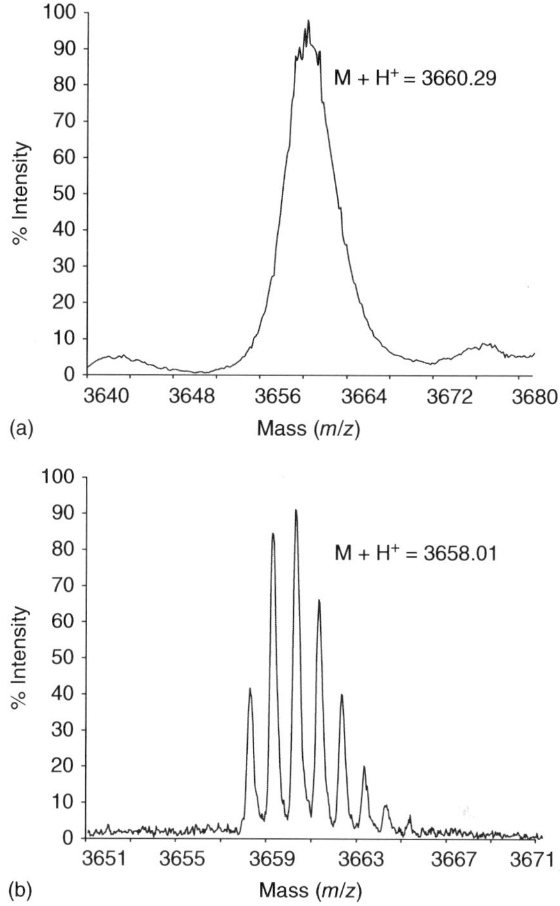


**Figure 2.20** MALDI-TOF spectra of Adrenocorticotropic Hormone (ACTH) (7-38) peptide acquired in a 4700 Proteomics Analyser (applied biosystems). (a) Average mass in linear positive acquisition mode at resolution 650. (b) Monoisotopic mass in reflector positive acquisition mode at resolution of 10,700 (FWHM). Theoretical average mass: $M + H^+ = 3660.19$ and theoretical monoisotopic mass: $M + H^+ = 3657.93$.

are resonantly activated and ejected by electronic manipulation of this field [275, 276]. Ion traps are very sensitive, because they can concentrate ions in the trapping field for varying lengths of time. Resolution limitation was recently improved by the development of linear ion traps (LITs) [277, 278].

The highest resolution mass analyser is the FTICR. It is an instrument that can determine exact mass of small molecules. Ion separation is done using high magnetic fields to trap the ions and cyclotron resonance to detect and excite the ions. Resolution is around 1,000,000; that is, $m/z$ 1000.000 can be separated from $m/z$ 1000.001 [279, 280].

*Tandem mass spectrometry* (MS/MS or $MS^2$) is the tool applied for structural determinations of organic compounds, peptides, and oligonucleotides. A tandem mass spectrometer configuration is necessary to perform MS/MS experiments because specific ions have to be selected first in the MS or $MS^1$ step. Once selected, such specific ions are fragmented and the mass of the product ions has to be measured in a second step (MS/MS or $MS^2$). Selected ions are named parent ions and the fragments or product ions are named daughter ions. Some tandem instrument configuration examples are

*Triple quadrupole mass spectrometers, QqQ.* In the first stage of analysis, the machine is operated in MS scan mode and all ions within a determined *m/z* ratio are transmitted to the third quadrupole for mass analysis. In the second stage, the mass spectrometer is operated in MS/MS mode and a particular peptide ion is selectively passed into the collision chamber (quadrupole 2, q). Inside the collision chamber, peptide ions are fragmented by interactions with an inert gas by a process known as collision-induced dissociation (CID) or collisionally activated dissociation (CAD). The peptide ion fragments are then resolved on the basis of their *m/z* ratio by the third quadrupole [265, 281].

*Quadrupole-TOF, QqTOF.* In the last years, several *hybrid* mass spectrometers have emerged from the combination of different ionization sources with different mass analysers. One example is the quadrupole-TOF mass spectrometer [282–285]. In this machine, the first quadrupole (Q1) and the quadrupole collision cell (*q*) of a triple-quadrupole machine have been combined with a TOF analyser. The main applications of a QqTOF mass spectrometer are protein identification by amino-acid sequencing and the characterization of protein modifications [286].

*MALDI-TOF/TOF.* The principal application of a MALDI-TOF is the determination of MM. Nevertheless, the improvement of MALDI-TOF machines by combining two *TOF* tubes, TOF/TOF, enabled this mass instrument to be used in structural elucidation [287]. The collision cell is located between the two TOF tubes. Similar to fragmentations in triple-quadrupoles, in a first step of analysis all the ions are transmitted to the second TOF in a scan mode mass analysis (MS). In the MS/MS mode, a determined *m/z* ion is selected by a time-ion-selector that allows only a narrow range of *m/z* ions to enter in the collision-cell where they are fragmented. Product ions are transmitted to the second TOF where they are resolved according to their *m/z* ratio [288].

*Ion traps* are the only mass spectrometer allowing $MS^n$ experiments without any other mass analyser. Different electronic modes are applied to trap, to select, to fragment, and to measure mass of product ions, all this analysis taking place in the same IT. Moreover, in an alternative scan mode ($MS^3$), the ion trap has the additional functionality of selecting one specific product ion and then inducing a subfragmentation spectrum. This approach can be extended to multiple isolation and fragmentation stages ($MS^n$).

*Fourier transform ion cyclotron resonance-ion trap, IT-FTICR.* This machine combines the advantage of ion traps, such as the good sensitivity and ability to perform experiments of $MS^n$, with the high resolution of FTICR mass analysers and the possibility of analysis of entire proteins, since the FTICR solve the limitation in mass range of ion traps [280]. More recently, the incorporation of high energy alternative fragmentation methods, such as electron capture dissociation (ECD) [289], has made this mass instrument the best alternative to analyze and fragment intact proteins [290].

### 2.12.3 Peptide Fragmentation

Peptide sequencing by *MS* involves the fragmentation of a peptide to produce smaller products. Fragmentation of the peptide backbone mainly occurs at the peptide amide bound. Nevertheless, peptides can fragment in different sites, multiple fragmentation of backbone and/or side chain can occur at the same time. Because of this, a nomenclature was created to indicate what type of ions are generated during the MS/MS analysis [291, 292]. Low energy collision activation method (CID) fragments peptide bonds and produces *b*/*y* ion pairs for single- and multiple-charged parent ions. Ions are named *b*-ions if the amino terminal fragment retains the charge, or *y*-ions, if the carboxy-terminal fragment retains the charge (Figure 2.21). On the contrary, ECD or electron transfer dissociation (ETD) are higher energy activation methods, producing *c*/*z* ion pairs with cleavage along the peptide backbone for multiple-charged precursor ions (>2 charges). Combination of these two types of peptide fragmentation improves the quality of peptide sequencing [293]. Mass values of fragment ions can be assembled to produce the original amino-acidic sequence, that is, differences in mass between two adjacent *b*- or *y*-ions should correspond to that of an amino acid (Figure 2.22). Identification of the amino acid and the peptide sequence can be done using MS/MS, with the exception of isoleucine and leucine, which are identical in mass, and then indistinguishable. Additional fragmentation along amino-acid side chains can be used to distinguish isoleucine and leucine [294]. Both *y*- and *b*-ions can suffer neutral loss of $NH_3$ (−17 Da), $H_2O$ (−18 Da), and CO (−28 Da), resulting in pairs of signals observed in the mass spectrum.

Immonium ions (*a1* ion in Figure 2.22) or fragments of immonium ions from individual amino acid residues in a peptide can also be detected. They are more frequently

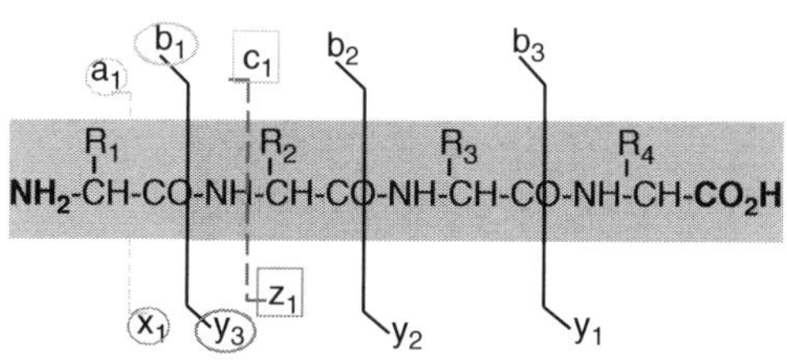


**Figure 2.21** Peptide fragmentation pattern and fragments nomenclature.

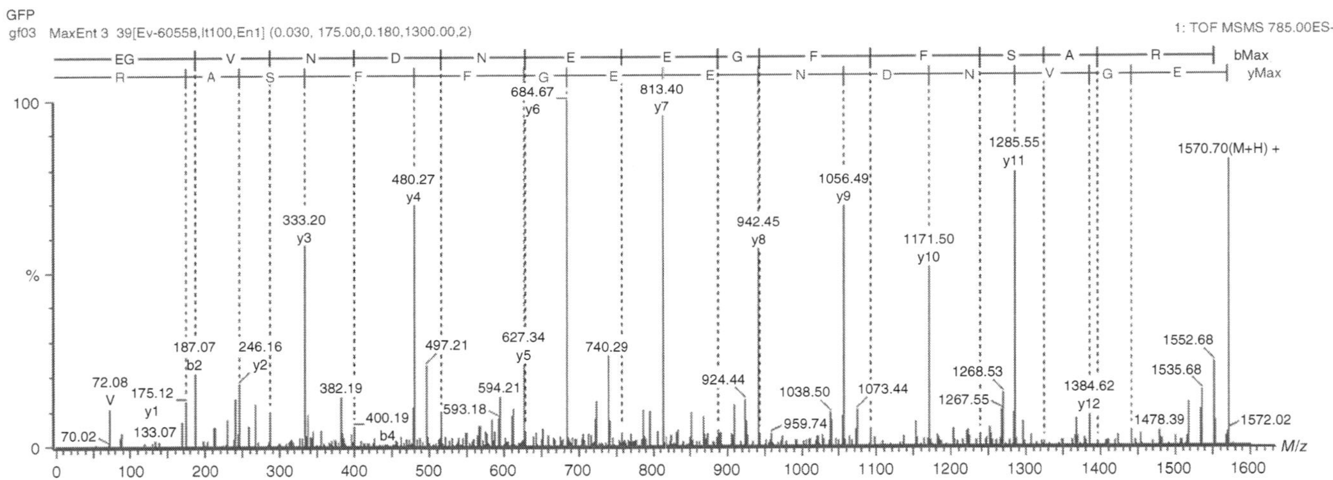


**Figure 2.22** Amino-acid sequence assigned to MSMS spectrum of Glu-fibrinogen peptide (theoretical amino acid sequence: EGVNDNEEGFFSAR; theoretical monoisotopic mass, $M + H^{+} = 1570.677$ Da; acquired in a nano-ESI-Q-TOF ultima (waters/micromass). (*See insert for color representation of this figure.*)

**TABLE 2.1 Immonium Ions Mass.**

| Amino Acid (Symbols) | Immonium Ion Mass (Related Ions) |
|---|---|
| Alanine (A) | 44 |
| Arginine (R) | 129 (112[a], 100, 87[a], 73, 70[a], 59) |
| Asparagine (N) | 87[a] (70) |
| Aspartic acid (D) | 88[a] |
| Cysteine (C) | 76 |
| Glutamic acid (E) | 102[a] |
| Glutamine (Q) | 101[a] (84[a], 129) |
| Glycine (G) | 30 |
| Histidine (H) | 110[a] (166, 138, 123, 121, 82) |
| Isoleucine (I) | 86[a] (72) |
| Leucine (L) | 86[a] (72) |
| Lysine (K) | 101[a] (129, 112, 84[a], 70) |
| Methionine (M) | 104[a] (61) |
| Phenylalanine (F) | 120[a] (91) |
| Proline (P) | 70[a] |
| Serine (S) | 60[a] |
| Threonine (T) | 74[a] |
| Tryptophan (W) | 159[a] |
| Tyrosine (Y) | 136[a] |
| Valine (V) | 72[a] |

[a]Major peaks according to Reference 63.

observed in some amino acid residues (His, Leu/Ile, Trp, Pro, Phe) than others. Detection of immonium ions depends on the mass range of the mass spectrometer. The characteristic *m/z* values of the immonium ions (Table 2.1) are useful for detecting and confirming some of the amino-acid residues in a peptide sequence, although no information regarding the position of these amino-acid residues in the peptide sequence can be inferred.

### 2.12.4 Quantification by MS

Pharmaceutical companies and proteomics are the most common fields that use MS as a quantitation tool. Quantification is done either by measuring the intensity (peak height) of a signal or by measuring the integrated area of the peak. In both cases, signal intensity is related to ion concentration, that is, mass intensity is proportional to the ion concentration. Signal intensity of different type of molecules cannot be compared as each type of molecules has different ionization capacity. Because of that, each type of ion has to be analyzed separately. Stable isotope labeling has been used in recent years in quantification experiments [295]. Analogs of the analyte to be tested are synthesized using stable isotopes $^{13}C$, $^{15}N$, or $^{2}H$ and known concentrations of the synthetic molecule are spiked into the solution being analyzed. Because the isotope analogs have the same ionization efficiency, intensities of mass signals can be compared. An example can be found in the recent publication of Nilsson et al. [296], where a lipophilic analyte was quantified in the blood plasma using an internal calibration through analyte/isotope standard ratio. The only difference between the pair of analogs is the difference in mass introduced by the stable or heavy isotopes. Proteomics MS-based quantitation uses different methods for protein labeling with stable isotopes resulting in peptide pairs of identical amino acid sequence with a characteristic mass difference (at least 4 Da). Peptides/proteins are quantified by measuring in MS mode the relative signal intensities for the pairs of peptide ions of identical sequence tagged with light and heavy isotope forms. Stable isotope label can be introduced into proteins or at peptide level using chemical, enzymatic, or metabolic methodologies (for a good review, see Reference 297). The methodology introduced in 2003 by Gerber et al. [298] named absolute quantification (AQUA) to perform absolute mass quantification consists of synthesizing stable isotopes labeled peptides in order to quantify proteins in complex mixtures. Isotopically labeled synthetic peptides that are used as internal standards have an amino-acid sequence identical to that of peptides formed by enzymatic digestion and are used to give an absolute quantitation of a protein in a complex sample.

## 2.13 CONCLUSIONS

This chapter illustrates the importance of recent developments in peptide chemistry, purification, and analysis, which have fueled research in both biomedicine and biomaterial areas using peptides as bricks. These developments have boosted the entry of peptides into clinical phases and therefore their appearance in the market.

Peptide science developed is causing a clear impact on the nature of peptides in drug discovery. As mentioned in the introduction, the oldest peptides described, which were evaluated for their therapeutic activities, contained natural sequences and had relatively low molecular weight. Nowadays, they show more sophisticated structures with longer amino-acid chains; sequences with aggregation tendency; cyclic peptides; containing nonnatural amino acids; presence of the nonpeptide moieties (pegylated, glycosylated, fatty acids, and chromophores); and hybrids with cell-penetrating peptides.

This is the result of the progress made by peptide scientists in last half a century, who have incessantly been developing novel strategies and chemical approaches. Those innovations have provided the academic community and pharmaceutical companies with significant tools to design and produce peptides as pharmaceutical ingredients that were difficult to produce in the past. Indeed, the new generation of peptide drugs launched recently to the pharmaceutical market, are more complex long peptides (up to 65 aminoacids), including multi-disulfide bridges [299]. In looking to the future, much remains to be accomplished since the requirements for peptide drugs from pharmaceutical market and the development of genomics and proteomics will continue demanding greater versatility of design and synthesis of target structures. Peptide science and scientists have a number of cases that remain unresolved, yet they are ready to find the right answers.

## ACKNOWLEDGMENTS

The work carried out into the laboratory of the authors was partially supported by CICYT (CTQ2012-30930), the *Generalitat de Catalunya* (2014 SGR 137), the Institute for Research in Biomedicine Barcelona (IRB Barcelona), AECYD (A/9846/07), ProteoRed (the Spanish Proteomics Network), (Spain); the National Research Foundation and the University of KwaZulu Natal (South Africa); and SENESCYT (Ecuador). J.T.-P. is a Juan de la Cierva fellow (Micinn).

## ABBREVIATIONS

| | |
|---|---|
| **ψPro** | pseudoproline |
| **2-CTC** | 2-chlorotrityl chloride resin |
| **6-Cl-HOBt** | 6-chloro-1-hydroxybenzotriazole |
| **Acm** | acetamidomethyl |
| **ACN** | acetonitrile |
| **Alloc** | allyloxycarbonyl |
| **AOP** | (7-azabenzotriazol-1-yl)oxytris-(dimethylamino) phosphonium hexafluorophosphate |
| **BAL** | backbone amide linker |
| **Bg** | *N*-benzydrylglycolamide |
| **Boc** | *t*-butyloxycarbonyl diazaphospholidinium hexafluorophosphate |
| **BOP** | benzotriazol-1-yloxytris(dimethylamino)-phosphonium hexafluorophosphate |
| **Bpoc** | 2-(4-biphenyl)isopropyoxycarbonyl |
| ***t*Bu** | *tert*-butyl |
| **Bzl** | benzyl |
| **Cam** | carboxyamidomethyl |
| **Cbz, Z** | benzyloxycarbonyl |
| **CD** | circular dichroism |
| **CHCA** | α-cyano-4-hydroxycinnamic acid |
| **CID** | collision-induced dissociation |
| **CM** | ChemMatrix |
| **COMU** | 1-[(1-(cyano-2-ethoxy-2-oxoethylideneaminooxy)-dimethylamino-morpholinomethylene)] methanaminium hexafluorophosphate |
| **DCC** | *N,N'*-dicyclohexylcarbodiimide |
| **Dde** | (1-(4,4-dimethyl-2,6-dioxocylohex-1-ylidene)-3-ethyl) |
| **Ddz** | α,α-dimethyl-3,5-dimethoxybezyloxy |
| **DEAE** | diethylamino ethyl |

| | |
|---|---|
| **DEPB** | diethyl phosphorobromidate |
| **DHB** | 2,5-dihydroxybenzoic acid |
| **Dhb** | didehydroaminobutyric acid |
| **DIC** | *N,N'*-diisopropylcarbodiimide |
| **DIEA (DIPEA)** | diisopropylethylamine |
| **DKP** | diketopiperazine |
| **DMAP** | 4-(*N,N*-dimethylamino)pyridine |
| **DMF** | *N,N*-dimethylformamide |
| **ECD** | electron capture dissociation |
| **EDC** | N-ethyl-N'-(3-dimethylaminopropyl)carbodimide |
| **ESI** | electrospray ionization |
| **Et** | ethyl ester |
| **ETD** | electron transfer dissociation |
| **FDA** | Food and Drug Administration |
| **Fmoc** | 9-fluorenylmethyloxcarbonyl |
| **FTICR** | fourier transform ion cyclotron resonance |
| **FWHM** | full width at half maximum |
| **HAPyU** | 1-(1-pyrrolidinyl-1*H*-1,2,3-triazolo[4,5-*b*]pyridin-1-ylmethylene) pyrrolidinium hexafluoro phosphate *N*-oxide |
| **HATU** | *N*-[(dimethylamino)-1*H*-1,2,3-triazolo[4,5-*b*]pyridin-1-ylmethylene]-*N*-methylmethanaminium hexafluorophosphate *N*-oxide |
| **HBTU** | *N*-[(1H-benzotriazol-1-yl)(dimethylamino)methylene]-*N*-methylmethanaminium hexafluorophosphate *N*-oxide |
| **HDMB** | 1-((dimethylamino)-(morpholino)methylene)-1*H*-benzotriazoliumhexafluorophosphate 3-oxide |
| **HDMC** | 6-chloro-1-((dimethylamino)-(morpholino)methylene)-1*H*-benzotriazolium hexafluorophosphate 3-oxide |
| **HFBA** | heptafluorobutyric acid |
| **Hmb** | *N*-(-2-hydroxy-4-methoxybenzyl) |
| **HMPA** | hexamethylphosphoramide |
| **HOAc** | acetic acid |
| **HOAt** | 1-hydroxy-7-azabenzotriazole |
| **HOBt** | 1-hydroxybenzotriazole |
| **HPLC** | high performance liquid chromatography |
| **cHx** | cyclohexyl |
| **IEC** | ion-exchange chromatography |
| **IT** | quadrupole ion trap |
| **LIT** | linear ion trap |

| | |
|---|---|
| **MALDI** | matrix assisted laser desorption/ionization |
| **MBHA** | *p*-methylbenzhydrylamine |
| **Me** | methyl ester |
| **MM** | molecular mass |
| **Mob** | *S*-4-Methoxybenzyl |
| **Moz** | 4-methoxy-benzyloxycarbonyl |
| **MS** | mass spectrometry |
| **MS/MS** | tandem mass spectrometry |
| **NCEs** | new chemical entities |
| **NMM** | *N*-methylmorpholine |
| **Nps** | nitrophenylsulfenyl |
| **Npys** | 3-nitro-2-pyridine-sulfenyl |
| **pNZ** | *p*-nitrobenzyloxycarbonyl |
| **Oxyma** | ethyl 2-cyano-2-(hydroxyimino)acetate |
| **Pac** | phenacyl ester |
| **PAL** | 5-(4-aminomethyl-3,4-dimethoxyphenoxy)valeric acid |
| **PyAOP** | (7-azabenzotriazol-1-yloxy)tris-(pyrrolidino) phosphonium hexafluorophosphate |
| **PyBOP** | benzotriazol-1-yloxytri(pyrrolidino)-phosphonium hexafluorophosphate |
| **PyBroP** | bromotri(pyrrolidino)phosphonium hexafluorophosphate |
| **PyCloP** | chlorotri(pyrrolidino)phosphonium hexafluorophosphate |
| **Pbf** | pentamethyl-2,3-dihydrobenzofuran-5-sulfonyl |
| **PBS** | phosphate buffered saline |
| **Ph** | phenyl ester |
| **PEGA** | poly(ethylene glycol)-poly(acrylamide) copolymer |
| **PEG-PS** | polyethylene glycol-polystyrene |
| **PyOxm** | *O*-[(cyano-(ethoxycarbonyl)methyliden)-amino] yloxytripyrrolidinophosphonium hexafluorophosphate |
| **Pns** | phenylnorstatine |
| **POEPOP** | polyoxyethylene cross-linked polyoxypropylene |
| **PS** | polystyrene |
| **RP-HPLC** | reverse-phase high performance liquid chromatography |
| **SA** | sinapinic acid |
| **Q** | quadrupole |
| **QqQ** | triple quadrupole |
| **Q-TOF** | quadrupole-time of flight |
| **SPOCC** | poly-oxyethylene-poly(3-methylene-3-methyloxethane) copolymer |

| | |
|---|---|
| **SPPS** | solid-phase peptide synthesis |
| **TBAF** | tetrabutyl ammonium fluoride |
| **Tce** | trichloroethyl ester |
| **TFA** | trifluoroacetic acid |
| **THP** | tetrahydropyranyl |
| **Tmob** | 2,4,6-trimethoxybenzyl |
| **TMP** | collidine |
| **Tmse** | 2-trimethylsilylethyl ester |
| **TOF** | time of flight |
| **Tos** | *p*-toluenesulfonyl |
| **Troc** | 2,2,2-trichloroethoxycarbonyl |
| **Trt** | trityl |
| **Z** | benzyloxycabonyl |

## REFERENCES


1. Bruckdorfer T, Marder O, Albericio F. From production of peptides in milligram amounts for research to multi-tons quantities for drugs of the future. Curr Pharm Biotechnol 2004;5:29–43.
2. Burrill & Company (2008) Analysis for pharmaceutical research and manufacturers of America.
3. Hughes B. 2008 FDA drug approvals. Nat Rev Drug Discovery 2009;8:93–96.
4. Reichert JM. Peptide Therapeutics: Trends in Global Clinical Development and Approval. Tufts Center for the Study of Drug Development. 2009.
5. García-Martin F, Albericio F. Solid supports for the synthesis of peptides. From the first resin used to the more sophisticated in the market. Chim Oggi 2008;26:29–34.
6. Merrifield RB. Solid-phase synthesis I. The synthesis of a tetrapeptide. J Am Chem Soc 1963;85:2149–2154.
7. Barany G, Merrifield RB. Solid-phase peptide synthesis. In: Gross E, Meienhofer J, editors. *The Peptides*. Vol. 2. New York: Academic Press; 1979. p 1–284.
8. Becker H, Lucas HW, Maul J, Pillai VN, Anzinger H, Mutter M. Makromol. Poly(ethylene glycols) grafted onto crosslinked polystyrenes: a new class of hydrophilic polymeric supports for peptide synthesis. Chem Rapid Commun 1982;3:217–223.
9. Zalipsky S, Albericio F Barany G. Preparation and use of an aminoethyl polyethylene glycol-crosslinked polystyrene graft resin support for solid-phase peptide synthesis. In: Peptides: structure and function. Proceedings of the ninth American peptide symposium, Rockford, Illinois; 1985. p 257–260.
10. Bayer E, Rapp W. New polymer supports for solid-liquid-phase peptide synthesis. Chem Pept Proteins 1986;3:3–8.
11. Adams JH, Cook RM, Hudson D, Jammalamadaka V, Lyttle MH, Songster MF. A reinvestigation of the preparation, properties, and applications of aminomethyl and 4-methylbenzhydrylamine polystyrene resins. J Org Chem 1998;63:3706–3716.

12. Gooding OW, Baudart S, Deegan TL, Heisler K, Labadie J, Newcomb WS, Porco JA, van Eikeren P. On the development of new poly(styrene-oxyethylene) graft copolymer resin supports for solid-phase organic synthesis. J Comb Chem 1999;1:113–122.
13. Meldal M. PEGA: a flow-stable polyethylene glycol-dimethylacrylamide copolymer for solid-phase synthesis. Tetrahedron Lett 1992;33:3077–3080.
14. Kempe M, Barany G. CLEAR: a novel family of highly cross-linked polymeric supports for solid-phase peptide synthesis. J Am Chem Soc 1996;118:7083–7093.
15. Meldal M. Properties of solid supports. Methods Enzymol 1997;289:83–104.
16. Rademann J, Meldal M, Bock K. Solid-phase synthesis of peptide isosteres by nucleophilic reactions with N-terminal peptide aldehydes on a polar support tailored for solid-phase organic chemistry. Chem Eur J 1999;5:1088–1095.
17. Rademann J, Meldal M, Bock K. In: Bajusz S, Hudecz F, editors. Proceedings of the 25th European Peptide Symposium Budapest: Akadémiai Kiadó, 1999, p 38–39.
18. Meldal M, Miranda LP. Matrix for Solid-Phase Organic Synthesis. WO 2003031489. 2003.
19. Rademann J, Grötli M, Meldal M, Bock K. SPOCC: a resin for solid-phase organic chemistry and enzymatic reactions on solid phase. J Am Chem Soc 1999;121:5459–5466.
20. Côté S. New Polyether Based Monomers, Crosslinkers, and Highly Crosslinked Amphiphile Polyether Resins. WO 2005012277. 2005.
21. García-Martín F, Quintanar-Audelo M, García-Ramos Y, Cruz LJ, Gravel C, Furic R, Côté S, Tulla-Puche J, Albericio F. ChemMatrix, a poly(ethylene glycol)-based support for the solid-phase synthesis of complex peptides. J Comb Chem 2006;8:213–220.
22. de la Torre BG, Jakab A, Andreu D. Polyethylene glycol-based resins as solid supports for the synthesis of difficult or long peptides. Int J Pept Res Ther 2007;13:265–270.
23. Mazzini S, García-Martin F, Alvira M, Aviñó A, Manning B, Albericio F, Eritja R. Synthesis of oligonucleotide derivatives using ChemMatrix supports. Chem Biodiversity 2008;5:209–218.
24. Patel H, Chantell CA, Fuentes G, Menakuru M, Park JH. Resin comparison and fast automated stepwise conventional synthesis of human SDF-1α. J Pept Sci 2008;14:1240–1243.
25. Kassem T, Sabatino D, Jia X, Zhu XX, Lubell WD. To Rink or not to Rink amide link, that is the question to address for more economical and environmentally sound solid-phase peptide synthesis. Int J Pept Res Ther 2009;15:211–218.
26. Meldal M, Svendsen I. Direct visualization of enzyme inhibitors using a portion mixing inhibitor library containing a quenched fluorogenic peptide substrate. Part 1. Inhibitors for subtilisin Carlsberg. J Chem Soc, Perkin Trans 1 1995:1591–1596.
27. Meldal M, Svendsen I, Juliano L, Juliano MA, Del Nery E, Scharfstein J. Inhibition of cruzipain visualized in a fluorescence quenched solid-phase inhibitor library assay. D-Amino acid inhibitors for cruzipain, cathepsin B and cathepsin L. J Pept Sci 1997;4:83–91.
28. Camperi SA, Marani MA, Iannucci NB, Côté S, Albericio F, Cascone O. An efficient strategy for the preparation of one-bead-one-peptide libraries on a new biocompatible solid support. Tetrahedron Lett 2005;46:1561–1564.
29. Meldal M. Smart assays in combinatorial chemistry. QSAR Comb Sci 2005;24:1125–1126.

30. Guillier F, Orain D, Bradley M. Linkers and cleavage strategies in solid-phase organic synthesis and combinatorial chemistry. Chem Rev 2000;100:2091–2158.
31. Songster MF, Barany G. Handles for solid-phase peptide synthesis. Methods Enzymol 1997;289:126–174.
32. Albericio F, Kneib-Cordonier N, Biancalana S, Gera L, Masada RI, Hudson D, Barany G. Preparation and application of the 5-(4-(9-fluoreny lmethyloxycarbonyl)aminomethyl-3,5-dimethoxyphenoxy)-valeric acid (PAL) handle for the solid-phase synthesis of C-terminal peptide amides under mild conditions. J Org Chem 1990;55:3730–3743.
33. Yraola F, Ventura R, Vendrell M, Colombo A, Fernandez J-C, de la Figuera N, Fernandez-Forner D, Royo M, Forns P, Albericio F. A re-evaluation of the use of Rink, BAL, and PAL resins and linkers. QSAR Comb Sci 2004;23:145–152.
34. Colombo A, De la Figuera N, Fernández JC, Fernández-Forner D, Albericio F, Forns P. A nonacid degradable linker for solid-phase synthesis. Org Lett 2007;9:4319–4322.
35. Stathopoulos P, Papas S, Tsikaris V. C-terminal N-alkylated peptide amides resulting from the linker decomposition of the Rink amide resin. A new cleavage mixture prevents their formation. J Pept Sci 2006;16:227–232.
36. Giraud M, Cavalier F, Martinez J. A side-reaction in the SPPS of Trp-containing peptides. J Pept Sci 1999;5:457–461.
37. Stanger KJ, Krchnack V. Incorporation of the Wang linker upon cleavage from polystyrene-based resin to form O-(4-hydroxy)benzyl derivatives. J Comb Chem 2006;8:652–654.
38. Cironi P, Tulla-Puche J, Barany G, Albericio F, Álvarez M. Solid-phase syntheses of furopyridine and furoquinoline systems. Org Lett 2004;6:1405–1408.
39. Gu W, Silverman RB. New stable backbone linker resins for solid-phase peptide synthesis. Org Lett 2003;5:415–418.
40. Isidro-llobet A, Álvarez M, Albericio F. Amino-acid protecting groups. Chem Rev 2009;109:2455–2504.
41. Carpino LA. Oxidative reactions of hydrazines. IV. Elimination of nitrogen from 1,1-disubstituted-2-arenesulfonhydrazides. J Am Chem Soc 1957;79:4427–4431.
42. Anderson GW, McGregor AC. *t*-Butyloxycarbonylaminoacids and their use in peptide synthesis. J Am Chem Soc 1957;79:6180–6183.
43. Merrifield RB. Solid-phase synthesis I. The synthesis of a tetrapeptide. J Am Chem Soc 1963;85:2149–2154.
44. Carpino LA, Han GY. 9-Fluorenylmethoxycarbonyl amino-protecting group. J Org Chem 1972;37:3404–3409.
45. Isidro-Llobet A, Guasch-Camell J, Álvarez M, Albericio F. *p*-Nitrobenzyloxycarbonyl (pNZ) as a temporary N.alpha.-protecting group in orthogonal solid-phase peptide synthesis - Avoiding diketopiperazine and aspartimide formation. Eur J Org Chem 2005;14:3031–3039.
46. Guibé F, Dangles O, Balavoine G. Palladium-catalyzed reaction of tributyltin hydride. Selective and very mild deprotection of allyl and allyloxycarbonyl derivatives of amino acids. Tetrahedron Lett 1986;27:2365–2368.
47. Dangles O, Guibé F, Balavoine G, Lavielle S, Marquet A. Selective cleavage of the allyl and (allyloxy)carbonyl groups through palladium-catalyzed hydrostannolysis with tributyltin hydride. Application to the selective protection-deprotection of amino acid derivatives and in peptide synthesis. J Org Chem 1987;52:4984–4993.

48. Patchornik A, Amit B, Woodward RB. Photosensitive protecting groups. J Am Chem Soc 1970;92:6333–6335.
49. Alsina J, Giralt E, Albericio F. Use of *N*-tritylaminoacids and PyAOP for the suppression of diketopiperazine formation in Fmoc/tBu solid-phase synthesis using alkoxybenzyl ester anchoring linkages. Tetrahedron Lett 1996;37:4195–4198.
50. Thieriet N, Alsina J, Giralt E, Guibé F, Albericio F. Use of alloc-amino acids in solid-phase peptide synthesis. Tandem deprotection-coupling under neutral conditions. Tetrahedron Lett 1997;38:7275–7278.
51. Thornton JM. Disulphide bonds in globular proteins. J Mol Biol 1981;151:261–287.
52. Creighton TE. Disulfide bonds and protein stability. Bioessays 1988;8:57–63.
53. Han Y, Albericio F, Barany G. Occurrence and minimization of cysteine racemization during stepwise solid-phase peptide synthesis. J Org Chem 1997;62:4307–4312.
54. Andreu D, Albericio F, Solé NA, Munson MC, Ferrer M, Barany G. Formation of disulfide bonds in synthetic peptides and proteins. In: Pennington MW, Dunn BM, editors. *Methods in Molecular Biology*. Peptide synthesis protocols. Vol. 35. Totowa, NJ, USA: Humana Press; 1994. p 91–169.
55. Annis I, Hargittai B, Barany G. Disulfide bond formation in peptides. Methods Enzymol 1997;289:198–221.
56. Munson MC, Garcia-Echevarria C, Albericio F, Barany G. *S*-2,4,6-trimethoxybenzyl (Tmob): a novel cysteine protecting group for the $N$-$^{\alpha}$-fluorenylcarbonyl (Fmoc) strategy of peptide synthesis. J Org Chem 1992;57:3013–3018.
57. Veber DF, Milkowski JD, Varga SL, Denkewalter RG, Hirschmann R. Acetamidomethyl. A novel thiol protecting group for cysteine. J Am Chem Soc 1972;94:5456–5461.
58. Albericio F, Hammer RP, García-Echevarría C, Molins MA, Chang JL, Munson MC, Pons M, Giralt E, Barany G. Cyclization of disulfide-containing peptides in solid-phase synthesis. Int J Pept Protein Res 1991;37:402–413.
59. Eritja R, Ziehler-Martin JP, Walker PA, Lee TD, Legesse K, Albericio F, Kaplan BE. Use of the S-tert-butylsulfenyl group for protection of cysteine in solid-phase peptide synthesis using Fmoc amino acids. Tetrahedron Lett 1987;43:2675–2680.
60. Matsueda R, Kimura T, Kaiser ET, Matsueda GR. 3-Nitro-2-pyridinesulfenyl group for protection and activation of the thiol function of cysteine. Chem Lett 1981;6:737–740.
61. Galanis AS, Albericio F, Grøtli M. Enhanced microwave-assisted method for on-bead disulfide bond formation: synthesis of alpha-conotoxin MII. Biopolymers Pept Sci 2009;92:23–34.
62. Albericio F, Andreu D, Giralt E, Novalpotro C, Pedroso E, Ponsati B, Ruiz-Gayo M. Use of the Npys thiol protection in solid phase peptide synthesis. Application to direct peptide-protein conjugation through cysteine residues. Int J Pept Protein Res 1989;34:124–128.
63. Tam JP, Wu C-R, Zhang J-W. Disulfide bond formation in peptides by dimethyl sulfoxide. Scope and applications. J Am Chem Soc 1991;113:6657–6662.
64. Munson MC, Barany G. Synthesis of α-conotoxin SI, a bycyclic tridecapeptide amide with two disulfide bridges: illustration of novel protection schemes and oxidation strategies. J Am Chem Soc 1993;115:10203–10210.
65. Moroder L, Besse D, Musiol H-J, Rudolph-Böhner S, Siedler F. Oxidative folding of cystine-rich peptides vs regioselective cysteine pairing strategies. Biopolymers 1996;40:207–234.

66. Dill KA, Ozkan SB, Shell MS, Weikl TR. The protein folding problem. Annu Rev Biophys 2008;37:289–316.

67. Sheehan JC, Hess GP. A new method of forming peptide bonds. J Am Chem Soc 1955;77:1067–1068.

68. Benoiton NL, Chen FMF. Not the alkoxycarbonylamino-acid *O*-acylisourea. J Chem Soc Chem Commun 1981:543–545.

69. Rebek R, Feitler D. Mechanism of the carbodiimide reaction. II. Peptide synthesis on the solid support. J Am Chem Soc 1974;96:1606.

70. Rich DH, Singh J. The carbodiimide method. In: Gross E, Meienhofer J, editors. *The Peptides: Analysis, Synthesis, Biology*. Vol. 1. New York: Academic; 1979. p 241–261.

71. Benoiton NL. Quantitation and the sequence dependence of racemization in peptide synthesis. In: Gross E, Meienhofer J, editors. *The Peptides: Analysis, Synthesis, Biology*. Vol. 5. New York: Academic; 1981. p 341–361.

72. Slebioda M, Wodecki Z, Kolodziejczyk AM. Formation of optically pure *N* -acyl-*N, N′*-dicyclohexylurea in *N,N′* -dicyclohexylcarbodiimide-mediated peptide synthesis. Int J Pept Protein Res 1990;35:539.

73. Benoiton NL, Chen FMF. 2-Alkoxy-5(4 *H* )-oxazolones from *N* –alkoxycarbonylamino acids and their implication in carbodiimide-mediated reactions in peptide synthesis. Can J Chem 1981;59:384.

74. Benoiton NL, Chen FMF. Reaction of *N* -t-butoxycarbonylamino acid anhydrides with tertiary amines and carbodiimides. New precursors for 2-*t*-butoxyoxazol-5(4 *H* )-one and *N* –acylureas. J Chem Soc Chem Commun 1981:1225.

75. Rich DH, Singh J. The carbodiimide method. In: Gross E, Meienhofer J, editors. *The Peptides: Analysis, Synthesis, Biology*. Vol. 1. New York: Academic; 1979. p 241–261.

76. König W, Geiger R. A new method for the synthesis of peptides: activation of the carboxyl group with dicyclohexylcarbodiimide and 1-hydroxybenzotriazoles. Chem Ber 1970;103:788–789.

77. Carpino LA. 1-Hydroxy-7-azabenzotriazole. An efficient peptide coupling additive. J Am Chem Soc 1993;115:4397–4398.

78. Carpino LA, El-Faham A, Minor CA, Albericio F. Advantageous applications of azabenzotriazole (triazolopyridine)-based coupling reagents to solid-phase peptide synthesis. J Chem Soc Chem Commun 1994:201–203.

79. Sabatino G, Mulinacci B, Alcaro MC, Chelli M, Rovero P, Papini AM. A comparative study of coupling reagents for automatic multiple peptide synthesizers. Peptides 2002, Proceedings of the European peptide symposium, 27th, Sorrento, Italy, 2002 Aug 31-Sept. 6; 2002. p 272–273.

80. Subirós-Funosas R, Prohens R, Barbas R, El-Faham A, Albericio F. Oxyma: an efficient additive for peptide synthesis to replace benzotriazole-based HOBt and HOAt with a lower risk of explosion. Chem Eur J 2009;15:9394–9403.

81. Wehrstedt KD, Wandrey PA, Heitkamp D. Explosive properties of 1-hydroxy benzotriazoles. J Hazard Mat 2005;A126:1–3.

82. Schüssler H, Zahn H. Contribution on the course of reaction of carbobenzoxyamino acids with dicyclohexylcarbodiimide. Chem Ber 1962;95:1076.

83. Weygand F, Huber P, Weiss K. The synthesis of peptides with symmetrical anhydrides I. Z Naturforsch 1967;22B:1084.

84. Hagenmeier H, Frank H. Increased coupling yields in solid phase peptide synthesis with a modified carbodiimide coupling procedure (1972). H-S Z Physiol Chem 1973;353.

85. For a review of the work on the use of protected amino acid chlorides see:Schröder E, Lübke K. *The Peptides*. Methods of peptide synthesis. Vol. I. New York: Academic Press; 1965. p 77.

86. Fischer E, Otto E. Synthesis of derivatives of some dipeptides. Machine translation. Ber Dtsch Chem Ges 1903;36:2106–2116.

87. Thern B, Rudolph J, Jung G. Triphosgene as highly efficient reagent for the solid-phase coupling of N-alkylated amino acids-total synthesis of cyclosporin O. Tetrahedron Lett 2002;43:5013–5016.

88. Carpino LA, Sadat-Aalaee D, Chao HG, DeSelms RH. [(9-Fluorenylmethyl) oxy]carbonyl (FMOC) amino acid fluorides. Convienient new peptide coupling reagents applicable to the FMOC/tert-butyl strategy for solution and solid-phase syntheses. J Am Chem Soc 1990;112:9651–9652.

89. Carpino LA, Mansour EME, Sadat-Aalaee D. tert-Butyloxycarbonyl and benzyloxycarbonyl amino acid fluorides. New, stable rapid-acting acylating agents for peptide synthesis. J Org Chem 1991;56:2611–2614.

90. Kim Y-A, Han S-Y. Comparative study of cyanuric fluoride and BOP-Cl as carboxyl activators in peptide coupling reactions. Bull Korean Chem Soc 2000;21:943–946.

91. Siavrda J, Chertanova L, Wakselman M. Activation of N,N-bis(alkoxycarbonyl) behavior of these amino acid derivatives. Tetrahedron 1994;50:5309–5322.

92. Carpino LA, Manssur EME, El-Faham A. Bis(BOC) amino acid fluorides as reactive peptide coupling reagents. J Org Chem 1993;58:4162–4164.

93. Carpino LA, Manssur EME, Sadat-Aalaee D. tert-Butyloxycarbonyl and benzyloxycarbonyl amino acid fluorides. New, stable rapid-acting acylating agents for peptide synthesis. J Org Chem 1991;56:2611–2614.

94. Olah GA, Nojima M, Kerekes I. Synthetic methods and reactions. IV. Fluorination of carboxylic acids with cyanuric fluoride. Synthesis 1973;1973:487–488.

95. Gawne G, Kenner G, Sheppard RC. Acyloxyphosphonium salts as acylating agents. Synthesis of peptides. J Am Chem Soc 1969;91:5669–5671.

96. Castro B, Dormoy JR, Evin G, Selve C. Peptide coupling reagents. IV. N-[Oxytris (dimethylamino) phosphonium]benzotriazole hexafluorophosphate. Tetrahedron Lett 1975:1219–1222.

97. Jakobsen MH, Buchardt O, Engdahl T, Holm A. A new facile one-pot preparation of pentafluorophenyl and 3,4-dihydro-4-oxo-1,2,3-benzotriazine-3-yl esters of Fmoc amino acids. Tetrahedron Lett 1991;32:6199–6202.

98. Frérot E, Coste J, Pantaloni A, Dufour M-N, Jouin P. PyBOP and PyBroP: two reagents for the difficult coupling of the α,α-dialklyl amino acid, Aib. Tetrahedron 1991;47:259–270.

99. Hudson D. Methodological implications of simultaneous solid-phase peptide synthesis. 1. Comparison of different coupling procedures. J Org Chem 1988;53:617–624.

100. Coste J, Frérot E, Jouin P, Castro B. Oxybenzotriazole free peptide coupling reagents for N-methylated amino acids. Tetrahedron Lett 1991;32:1967–1970.

101. Coste J, Frérot F, Jouin P. Coupling N-methylated amino acids using PyBroP and PyCloP halogenophosphonium salts: mechanism and fields of application. J Org Chem 1994;59:2437–1446.

102. Kim MH, Patel DV. BOP as a reagent for mild and efficient preparation of esters. Tetrahedron Lett 1994;35:5603–3506.
103. Coste J, Campagne JM. Esterification of carboxylic acids by BOP or PyBOP. Tetrahedron Lett 1995;36:4253–4256.
104. Campagne JM, Coste J, Joiun P. (1H-Benzotriazol-1-yloxy)tris(dimethylamino)phos phonium Hexafluorophosphate- and (1H-Benzotriazol-1-yloxy)tripyrrolidinophos phonium hexafluorophosphate-mediated activation of monophosphonate esters: synthesis of mixed phosphonate diesters, the reactivity of the benzotriazolyl phosphonic esters vs the reactivity of the benzotriazolyl carboxylic esters. J Org Chem 1995;60:5214–5223.
105. Carpino LA, Ionescu D, El-Faham A. Peptide coupling in the presence of highly hindered tertiary amine. J Org Chem 1996;61:2460–2465.
106. Kates SA, Diekmann E, El-Faham A, Herman LW, Ionescu D, McGuinness BF, Triolo SA, Albericio F, Carpino LA. On the use of novel coupling reagents for solid-phase peptide synthesis. In: Marshak DR, editor. *Techniques in Protein Chemistry VII*. New York: Academic Press; 1996. p 515.
107. Ehrlich A, Heyne H-U, Winter R, Beyermann M, Haber H, Carpino LA, Bienert M. Cyclization of all-L-pentapeptides by means of 1-hydroxy-7-azabenzotriazole-derived uronium and phosphonium reagents. J Org Chem 1996;61:8831–8838.
108. Jou G, Gonzalez I, Albericio F, Lloyd PW, Giralt E. Total synthesis of dehydrodidemnin B. Use of uronium and phosphonium salt coupling reagents in peptide synthesis in solution. J Org Chem 1997;62:354–366.
109. Han Y, Albericio F, Barany G. Occurrence and minimization of cysteine racemization during stepwise solid-phase peptide synthesis. J Org Chem 1997;62:4307–4312.
110. Albericio F, Cases M, Alsina J, Triolo SA, Carpino LA, Kates SA. On the use of PyAOP, a phosphonium salt derived from HOAt, in solid-phase peptide synthesis. Tetrahedron Lett 1997;38:4853–4856.
111. El-Faham A, Subirós-Funosas R, Albericio F. PyOxP and PyOxB: the Oxyma-based novel family of phosphonium salts. Org. Biomol. Chem 2010;8:3665–3673.
112. Gairí M, Lloyd-Williams P, Albericio F, Giralt E. Use of BOP reagent for the suppression of diketopiperazine formation in Boc/Bzl solid-phase peptide synthesis. Tetrahedron Lett 1990;31:7363–7366.
113. Abdelmoty I, Albericio F, Carpino LA, Forman BM, Kates SA. Structural studies of reagents for peptide bond formation: crystal and molecular structures of HBTU and HATU. Lett Pept Sci 1994;1:57–67.
114. Bofill JM, Albericio F. Theoretical and experimental structural studies of reagents for peptide bond formation: molecular structure and mechanism of HATU. J Chem Res-S 1996:302–303.
115. Carpino LA, Henklein P, Foxman BM, Abdelmoty I, Costisella B, Wray V, Domke T, El-Faham A, Mugge C. The solid state and solution structure of HAPyU. J Org Chem 2001;66:5245–5247.
116. del Fresno M, El-Faham A, Carpino LA, Royo M, Albericio F. Substituted guanidines: introducing diversity in combinatorial chemistry. Org Lett 2000;2:3539–3542.
117. Li P, Xu JC. New and highly efficient immonium-type peptide coupling reagents: synthesis, mechanism, and application. Tetrahedron 2000;56:4437–4445.
118. Dourtoglou V, Ziegler J-C, Gross B. O-Benzotriazolyl-N,N-tetramethyluronium hexafluorophosphate: a new and effective reagent for peptide coupling. Tetrahedron Lett 1978:1269–1272.

119. Dourtoglou V, Gross B, Lambropoulou V, Zioudrou C. O-Benzotriazolyl-N,N, N',N'-tetramethyluronium hexafluorophosphate as coupling reagent for the synthesis of peptides of biological interest. Synthesis 1984;1984:572–574.

120. Knorr R, Trzeciak A, Bannwarth W, Gillessen D. New coupling reagents in peptide chemistry. Tetrahedron Lett 1989;30:1927–1930.

121. Carpino LA, El-Faham A, Albericio F. Efficiency in peptide coupling: 1-hydroxy-7-azabenzotriazole vs 3,4-dihydro-3-hydroxy-4-oxo-1,2,3-benzotriazine. J Org Chem 1995;50:3561–3564.

122. El-Faham A. New syntheses of bis(tetramethylene)fluoroformamidinium hexafluorophosphate (BTFFH) and 1,3-dimethyl-2-fluoro-4,5-dihydro-1H-imidazolium hexafluorophosphate (DFIH). Utility in peptide coupling reactions. Org Prep Proced Int 1998;30:477–481.

123. El-Faham A, Albericio F. Novel proton acceptor immonium-type coupling reagents: application in solution and solid-phase peptide synthesis. Org Lett 2007;9:4475–4477.

124. El-Faham A, Albericio F. Morpholine-based immonium and halogenoamidinium salts as coupling reagents in peptide synthesis. J Org Chem 2008;73:2731–2737.

125. El-Faham A, Subirós-Funosas R, Prohens R, Albericio F. COMU, a safer and more effective replacement for benzotriazole-based uronium coupling reagents. Chem Eur J 2009;15:9404–9416.

126. (a) Shioiri T, Ninomiya K, Yamada S. Diphenylphosphoryl azide. New convenient reagent for a modified curtius reaction and for peptide synthesis. J Am Chem Soc 1972;94:6203–6205; (b) Shioiri T, Yamada S-I. Amino acids and peptides. XI. Phosphorus in organic synthesis. VI. Application of diphenyl phosphorazidate to the synthesis of peptides containing various functions. Chem Pharm Bull 1974;22:859–863.

127. Hyde C, Johnson T, Sheppard RC. Internal aggregation during solid phase peptide synthesis. Dimethyl sulfoxide as a powerful dissociating solvent. J Chem Soc Chem Commun 1992:1573–1575.

128. Zhang L, Goldammer C, Henkel B, Zühl F, Panhaus G, Jungs G, Bayer E. "Magic Mixture", a powerful solvent system for solid-phase synthesis of "difficult sequences". In: Epton R, editor. *Innovation and Perspectives in Solid Phase Synthesis*. Birmingham: Mayflower; 1994. p 711–716.

129. Seebach D, Thaler A, Back AK. Solubilization of peptides in non-polar organic solvents by the addition of inorganic salts: facts and implications. Helv Chim Acta 1989;72:857–867.

130. Toniolo C, Bonora GM, Mutter M. Linear oligopeptides. 78. Effect of the insertion of a proline residue on the solution conformation of host peptides. Makromol Chem 1981;182:2007–2014.

131. García-Martín F, White P, Steinauer R, Côté S, Tulla-Puche J, Albericio F. The synergy of ChemMatrix Resin and pseudoproline building blocks renders RANTES, a complex aggregated chemokine. Biopolymers Pept Sci 2006;84:566–575.

132. Frutos S, Tulla-Puche J, Albericio F, Giralt E. Chemical synthesis of 19F-labeled HIV-1 protease using Fmoc-chemistry and chemmatrix resin. Int J Pept Res Ther 2007;13:221–227.

133. Johnson T, Quibell M, Owen D, Sheppard RC. A reversible protecting group for the amide bond in peptides: use in the synthesis of difficult sequences. J Chem Soc Chem Commun 1993;4:369–372.

134. Zeng W, Regamey P-O, Rose K, Wang Y, Bayer E. Use of Fmoc-N-(2-hydroxy-4-methoxybenzyl)amino acids in peptide synthesis. J Pept Res 1997;49:273–279.
135. Simmonds RG. Use of the Hmb backbone-protecting group in the synthesis of difficult sequences. Int J Pept Protein Res 1996;47:36–41.
136. Cardona V, Eberle I, Barthelemy S, Beythien J, Doerner B, Schneeberger P, Keyte J, White PD. Application of Dmb-dipeptides in the Fmoc SPPS of difficult and aspartimide-prone sequences. Int J Pept Res Ther 2008;14:285–292.
137. Carpino LA, Nasr K, Abdel-Maksoud AA, El-Faham A, Ionescu D, Henklein P, Wenschuh H, Beyermann M, Krause E, Bienert M. Dicyclopropylmethyl peptide backbone protectant. Org Lett 2009;11:3718–3721.
138. Mutter M, Nefzi A, Sato T, Sun X, Wahl T, Wuhr T. Pseudo-prolines (psi Pro) for accessing "inaccessible" peptides. Pept Res 1995;8:145–153.
139. Wöhr T, Wahl F, Nefzi A, Rohwedder B, Sato T, Sun X, Mutter M. Pseudo-prolines as a solubilizing, structure-disrupting protection technique in peptide synthesis. J Am Chem Soc 1996;118:9218–9227.
140. White P, Keyte JW, Bailey K, Bloomberg G. Expediting the Fmoc solid phase synthesis of long peptides through the application of dimethyloxazolidine dipeptides. J Pept Sci 2004;10:18–26.
141. Sampson WR, Patsiouras H, Ede NJ. The synthesis of "difficult" peptides using 2-hydroxy-4-methoxybenzyl or pseudoproline amino acid building blocks: a comparative study. J Pept Sci 1999;5:403–409.
142. Keller M, Miller AD. Access to the inaccessible sequence of Cpn 60.1 (195-217) by temporary oxazolidine protection of selected amide bonds. Bioorg Med Chem Lett 2001;11:857–859.
143. Abedini A, Raleigh DP. Incorporation of pseudoproline derivatives allows the facile synthesis of human IAPP, a highly amyloidogenic and aggregation-prone polypeptide. Org Lett 2005;7:693–696.
144. Cremer G-A, Tariq H, Delmas AF. Combining a polar resin and a pseudo-proline to optimize the solid-phase synthesis of a 'difficult sequence'. J Pept Sci 2006;12:437–442.
145. Sohma Y, Sasaki M, Hayashi Y, Kimura T, Kiso Y. Novel and efficient synthesis of difficult sequence-containing peptides through O-N intramolecular acyl migration reaction of *O*-acyl isopeptides. Chem Commun 2004;1:124–125.
146. Sohma Y, Yoshiya T, Taniguchi A, Kimura T, Hayashi Y, Kiso Y. Development of *O*-acyl isopeptide method. Biopolymers Pept Sci 2007;88:253–262.
147. Carpino LA, Krause E, Sferdean CD, Schümann M, Fabian H, Bienert M, Beyermann M. Synthesis of "difficult" peptide sequences: application of a depsipeptide technique to the Jung-Redemann 10- and 26-mers and the amyloid peptide Aβ.(1-42). Tetrahedron Lett 2004;45:7519–7523.
148. Mutter M, Chandravarkar A, Boyat C, Lopez J, Dos Santos S, Mandal B, Mimna R, Murat K, Patiny L, Saucède L, Tuchscherer G. Switch peptides in statu nascendi: induction of conformational transitions relevant to degenerative diseases. Angew Chem Int Ed 2004;43:4172–4178.
149. Oliyai R, Stella VJ. Prodrugs of peptides and proteins for improved formulation and delivery. Annu Rev Pharmcol Toxicol 1993;33:521–544.
150. Kimura T, Ohtake J, Nakata S, Enomoto H, Moriwaki H, Akaji K, Kiso Y. Synthesis of prodrugs of HIV protease inhibitors. Pept Chem 1995;32:157–160.

151. Dos Santos S, Chandravarkar A, Mandal B, Mimna R, Murat K, Saucède L, Tella P, Tuchscherer G, Mutter M. Switch-peptides: controlling self-assembly of amyloid β-derived peptides in vitro by consecutive triggering of acyl migrations. J Am Chem Soc 2005;127:11888–11889.

152. Tuchscherer G, Chandravarkar A, Camus M-S, Bérard J, Murat K, Schmid A, Mimna R, Lashuel HA, Mutter M. Switch-peptides as folding precursors in self-assembling peptides and amyloid fibrillogenesis. Biopolymers Pept Sci 2006;88:239–251.

153. Mimna R, Camus M-S, Schmid A, Tuchscherer G, Lashuel HA, Mutter M. Disruption of amyloid-derived peptide assemblies through the controlled induction of a β-sheet to a-helix transformation: application of the switch concept. Angew Chem Int Ed 2007;46:2681–2684.

154. Camus M-S, Dos Santos S, Chandravarkar A, Mandal B, Schmid AW, Tuchscherer G, Mutter M, Lashuel HA. Switch-peptides: design and characterization of controllable super-amyloid-forming host-guest peptides as tools for identifying anti-amyloid agents. Chembiochem 2008;9:2104–2112.

155. Sohma Y, Hayashi Y, Kimura M, Chiyomori Y, Taniguchi A, Sasaki M, Kimura T, Kiso Y. The '*O*-acyl isopeptide method' for the synthesis of difficult sequence-containing peptides: application to the synthesis of Alzheimer's disease-related amyloid β peptide (Aβ) 1-42. J Pept Sci 2005;11:441–451.

156. Sohma Y, Yoshiya T, Taniguchi A, Kimura T, Hayashi Y, Kiso Y. Development of *O*-acyl isopeptide method. Biopolymers Pept Sci 2007;88:253–262.

157. Sohma Y, Taniguchi A, Skwarczynski M, Yoshiya T, Fukao F, Kimura T, Hayashi Y, Kiso Y. *O*-Acyl isopeptide method' for the efficient synthesis of difficult sequence-containing peptides: use of '*O*-acyl isodipeptide unit. Tetrahedron Lett 2006;47:3013–3017.

158. Yoshiya T, Sohma Y, Kimura T, Hayashi Y, Kiso Y. '*O*-Acyl isopeptide method': racemization-free segment condensation in solid phase peptide synthesis. Tetrahedron Lett 2006;47:7905–7909.

159. Coin I, Dölling R, Krause E, Bienert M, Beyermann M, Sferdean CD, Carpino LA. Depsipeptide methodology for solid-phase peptide synthesis: circumventing side reactions and development of an automated technique via depsidipeptide units. J Org Chem 2006;71:6171–6177.

160. Carpino LA, Philbin M, Ismail M, Truran GA, Mansour EME, Iguchi S, Ionescu D, El-Faham A, Warass R, Weiss MS. New family of base- and nucleophile-sensitive amino-protecting groups. A michael-acceptor-based deblocking process. Practical utilization of the 1,1-dioxobenzo[*b*]thiophene-2-ylmethyloxycarbonyl (Bsmoc) Group. J Am Chem Soc 1997;19:9915–9916.

161. Carpino LA, Ismail M, Truran GA, Mansour EME, Iguchi S, Ionescu D, El-Faham A, Riemer C, Warass R. The 1,1-dioxobenzo[*b*]thiophene-2-ylmethyloxycarbonyl (Bsmoc) amino-protecting group. J Org Chem 1999;64:4324–4338.

162. Hentschel J, Borner HG. Peptide-directed microstructure formation of polymers in organic media. J Am Chem Soc 2006;128:14142–14149.

163. Werbitzky O, Oehlers D. Manufacturing peptide APIs in the scope of upcoming new challenges higher complexity and industrialization. Chem Today 2008;26:26–28.

164. Bergmann M, Zervas L. Über ein allgemeines merfahren der peptid-synthese. Ber Dtsch Chem Ges 1932;65:1192–1201.

165. Kenner GW. Recent progress in the chemistry of peptides. J Chem Soc 1956:3689–3700.

166. O'Ferrall RAM, Slae S. β-elimination of 9-fluorenylmethanol in aqueous solution: an ElcB mechanism. J. Chem. Sot. (B) 1970;4:260–268.

167. Rosenmund KW, Zetzsche F, Heise F. Über die beeinflussung der wirksamkeit von katalysatoren, V. Mitteilung; Karl W. Rosenmund und F. Heise: zur kenntnis der katalytischen reduktion von estern und aldehyden. Ber Dtsch Chem Ges 1921;54:2038–2042.

168. Ben-Ishai D, Berger A. Cleavage of N-carbobenzoxy groups by dry hydrogen bromide and hydrogen chloride. J Org Chem 1952;17:1564–1570.

169. Sieber P, Iselin B. Selektive acidolytische spaltung von aralkyloxy carbonyl-aminoschutzgruppen. Helv Chim Acta 1968;51:614–622.

170. Weygand F, Hunger K. Acylierung von aminosäuren mitp HT-methoxybenzy loxycarbonyl-azid. Chem Ber 1962;95:1–6.

171. Birr C, Lochinger W, Stahnke G, Lang P. Der .-Dimethyl-3.5-dimethoxy benzyloxycarbonyl (Ddz)-Rest, eine photo- und säurelabile stickstoff-schutzgruppe für die peptidchemie. Justus Liebigs Ann Chem 1972;763:162–172.

172. Chamberlin JW. Use of the 3,5-Dimethoxybenzyloxycarbonyl Group as a Photosensitive N-Protecting Group. J Org Chem 1966;31:1658–1660.

173. Inamura S, Fukase K, Kusumoto S. Synthetic study of peptidoglycan partial structures. Synthesis of tetrasaccharide and octasaccharide fragments. Tetrahedron Lett 2001;42:7613–7616.

174. Somsák L, Czifrák K, Veres E. Selective removal of 2,2,2-trichloroethyl- and 2,2,2-trichloroethoxycarbonyl protecting groups with Zn–N-methylimidazole in the presence of reducible and acid-sensitive functionalities. Tetrahedron Lett 2004;45:9095–9097.

175. Zervas L, Borovas O, Gazis E. New methods in peptide synthesis. I. tritylsulfenyl and o-nitrophenylsulfenyl groups as N-protecting groups. J Am Chem Soc 1963;90:3660–3666.

176. Goerdeler J, Holst A. N-o-nitrobenzolsulfenyl-aminosäuren und –aminosäureester. Angew Chem 1959;71:775.

177. Brandenburg D. Zur spaltung der sulfenamidbindung in o-nitrophenylsulfeny laminosäuren und -peptiden. Tetrahedron Lett 1966;7:6201–6207.

178. Ekstrom MB, Sjoberg B. Semisynthetic penicillins. VI. The use of the o-nitrophenyl sulphenyl protecting group in the preparation of aminopenicillins. Acta Chem Scand 1965;19:1245–1247.

179. Tun-Kyi A. Selective removal of the o-nitrophenylsulfenyl protecting group in peptide synthesis. Helv Chim Acta 2004;61:1086–1090.

180. du Vegneaud V, Behrens OK. A method for protecting the imidazole ring of histidine during certain reactions and its application to the preparation of *L*-amino-N-methylhistidine. J Biol Chem 1937;117:27–36.

181. Greene TW, Wuts PGM, editors. *Protective Groups in Organic Synthesis*. 3rd ed. New York: John Wiley & Sons, Inc.; 1999.

182. McDermott JR, Benoiton NL. N-methylamino acids in peptide synthesis. III. racemization during deprotection by saponification and acidolysis. Can J Chem 1973;51:2555–2561.

183. Meienhofer J, Dzieduszycka M, Smulkowski M, Makofske R, Danho W. Preparative oxidative conversion of protected peptide Cα-hydrazides into the corresponding acids by N-bromosuccinimide. Int J Pept Protein Res 1981;1i8:6–11.

184. Woodward RB. Neueste fortschritte in der chemie der naturstoffe. Die totalsynthese des cephalosporins C. Nobel-vortrag am 11. Dezember 1965. Angew Chem 1966;78:557–564.

185. Martinez J, Laur J, Castro B. Carboxamidomethyl esters (CAM esters) as carboxyl protecting groups. Tetrahedron Lett 1983;24:5219–5222.

186. Amblard M, Rodriguez M, Martinez J. N-benzhydryl-glycolamide esters (OBg esters) as carboxyl protecting groups in peptide synthesis. Tetrahedron 1988;44:5101–5108.

187. Sieber P. Der 2-trimethylsilyläthyl-rest als selektiv abspaltbare carboxy-schutzgruppe. Helv Chim Acta 1977;60:2711–2716.

188. Gerlach H. 2-(Trimethylsilyl)äthylester als carboxylschutzgruppe; anwendung bei der synthese des (-)-(S)-curvularins. Helv Chim Acta 1977;60:3039–3044.

189. Galpin IJ, Hardy PM, Kenner GW, McDermott JR, Ramage R, Seely JH, Tyson RG. Peptides—XXXII: the use of phenyl esters in peptide synthesis. Tetrahedron 1979;35:2577–2582.

190. Galpin IJ, Handa BK, Hudson D, Jackson AG, Kenner GW, Ohlsen SR, Ramage R, Singh ZB, Tyson RG. Studies on the synthesis of 1-75 fragment of a lysozyme analog. In: Loffet A, editor. *Peptides*. Bruxelles: Edition de l'Universite de Bruxelles; 1976. p 247–256.

191. Hofmann K, Lindenmann A, Magee MZ, Khan NH. Studies on polypeptides. III.1 novel routes to α-amino acid and polypeptide hydrazides. J Am Chem Soc 1952;74:470–476.

192. Wolman Y, Gallop PM, Patchornik A. Peptide synthesis via oxidation of hydrazides. J Am Chem Soc 1961;83:1263–1264.

193. Milne HB, Kilday W. Peptide synthesis via oxidation of N-acetyl-α-amino acid phenylhydrazides. J Org Chem 1965;30:64–66.

194. Sheehan J, Cruickshank P, Boshart G. A convenient synthesis of water-soluble carbodiimides. J Org Chem 1961;26:2525–2528.

195. Bodanszky M, editor. *Peptide Chemistry: A Practical Textbook*. Berlin-Germany: Springer-Verlag; 1988.

196. Dawson PE, Muir TW, Clark-Lewis I, Kent SB. Synthesis of proteins by native chemical ligation. Science 1994;266:776–779.

197. Macmillan D. Evolving strategies for protein synthesis converge on native chemical ligation. Angew Chem Int Ed 2006;45:7668–7672.

198. Beligere GS, Dawson PE. Conformationally assisted protein ligation using C-terminal thioester peptides. J Am Chem Soc 1999;121:6332–6333.

199. Canne LE, Bark SJ, Kent SBH. Extending the applicability of native chemical ligation. J Am Chem Soc 1996;118:5891–5896.

200. Offer J, Boddy C, Dawson PE. Extending synthetic access to proteins with a removable acyl transfer auxiliary. J Am Chem Soc 2002;124:4642–4646.

201. Nilsson BL, Kiessling LL, Raines RT. Staudinger ligation: a peptide from a thioester and azide. Org Lett 2000;2:1939–1941.

202. Saxon E, Armstrong JI, Bertozzi CR. A "Traceless" staudinger ligation for the chemoselective synthesis of amide bonds. Org Lett 2000;2:2141–2143.

203. Tam A, Soellner MB, Raines RT. Water-soluble phosphinothiols for traceless staudinger ligation and integration with expressed protein ligation. J Am Chem Soc 2007;129:11421–11430.

204. Tam JP, Yu Q. Methionine ligation strategy in the biomimetic synthesis of parathyroid hormones. Biopolymers 1998;46:319–327.

205. Yan LZ, Dawson PE. Synthesis of peptides and proteins without cysteine residues by native chemical ligation combined with desulfurization. J Am Chem Soc 2001;123:526–533.

206. Haase C, Rohde H, Seitz O. Native chemical ligation at valine. Angew Chem Int Ed 2008;47:6807–6810.

207. Crich D, Banerjee A. Native chemical ligation at phenylalanine. J Am Chem Soc 2007;129:10064–10065.

208. Chen J, Wan Q, Yuan Y, Zhu J, Danishefsky SJ. Native chemical ligation at valine: a contribution to peptide and glycopeptide synthesis. Angew Chem Int Ed 2008;47:8521–8524.

209. Ficht S, Payne RJ, Brik A, Wong CH. Second-generation sugar-assisted ligation: a method for the synthesis of cysteine-containing glycopeptides. Angew Chem Int Ed 2007;46:5975–5979.

210. Payne RJ, Ficht S, Tang S, Brik A, Yang Y-Y, Case DA, Wong C-H. Extended sugar-assisted glycopeptide ligations: development, scope, and applications. J Am Chem Soc 2007;129:13527–13536.

211. Kumar AKS, Harpaz Z, Haj-Yahya M, Brik A. Side-chain assisted ligation in protein synthesis. Bioorg Med Chem Lett 2009;19:3870–3874.

212. Kates SA, Solé NA, Albericio F, Barany G. In: Basava C, Anantharamaiah GM, editors. *Peptides: Design, Synthesis and Biological Activity*. Berlin: Birkhauser; 1994. p 39–58.

213. Wipf P. Synthetic studies of biologically active marine cyclopeptides. Chem Rev 1995;95:2115–2134.

214. Davies JS. The cyclization of peptides and depsipeptides. J Pept Sci 2003;9:471–501.

215. Kessler H. Peptide conformations. Part 19. Conformation and biological effects of cyclic peptides. Angew Chem Int Ed Engl 1982;21:512–523.

216. Rizo J, Gierasch LM. Constrained peptides: models of bioactive peptides and protein substructures. Annu Rev Biochem 1992;61:387–418.

217. Blackburn C, Kates SA. Solid-phase synthesis of cyclic homodetic peptides. Methods Enzymol 1997;289:175–198.

218. Lambert JN, Mitchell JP, Roberts KD. The synthesis of cyclic peptides. J Chem Soc, Perkin Trans 1 2001;5:471–484.

219. Gilon C, Mang C, Lohof E, Friedler A, Kessler H. In: Goodman M, Felix AM, Moroder L, Toniolo C, editors. *Synthesis of Peptides and Peptidomimetics (Houben-Weyl E22: Methods of Organic Chemistry)*. Stuttgart and New York: Georg Thieme Verlag; 2002. p 461–542.

220. McMurray JS. Solid phase synthesis of a cyclic peptide using Fmoc chemistry. Tetrahedron Lett 1991;32:7679–7682.

221. Rovero P, Quartara L, Fabbri G. Synthesis of cyclic peptides on solid support. Tetrahedron Lett 1991;32:2639–2642.

222. Trzeciak A, Bannwarth W. Synthesis of head-to-tail cyclized peptides on solid support by Fmoc [9-fluorenylmethoxycarbonyl] chemistry. Tetrahedron Lett 1992;33:4557–4560.

223. Kates SA, Solé NA, Johnson CR, Hudson D, Barany G, Albericio F. A novel, convenient, three-dimensional orthogonal strategy for solid-phase synthesis of cyclic peptides. Tetrahedron Lett 1993;34:1549–1552.

224. Romanovskis P, Spatola AF. Preparation of head-to-tail cyclic peptides via side-chain attachment: implications for library synthesis. J Pept Res 1998;52:356–374.

225. Valero ML, Giralt E, Andreu D. A comparative study of cyclization strategies applied to the synthesis of head-to-tail cyclic analogs of a viral epitope. J Pept Res 1999;53:56–67.

226. Büttner F, Norgren AS, Zhang S, Prabpai S, Kongsaeree P, Arvidsson PI. Cyclic-β-tetra- and pentapeptides: synthesis through on-resin cyclization and conformational studies by X-ray, NMR and CD spectroscopy and theoretical calculations. Chem Eur J 2005;11:6145–6158.

227. Alsina J, Rabanal F, Giralt E, Albericio F. Solid-phase synthesis of "head-to-tail" cyclic peptides via lysine side-chain anchoring. Tetrahedron Lett 1994;35:9633–9635.

228. Alsina J, Chiva C, Ortiz M, Rabanal F, Giralt E, Albericio F. Active carbonate resins for solid-phase synthesis through the anchoring of a hydroxyl function. Synthesis of cyclic and alcohol peptides. Tetrahedron Lett 1997;38:883–886.

229. Cabrele C, Langer M, Beck-Sickinger AG. Amino acid side chain attachment approach and its application to the synthesis of tyrosine-containing cyclic peptides. J Org Chem 1999;64:4353–4361.

230. Isied SS, Kuehn CG, Lyon JM, Merrifield RB. Specific peptide sequences for metal ion coordination. 1. Solid-phase synthesis of cyclo-$(Gly\text{-}His)_3$. J Am Chem Soc 1982;104:2632–2634.

231. Alcaro MC, Orfei M, Chelli M, Ginanneschi M, Papini AM. Solid-phase approach to the synthesis of cyclen scaffolds from cyclotetrapeptides. Tetrahedron Lett 2003;44:5217–5219.

232. Rietman BH, Smulders RHPH, Eggen IF, Van Vliet A, Van de Werken G, Tesser GI. Protected peptide disulfides by oxidative detachment from a support. Int J Pept Protein Res 1994;44:199–206.

233. Barany G, Han Y, Hargittai B, Liu R-q, Varkey JT. Side-chain anchoring strategy for solid-phase synthesis of peptide acids with C-terminal cysteine. Biopolymers Pept Sci 2004;71:652–666.

234. Jensen KJ, Alsina J, Songster MF, Vágner J, Albericio F, Barany G. Backbone amide linker (BAL) strategy for solid-phase synthesis of *C*-terminal-modified and cyclic peptides. J Am Chem Soc 1998;120:5441–5452.

235. Alsina J, Jensen KJ, Albericio F, Barany G. Solid-phase synthesis with tris(alkoxy)benzyl backbone amide linkage (BAL). Chem Eur J 1999;5:2787–2795.

236. Alsina J, Albericio F. Solid-phase synthesis of *C*-terminal modified peptides. Biopolymers Pept Sci 2003;71:454–477.

237. Boas U, Brask J, Jensen KJ. Backbone amide linker in solid-phase synthesis. Chem Rev 2009;109:2092–2118, and references therein.

238. Lemmens-Gruber R, Kamyar MR, Dornetshuber R. Cyclodepsipeptides – potential drugs and lead compounds in the drug development process. Curr Med Chem 2009;16:1122–1137.

239. Hamada Y, Shioiri T. Recent progress of the synthetic studies of biologically active marine cyclic peptides and depsipeptides. Chem Rev 2005;105:4441–4482.

240. López-Macià A, Jiménez JC, Royo M, Giralt E, Albericio F. Synthesis and structure determination of kahalalide F. J Am Chem Soc 2001;123:11398–11401.

241. Jiménez JC, López-Macià A, Gracia C, Varón S, Carrascal M, Caba JM, Royo M, Francesch AM, Cuevas C, Giralt E, Albericio F. Structure-activity relationship of kahalalide F synthetic analogues. J Med Chem 2008;51:4920–4931.

242. Tulla-Puche J, Bayó-Puxan N, Moreno JA, Francesch AM, Cuevas C, Álvarez M, Albericio F. Solid-phase synthesis of oxathiocoraline by a key intermolecular disulfide dimer. J Am Chem Soc 2007;129:5322–5323.

243. Bayó-Puxan N, Tulla-Puche J, Albericio F. Oxathiocoraline: lessons to be learned from the synthesis of complex *N*-methylated depsipeptides. Eur J Org Chem 2009;18:2957–2974.

244. Tulla-Puche J, Marcucci E, Fermin M, Bayó-Puxan N, Albericio F. Protection by conformationally testricted mobility: first solid-phase synthesis of triostin A. Chem Eur J 2008;14:4475–4478.

245. Coin I, Beerbaum M, Schmieder P, Bienert M, Beyermann M. Solid-phase synthesis of a cyclodepsipeptide: cotransin. Org Lett 2008;10:3857–3860.

246. Spengler J, Koksch B, Albericio F. Simple machine-assisted protocol for solid-phase synthesis of depsipeptides. Biopolymers Pept Sci 2007;88:823–828.

247. Stellwagen E. Gel filtration. Methods Enzymol 1990;182:317–328.

248. Eriksson K-O. Gel filtration. In: Vijayalakshmi MA, editor. *Theory and Practice of Biochromatography*. London: Taylor and Francis; 2002. p 9–23.

249. Bollag DM. Ion-exchange chromatography. In: Dunn BM, Pennigton MW, editors. *Methods in Molecular Biology*. Peptide analysis protocols. Vol. 36. New Jersey: Humana Press Inc.; 1994. p 11–22.

250. Lee T-H, Aguilar M-I. High performance liquid chromatography (HPLC) of peptides and proteins. In: Rapley JWR, editor. *Molecular Biomethods Handbook*. New Jersey: Humana Press Inc.; 2008. p 767–792.

251. Geng X, Regnier FE. Retention model for proteins in reversed-phase liquid chromatography. J Chromatogr 1984;296(27):15–30.

252. Mant CT, Hodges RS. The effects of anionic ion-pairing reagents on the peptide retention in reversed-phase chromatography. In: Mant CT, Hodges RS, editors. *High-Performance Liquid Chromatography of Peptides and Proteins: Separation, Analysis and Conformation*. Florida: CRC Press; 1991. p 327–341.

253. Horvath C, Melander W, Molnar I. Liquid chromatography of ionogenic substances with nonpolar stationary phases. Anal Chem 1977;49:142–154.

254. Goldberg AP, Nowakowska AE, Antle PE, Snyder LR. Retention-optimization strategy for the high-performance liquid chromatographic ion-pair separation of samples containing basic compounds. J Chromatogr 1984;316:241–260.

255. Cazes J. *Encyclopedia of Chromatography*. Marcel Dekker Inc.: New York-Basel; 2001. p 997.

256. Simpson RJ. *Purifying Proteins for Proteomics: A Laboratory Manual*. New York: Cold Spring Harbor Laboratory Press; 2004.

257. Hearn MTW. Physico-chemical factors in polypeptide and protein purification and analysisby high-performance liquid chromatographic techniques: current status and future challenges. In: Ahuja S, editor. *Handbook of Bioseparation*. New York: Academic Press; 2000. p 72–235.

258. Chloupek RC, Hancock WS, Marchylo BA, Kirkland JJ, Boyes BE, Snyder LR. Temperature as a variable in reversed-phase high-performance liquid chromatographic separations of peptide and protein samples II. Selectivity effects observed in the separation of several peptide and protein mixtures. J Chromatogr A 1994;686:45–59.

259. Haidacher D, Vailaya A, Horvath C. Temperature effects in hydrophobic interaction chromatography. Proc Natl Acad Sci U S A 1996;93:2290–2295.

260. Campbell ID, Dwek RA. *Biological Spectroscopy*. New Jersey: Benjamin/Cummings publishing Co., Inc.; 1984. p 255–277.

261. Ching W. Separation of isomeric peptides using electrospray ionization/high-resolution ion mobility spectrometry. Anal Chem 2000;72:391–395.

262. Karas M, Bachmann D, Bahr U. HIllenkamp, F. Int J Mass Spectrom Ion Processes 1987;78:53–68.

263. Fenn JB, Mann M, Meng CK, Wong SF, Whitehouse CM. Electrospray ionization for mass spectrometry of large biomolecules. Science 1989;246:64–71.

264. Hunt DF, Yates JR III, Shabanowitz J, Winston S, Hauer CR. Protein sequencing by tandem mass spectrometry. Proc Natl Acad Sci U S A 1986;83(17):6233–7.

265. Bakhtiar R, Hofstadler SA, Smith RD. Electrospray ionization mass spectrometry. Part II applications in characterization of peptides and proteins. J Chem Educ 1996;73:A118–A123.

266. Karas M, Hillenkamp F. Laser desorption ionization of proteins with molecular masses exceeding 10,000 daltons. Anal Chem 1988;60:2299–2301.

267. Hillenkamp F, Karas M. Matrix-assisted laser desorption/ionisation, an experience. Int J Mass Spectrom 2000;200:71–77.

268. Muddiman DC, Bakhtiar R, Hofstadler SA, Smith RD. Matrix-assisted laser desorption/ionization mass spectrometry. instrumentation and applications. J Chem Educ 1997;74:1288–1292.

269. Laugesen S, Roepstorff P. Combination of two matrices results in improved performance of MALDI MS for peptide mass mapping and protein analysis. J Am Soc Mass Spectrom 2003;14:992–1002.

270. Gross JH. *Mass Spectrometry*. Berlin, Heidelberg: Springer-Verlag; 2004. p 88–99.

271. Burlingame AL, Boyd RK, Gaskell SJ. Mass spectrometry. Anal Chem 1998;70: 647R–716R.

272. Wysockia VH, Resingb KA, Zhanga Q, Cheng G. Mass spectrometry of peptides and proteins. Methods 2005;35:211–222.

273. Miller PE, Denton BM. The quadrupole mass filter: basic operating concepts. J Chem Educ 1986;63:617–622.

274. Motshwene PG, Moncrieffe MC, Grossmann JG, Kao C, Ayaluru M, Sandercock AM, Robinson CV, Latz E, Gay NJ. An oligomeric signaling platform formed by the toll-like receptor signal transducers MyD88 and IRAK-4. J Biol Chem 2009;284(37):25404–25411.

275. Cooks RG, Glish GL, McLuckey SA, Kaiser RE. Ion trap mass spectrometry. Chem Eng News 1991;25:26–41.

276. Jonscher KR, Yates JR III. The quadrupole ion trap mass spectrometer - a small solution to a big challenge. Anal Biochem 1997;244:1–15.

277. Schwartz JC, Senko MW, Syka JEP. A two-dimensional quadrupole ion trap mass spectrometer. J Am Soc Mass Spectrom 2002;13:659.

278. Hopfgartner G, Varesio E, Tschappat V, Grivet C, Bourgogne E, Leuthold LA. Triple quadrupole linear ion trap mass spectrometer for the analysis of small molecules and macromolecules. J Mass Spectrom 2004;39:845–855.

279. Amster J. Fourier transform mass spectrometry. J Mass Spectrom 1996;31:1325–1337.

280. Belov ME, Nikolaev EN, Alving K, Smith RD. A new technique for unbiased external ion accumulation in a quadrupole two-dimensional ion trap for electrospray ionization Fourier transform ion cyclotron resonance mass spectrometry. Rapid Commun Mass Spectrom 2001;15:1172.

281. Thomson BA, Douglas DJ, Cow JJ, Hager JW, Jolliffe CL. Improved collisionally activated dissociation efficiency and mass resolution on a triple quadrupole mass spectrometer system. Anal Chem 1995;67:1696–1704.

282. Morris HR, Paxton T, Dell A, Langhorne J, Berg M, Bordoli RS, Hoyes J, Bateman RH. High sensitivity collisionally-activated decomposition tandem mass spectrometry on a novel quadrupole/orthogonal-acceleration time-of-flight mass spectrometer. Rapid Commun Mass Spectrom 1996;10:889–896.

283. Morris HR, Paxton T, Panico M, McDowell R, Dell A. A novel geometry mass spectrometer, the Q-TOF, for low-femtomole/attomole-range biopolymer sequencing. J Protein Chem 1997;16:469–479.

284. Verentchikov AN, Ens W, Standing KG. Reflecting time-of-flight mass spectrometer with an electrospray ion source and ortogonal extraction. Anal Chem 1994;66:126–133.

285. Chernushevich IV, Loboda AV, Thomson BA. An introduction to quadrupole-time-of-flight mass spectrometry. J Mass Spectrom 2001;36:849–865.

286. Shevchenko A, Chernushevich I, Ens W, Standing KG, Thomson B, Wilm M, Mann M. Rapid 'de novo' peptide sequencing by a combination of nanoelectrospray, isotopic labeling and a quadrupole/time-of-flight mass spectrometer. Rapid Commun Mass Spectrom 1997;11:1015–1024.

287. Gogichaeva NV, Williams T, Alterman MA. Tandem mass spectrometry as a new tool for amino acid analysis. J Am Soc Mass Spectrom 2007;18:279–284.

288. Medzihradszky KF, Campbell JM, Baldwin MA, Falick AM, Juhasz , Vestal ML, Burlingame AL. The Characteristics of peptide collision-induced dissociation using a high-performance MALDI-TOF/TOF tandem mass spectrometer. Anal Chem 2000;72:552–558.

289. Zubarev RA, Kelleher NL, McLafferty FW. Electron capture dissociation of multiply charged protein cations. A nonergodic process. J Am Chem Soc 1998;120:3265–3266.

290. Mackay CL, Ramsahoye B, Burgess K, Cook K, Weidt S, Creanor J, Harrison D, Langridge-Smith P, Hupp T, Hayward L. Sensitive, specific, and quantitative FTICR mass spectrometry of combinatorial post-translational modifications in intact histone H4. Anal Chem 2008;80:4147–4153.

291. Hunt DF, Yates JR 3rd, Shabanowitz J, Winston S, Hauer CR. Protein sequencing by tandem mass spectrometry. Proc Natl Acad Sci U S A 1986;83:6233–6237.

292. Roepstorff P. Proposal for a common nomenclature for sequence ions in mass spectra of peptides. Biomed Mass Spectrom 1984;11:601.

293. Chowdhury SM, Du X, Tolic N, Wu S, Moore RJ, Mayer MU, Smith RD, Adkins JN. Identification of cross-linked peptides after click-based enrichment using sequential collision-induced dissociation and electron transfer dissociation tandem mass spectrometry. Anal Chem 2009;81:5524–5532.

294. Savitski MM, Nielsen ML, Zubarev RA. Side-chain losses in electron capture dissociation to improve peptide identification. Anal Chem 2007;79:2296–2302.

295. de Leenheer AP, Thienpont LM. Applications of isotope dilution-mass spectrometry in clinical chemistry, pharmacokinetics, and toxicology. Mass Spectrom Rev 1992;11(4):249–307.

296. Nilsson LB, Eklund G. Direct quantification in bioanalytical LC–MS/MS using internal calibration via analyte/stable isotope ratio. J Pharm Biomed Anal 2007;43:1094–1099.

297. Ong SE, Mann M. Mass spectrometry-based proteomics turns quantitative. Nat Chem Biol 2005;1:252–262.

298. Gerber SA, Rush J, Stemman O, Kirschner MW, Gygi SP. Absolute quantification of proteins and phosphoproteins from cell lysates by tandem MS. PNAS 2003;100:6940–6945.

299. Zompra AA, Galanis AS, Werbitzky O, Albericio F. Manufacturing peptides as active pharmaceutical ingredients (API). Future Med Chem 2009;1:361–377.

300. Volkmer R. Synthesis and application of peptide arrays: quo vadis SPOT technology. ChemBioChem 2009;10:1431–1442.

# 3

# PEPTIDE DESIGN STRATEGIES FOR G-PROTEIN COUPLED RECEPTORS (GPCRs)

Anamika Singh and Carrie Haskell-Luevano

*Department of Medicinal Chemistry, University of Minnesota, Minneapolis, MN, USA*

## 3.1 INTRODUCTION

The plasma membrane of mammalian cells separates the extracellular and intracellular environments and can function as a messenger to carry information across a cell membrane to the interior of the cell, thus providing a mechanism of communication. Membrane proteins represent a large and versatile group of protein sensors that are involved in diverse physiological processes, such as neurotransmission, cellular metabolism, secretion, cellular differentiation, growth, inflammation, and immune responses, and are thus primary targets for drug discovery [1]. Peptides act as a primary source of intercellular communication in many diverse biological systems by interacting with their corresponding receptors. A large subset of these receptors couple with guanine-nucleotide binding proteins (G-proteins) to produce a signal transduction cascade of cellular actions, hence called G-protein coupled receptors (GPCRs).

GPCRs consist of seven α-helical transmembrane (TM) segments and represent the largest family of membrane proteins in the human genome. Molecular cloning studies and genome data analysis have revealed nearly 1000 members of the GPCR

*Peptide Chemistry and Drug Design*, First Edition. Edited by Ben M. Dunn.

super family, which is the largest group of membrane-spanning surface receptors on human cells [2, 3]. GPCRs respond to a diverse range of extracellular stimuli, such as neurotransmitters, peptide hormones, amino acids, amines, lipids, sugars, and even photons and cations, and thus assist in variety of different functions [1]. GPCRs represent major targets for the development of novel drug candidates in all clinical areas. It is estimated that 30% of clinically prescribed drugs function as either agonists or antagonists at GPCRs, which highlights their immense therapeutic potential [2, 4–6].

With the exception of the thyroid hormone receptor, the receptors for peptide hormones are located in the plasma membrane. The vast majority of endogenous hormones and neuroactive peptides utilize GPCRs to alter the physiology of their target cells. Irrespective of the nature of ligand and receptor structure, the basic principle that has been established is that the ligand–receptor interaction drives a conformational change of the GPCR, which transfers information to inside the cell through guanine-nucleotide binding proteins (G-protein).

## 3.2 CLASSIFICATION OF GPCRs

GPCRs generally consist of a single polypeptide chain of 400–3000 residues that form a variable extracellular N-terminus (7–3000 amino acids), seven TM spanning segments consisting α-helices (20–27 residues each), an intracellular C-terminus (12–400 amino acids), and three endoloops and three exoloops (5–250 amino acids each). The TM segment is highly conserved structurally among the GPCRs but they differ considerably in amino acid sequences. Peptide GPCRs proteins vary especially in the size of the extracellular amino-terminal tails, cytoplasmic loops, and carboxy-terminal tails, which account for their diverse structure and function [1]. Based on the sequential similarity in the TM regions, GPCRs can be divided into three main families: *Class A* receptors share sequence similar to rhodopsin and the calcitonin receptor and are the largest group, which constitutes 90% of all GPCRs, most of which have short N-terminal segments and highly conserved TM regions. These GPCRs are structurally related to rhodopsin or adrenergic receptors, and bind to many amine, purine, and peptide ligands [7]. *Class B* is the secretin/glucagon-like receptors that share little structure similarity to the other classes of GPCRs. These receptors tend to have six conserved cysteines and a hormone-binding domain in their long N-terminus. These receptors bind large peptide ligands such as glucagon, corticotropin-releasing hormone (CRH), parathyroid hormone, vasoactive intestinal peptide (VIP), growth hormone releasing hormone, calcitonin, gastric inhibitory polypeptide, and adenylate cyclase activating polypeptide receptor (PACAP) [8]. *Class C* GPCRs are related in structure to the metabotropic receptors and are the neurotransmitter receptors with long N-terminal tails (500–600 residues), comprising of a separately folded ligand-binding domain, and no TM homology with other GPCR families [9]. This family includes the $Ca^{2+}$ receptor, γ-aminobutyric acid (GABA) receptors, and metabotropic glutamate receptors (mGluRs) [8]. Other than these three families, there are classes such as adhesion and frizzled/TAS2-GPCRs, which are not classified with the above families [10]. Also, there are orphan receptors in which the endogenous ligands remain to be identified.

## 3.3 CATALOG OF PEPTIDE-ACTIVATED G-PROTEIN COUPLED RECEPTORS

Table 3.1 is a list of some GPCRs that bind to peptide ligands including class A and B receptor subtypes. The table illustrates characteristic properties of these receptors and enlists the number of amino acids in sequence, endogenous ligands, primary signal transduction pathway, tissue expression and function, knockout phenotype, if any, and disease relevance of specific receptor. Specialized databases of GPCRs can be found at http://www.gpcr.org/7tm; http://www.iuphar-db.org and http://www.ncbi.nlm.nih.gov/sites/entrez [12].

## 3.4 STRUCTURE OF GPCRs: COMMON FEATURES

Like other membrane proteins, GPCRs are partially buried in the nonpolar environment of the lipid bilayer, forming a compact bundle of TM helices. Before crystal structures of GPCRs became available, attempts to identify the orientation of helices, respectively, with each other in 3D space, chimeric receptors were generated. Studies with chimeric muscarinic receptors provided evidence that TM1 and TM7 are oriented relative to each other [13]. Other studies, such as functional analysis of engineered metal ion-binding sites [14], disulfide bonds [15], as well as spectroscopic approaches [16], aided in the identification of amino acids that are involved in the TM helix–helix interactions and their respective orientations. Despite the remarkable diversity in the structure of GPCR ligands, all GPCRs share some highly conserved structural features (Figure 3.1a): (i) insertion into the plasma membrane of the cell, (ii) the presence of seven α-helical TM segments, (iii) three extracellular and three intracellular loops, (iv) extracellular amino terminus, (v) intracellular carboxyl terminus. As described earlier, most conserved structural region among GPCRs is the TM spanning segments and most variable are the carboxyl terminus, the intracellular loop spanning TM5 and TM6, and the amino terminus. Some peptide hormones and proteins bind to the N-terminus and extracellular loops joining the TM segments (Figure 3.1b), whereas the others are postulated to bind to the TM region of the GPCR [1].

### 3.4.1 Crystal Structures

General features of the GPCRs were defined after the successful X-ray structure of bacteriorhodopsin, bovine rhodopsin, and a few other members of the super family. The first insight came from the two-dimensional crystals of rhodopsin that revealed the general arrangement of the TM helices [17]. The first three-dimensional crystal structure of dark rhodopsin was reported in 2000 [18] at 2.8 Å, and subsequent

**TABLE 3.1 Catalog of the Peptide Activated G-Protein Coupled Receptors.**

| Name | Receptor Subtypes | Amino Acids | Agonist (endogenous) | 1° Signal Transduction | Tissue Expression | Tissue Function | Knockout Phenotype | Disease Relevance |
|---|---|---|---|---|---|---|---|---|
| Angiotensin receptors | AT1 | 359 | ang 1–7, ang I, ang II, ang III | $G_i/G_0$, $Gq/G_{11}$; adenylate cyclase inhibition, phospholipase C stimulation, calcium channel, phospholipase $A_2$ stimulation | Liver, kidney, adrenal, lung, heart, thymus, uterus, ovary, aorta | Regulation of blood pressure and volume, regulator of aldosterone secretion | Normal phenotype | Increased cardiovascular risk |
| | AT2 | 363 | ang I, ang II, [*p*-aminoPhe6] ang II, ang III | $G_i/G_0$; phospholipase $A_2$ stimulation | Non pregnant uterus, heart, lung, kidney | Cardiovascular functions, Inhibition of cell growth and stimulation of apoptosis *in vitro* and *in vivo* | Normal phenotype | Mental retardation, coronary risk |
| Apelin receptor | APJ | 380 | Apelin-13 apelin-36 | $G_i/G_0$; adenylate cyclase inhibition | Heart, coronary artery, aorta, internal mammary artery, pulmonary artery and saphenous vein | Vasoconstriction | Normal phenotype | Cardiovascular |

| | | | | | | | | |
|---|---|---|---|---|---|---|---|---|
| Bombesin receptors | BB1 | 390 | Gastrin-releasing peptide (GRP) Neuromedin B (NMB) | $G_q/G_{11}$; phospholipase C stimulation | Brain, esophagus | Growth regulation | Increased stress vunerability, decreased marble burying behavior | Disruption in thermoregulation |
| | BB2 | 384 | [Phe13]bombesin, GRP, litorin, NMB, ranatensin | $G_q/G_{11;}$ phospholipase C, phospholipase A2 and phospholipase D stimulation | Pancreas, stomach, adrenal cortex , brain | Hypothermia, meal-related satiety signal | Normal phenotype | Related with cancers such as those of the lung, colon, and prostate |
| | BB3 | 399 | none | $G_q/G_{11;}$ phospholipase C and phospholipase D stimulation | Central nervous system (CNS), gastrointestinal tract | Modulate smooth-muscle contraction, exocrine and endocrine processes, metabolism, and behavior, modulate smooth-muscle contraction | Mild obesity, hyperphagia | |
| Bradykinin receptors | B1 | 353 | Bradykinin, [des-Arg9] bradykinin, kallidin, [des-Arg10] kallidin | $G_i/G_0$, $Gq/G_{11}$; adenylate cyclase inhibition, phospholipase C stimulation | Thalamus and hypothalamus, prostatic intraepithelial neoplasia and malignant lesions, T-lymphocytes, Lung fibroblasts, Lung mucosa | Modulation of blood pressure, inhibition of arterial smooth muscle cell migration, control of inflammatory pain conditions | Higher glycemia, pancreas abnormalities, hypoinsulinemia | Associated with renal disease |

(*continued*)

**TABLE 3.1** ***(Continued)***

| Name | Receptor Subtypes | Amino Acid | Agonist (endogenous) | 1° Signal Transduction | Tissue Expression | Tissue Function | Knockout Phenotype | Disease Relevance |
|---|---|---|---|---|---|---|---|---|
| | B2 | 391 | Bradykinin, kallidin | Gs, $G_i/G_0$, $Gq/G_{11}$; adenylate cyclase inhibition, phospholipase C and $A_2$ stimulation | Endothelium and tunica media of aorta and endocardium, lung fibroblasts, gastric mucosa, medulla | Mediation of vasolidation in forearm vasculature, mediation of normal vasomotor responses in resistance and epicardial coronary vessels | Insulin resistance | Associated with cardiovascular disease, renal disease, airway disease and insulin resistance |
| Calcitonin receptors | CT | 490 | Calcitonin, amylin, α-CGRP, β-CGRP | Gs, $Gq/G_{11}$; adenylate cyclase stimulation | Brain, testes, stomach, duodenum, jejunum, kidney, bone | Inhibits food intake when injected into hypothalamic areas, inhibitor of bone resorption acting directly on osteoclasts | High bone mass | |
| Chemokine receptors | CCR5 | 352 | CCL2, CCL3, CCL4, CCL5, CCL8, CCL11, CCL13, CCL14 | $G_i/G_0$; calcium channel | Testes, blood and cerebrospinal fluid, tonsil B and Th1-type lymphocytes, microglia, tonsil B lymphocytes, basophils, blood dendritic cells | | | HIV-1-AIDS |

| | | | | | | | | |
|---|---|---|---|---|---|---|---|---|
| Cholecystokinin receptors | CCK1 | 428 | CCK-33, CCK-4, CCK-58, CCK-8 (sulfated, desulfated), gastrin, pentagastrin | $G_q/G_{11}$; phospholipase C stimulation | Stomach, exocrine and endocrine pancreas, gallbladder, adrenal gland, blood cells, kidney | Stimulate the secretion of the digestive enzyme pancreatic amylase, stimulate gallbladder contraction | Increased intestinal cholesterol absorption, increased gastric acid secretion | |
| | CCK2 | 447 | CCK-33, CCK-4, CCK-58, CCK-8 (sulfated, desulfated), gastrin, pentagastrin | $G_q/G_{11}$; phospholipase C stimulation | Stomach, exocrine and endocrine pancreas, adipocytes, adrenal gland, blood cells | Hyperphagia and increased fat deposition improvement of memory processes | Impaired learning and memory | Role in the growth of pancreatic cancer |
| Corticotropin -releasing factor receptors | CRF1 | 415 | CRF, urocortin | Gs; adenylate cyclase stimulation | Neocortex, limbic structures, cerebellum and pituitary | Define the behavioral stress response | Reduce anxiety-like behavior under stress conditions | |
| | CRF2 | 411 | CRF, urocortin | Gs; adenylate cyclase stimulation | Limbic structures that are involved in stress responses | Delayed anxiolytic effect | Hypersensitive to stress | |
| Endothelin receptors | $ET_A$ | 427 | ET-1, ET-2 | $G_q/G_{11}$; phospholipase C, phospholipase $A_2$ and phospholipase D stimulation | Vascular smooth muscle cells | Vasoconstrictor, release of endothelium -derived relaxing factors such as nitric oxide (NO) and prostanoids | Velocardiofacial syndrome | |

*(continued)*

**TABLE 3.1 (*Continued*)**

| Name | Receptor Subtypes | Amino Acid | Agonist (endogenous) | 1° Signal Transduction | Tissue Expression | Tissue Function | Knockout Phenotype | Disease Relevance |
|---|---|---|---|---|---|---|---|---|
| | $ET_B$ | 442 | [$^{125}$I]ET-1, ET-3 | Gs, $G_i/G_0$, $Gq/G_{11}$; phospholipase C, $A_2$ and D stimulation | Endothelial cells lining wall of all blood vessels | Endothelium dependent vasodilatation, broncoconstriction | Aganglioic megacolon (resembling Hirschsprung's disease), associated with coat color spotting. | Hirschsprung's disease |
| Galanin receptors | $GAL_1$ | 349 | Galanin | Gi/Go; inhibits adenylyl cyclase | Brain, spinal cord, as well as in peripheral sites such as the small intestine and heart | Stimulates cortisol secretion , cartilage growth plate physiology and fracture repair | | |
| Gonadotrophin -releasing hormone receptors | GnRHR | 328 | GnRH | $Gq/G_{11}$; phospholipase C stimulation | Pituitary, breast, ovary, prostate | Gonadotropin synthesis and secretion, ovarian steroidogenesis, cell proliferation, apoptosis | Deficiency in Luteinizing hormone (LH), follicle- stimulating hormone and gonadal steroids, failure of the sex organs to develop postnatally | Hypogonadotropic hypogonadism |
| Growth hormone -releasing receptor | GHRHR | 423 | GHRH(1–29) NH2 (rat) | Gs; adenylate cyclase stimulation, calcium channel | Pituitary, kidney | Stimulation of GH release | short stature, delayed bone age | Pancreatic and pituitary tumors |
| Glucagon | GcgR | 477 | Glucagon | Gs; adenylate cyclase stimulation | Liver, kidney | Glycogenolysis | Increase of glucagon content in the pancreas and blood | Pancreatic cancer |

| | | | | | | | | |
|---|---|---|---|---|---|---|---|---|
| Ghrelin | GHSR | 366 | Ghrelin, des-octanoyl ghrelin | $Gq/G_{11}$; adenylate cyclase stimulation, phospholipase C stimulation, calcium channel | Heart, large conduit vessels, brain | Stimulate food intake and lead to body weight gain | | Atherosclerosis |
| KiSS1-derived peptide receptor | KISS1R | 398 | Kisspeptin-9,10, 13,14,15,28, 54 | $Gq/G_{11}$; phospholipase C stimulation | Pancrease, brain, placenta, pituitary | Metastasis inhibitor, Inhibition of trophoblast migration | Reduced FSH and LH secretion | Idiopathic hypog-onadotrophic hypogonadism |
| Melanin- con-centrating hormone receptors | $MCH_1$ | 422 | | | | | | |
| Melanocortin receptors | MC1R | 317 | α-MSH | Gs; adenylate cyclase stimulation | Dermal papilla cells, endothelial cells, melanoma cells, pituitary and testes | Regulates the synthesis of eumelanin (black-brown pigment) versus phaeomelanin (yellow-red pigment), anti-inflammatory actions | Responsible for pigmentation phenotypes such as the red fox or yellow mouse | Melenoma |
| | MC2R | 297 | Adrenocorticotropic hormone (ACTH) (1–39) | Gs; adenylate cyclase stimulation | Adrenal gland and skin tissues | Adrenal cortical steroidogenesis | Cushing's syndrome | Glucocorticoid deficiency |

(*continued*)

**TABLE 3.1 (*Continued*)**

| Name | Receptor Subtypes | Amino Acid | Agonist (endogenous) | 1° Signal Transduction | Tissue Expression | Tissue Function | Knockout Phenotype | Disease Relevance |
|---|---|---|---|---|---|---|---|---|
| | MC3R | 360 | α-MSH, γ-MSH | Gs; adenylate cyclase stimulation | Hypothalamus, stomach, duodenum and pancreas, heart, placenta | Energy homeostasis, naturesis | Increased fat mass, reduced lean mass and a higher feed efficiency, hyperleptinaemic | Feeding disorder |
| | MC4R | 332 | α-MSH | Gs; adenylate cyclase stimulation | Hypothalamus, pituitary gland, dermal papilla cells | Inhibition of feeding behavior and the regulation of metabolism, involved in sexual behavior and male erectile function | Obese, hyperphagic, hyperinsulinemic, and hyperglycemic | Obesity and related diseases |
| | MC5R | 325 | α-MSH | Gs; adenylate cyclase stimulation | Adrenal gland, fat cells, kidney, leukocytes, lung, lymph node, mammary gland, ovary, pituitary, testis and uterus | Sebum production | Decreased production of sebaceous lipids | |
| Motilin | | | | | | | | |
| Neuropeptide Y receptors | Y1 | 384 | NPY, PP, PYY | $G_i/G_0$; adenylate cyclase inhibition | Colon, kidney, adrenal gland, heart, placenta | Modulation of food intake, Inhibition of electrogenic ion transport | Obesity and hyperinsulinemia | Obesity and related diseases |

| | | | | | | | |
|---|---|---|---|---|---|---|---|
| Neurotensin receptors | NTS1 | 418 | Neurotensin, neuromedin N, large neurotensin | $Gq/G_{11}$; phospholipase C stimulation | Brain, colon | Smooth muscle contraction, modulation of turning behavior | Hyperthermia, a small increase in body weight and an increase in food intake |
| | NTS2 | 410 | Neurotensin, neuromedin N | $Gq/G_{11}$; phospholipase C stimulation | Brain, spinal cord | Analgesia | Normal phenotype |
| Neuromedin U | NMU1 | 426 | NMS, NMU-25 | $Gq/G_{11}$; phospholipase C stimulation | Small intestines and stomach | | |
| Opioid receptors | $\delta$ | 372 | β-Endorphin, [Leu]-enkephalin, [Met]-enkephalin, dynorphins, α-neoendorphin | $Gq/G_{11}$; adenylate cyclase inhibition, phospholipase C stimulation, potassium and calcium channel | Forebrain, skin, immune cells | Analgesia | Anxiety, depression, increased sensitivity to inflammatory pain |
| | $\kappa$ | 380 | β-Endorphin, [Leu]-enkephalin, [Met]-enkephalin, dynorphins, α-neoendorphin, β-neoendorphin | $Gq/G_{11}$; adenylate cyclase inhibition, potassium and calcium channel | CNS, skin, immune cells | Water diuresis, potentially by modulation of vasopressin | Enhanced sensitivity to chemical visceral pain, abolished hypolocomotor |
| | $\mu$ | 400 | β-Endorphin, [Leu]-enkephalin, [Met]-enkephalin | $Gq/G_{11}$; adenylate cyclase inhibition, adenylate cyclase stimulation, phospholipase C stimulation, potassium and calcium, phospholipase A2 and D stimulation | CNS, skin, immune cells | Constriction of the pupil. | |

*(continued)*

**TABLE 3.1 (*Continued*)**

| Name | Receptor Subtypes | Amino Acid | Agonist (endogenous) | 1° Signal Transduction | Tissue Expression | Tissue Function | Knockout Phenotype | Disease Relevance |
|---|---|---|---|---|---|---|---|---|
| Orexin | OX1 | 425 | Orexin-A, orexin-B | $G_i/G_0$, $Gq/G_{11}$; phospholipase C stimulation | Pituitary, retina, lung, skeletal muscle, kidney, testis , CNS | Stimulation of glucocorticoid secretion, modulation of feeding, Facilitation of learning and memory | | |
| | OX2 | 444 | Orexin-A, orexin-B | $Gq/G_{11}$; phospholipase C stimulation | Pituitary, CNS, brain, lung, spleen, testis, olfactory system | Stimulation of food intake, increase in wake duration and decrease in Rapid eye movement (REM) and non-REM sleep | Disrupted wakefulness, abnormal attacks of non-REM sleep | Excessive daytime sleepiness, risk of developing cluster headaches |
| Parathyroid hormone receptor | PTH1 | 593 | PTH 1–34 (human), PTHrP 1–36 (human) | Gs; adenylate cyclase stimulation | Kidney, bone, CNS | Bone formation and resorption, reduced promoter activity | Altered bone development | Eiken syndrome, blomstrand chondrodys-plasia |
| | PTH2 | 550 | PTH 1–34 (rat), PTHrP 1–34, PTHrP 1–36 (human), TIP39 (human/bovine) | Gs; adenylate cyclase stimulation | Pancrease, Thyroid, brain, CNS, Renal vessels | Stimulation of hypothalamic releasing factor secretion, positive inotropy, modulation of renal vessel tone | | Osteoarthritis |

| | | | | | | | | |
|---|---|---|---|---|---|---|---|---|
| Prokineticin receptors | PKR1 | 393 | Prokineticin 1, prokineticin 2, prokineticin 2$\beta$ | $Gq/G_{11}$; phospholipase C stimulation | Testis, medulla oblongata, skin, skeletal muscle, CNS | Macrophage migration and the production of pro-inflammatory cytokines | Lack macrophage chemotaxis and cytokine production | |
| | PKR2 | 384 | Prokineticin 1, prokineticin 2 | $G_i/G_0$, $Gq/G_{11}$ | CNS, endocrine tissues, testis, brain, digestive tract | Inhibition of food intake, stimulation of water intake | | |
| Prolactin -releasing peptide receptor | PrRP | 370 | NPY 18–36 (human), PrRP-20 (human), PrRP-31 (human), PrRP-24-31 (human) | $Ca^{2+}$ mobilization | Pituitary gland, CNS | Stimulation of corticotropin -releasing hormone (CRH)-mediated adrenocorticotropin (ACTH) release, reduction in food intake | Increased body fat, increased levels of insulin and leptin and decreased glucose tolerance | Severe early-onset obesity, diabetes mellitus OLETF type I |
| Somatostatin receptors | sst1 | 391 | CST-14, CST-17, SRIF-28, SRIF-14 | $G_i/G_0$; adenylate cyclase inhibition | Hypthalamus, pancreatic islets, spleen, Whole placenta, pituitary, blood vessels | Antiproliferation, modulates somatostatin levels, inhibition of GH level | Inability to regulate basal Growth hormone levels | Growth control |
| | sst2 | 369 | CST-14, CST-17, SRIF-28, SRIF-14 | $G_i/G_0$; adenylate cyclase inhibition | Hypothalamus, pituitary, spleen, pancreas | Inhibition of GH release, gastric acid secretion, histamine release and glucagon release | Hyperplasia of thyrotropes, increased release of ACTH, lower gastric pH values | Prostate cancer |

(*continued*)

**TABLE 3.1** ***(Continued)***

| Name | Receptor Subtypes | Amino Acid | Agonist (endogenous) | 1° Signal Transduction | Tissue Expression | Tissue Function | Knockout Phenotype | Disease Relevance |
|---|---|---|---|---|---|---|---|---|
| | sst3 | 418 | CST-14, CST-17, SRIF-28, SRIF-14 | $G_i/G_0$; adenylate cyclase inhibition | Cerebellum, amygdala, hippocampus, striatum, olfactory bulb and cortex, spleen, kidney, liver | Inhibition of IL-2 secretion and cell proliferation and GH release | | Gastric cancers |
| | sst4 | 388 | CST-14, CST-17, SRIF-28, SRIF-14 | $G_i/G_0$; adenylate cyclase inhibition | Hippocampus, brain and lung tissues, pituitary, blood vessels | Antiproliferation | | |
| | sst5 | 364 | CST-14, CST-17, SRIF-28, SRIF-14 | $G_i/G_0$; adenylate cyclase inhibition | Pancreatic islets, Small intestine, heart, adrenal, cerebellum, pituitary, placenta, skeletal muscle | Inhibition of GH release and insulin release, antiproliferation | Increased insulin, leptin and glucagon concentra-tions | Grave's disease |
| Tachykinin receptors | NK1 | 407 | neurokinin 1 | $Gq/G_{11}$; phospholipase C stimulation | Native human airway smooth muscle cells, colon, amygdala | Mood and stress regulation, cognitive functions | Normal phenotype | Laryngeal cancer, inflammatory bowel disease, psoriasis, blood vessels |
| | NK2 | 398 | Neurokinin A | $Gq/G_{11}$; phospholipase C stimulation | Colon, native human airway smooth muscle cells | Colonic functions | Normal phenotype | Laryngeal cancer, Chronic obstructive pulmonary disease (COPD) |

| | | | | | | | | |
|---|---|---|---|---|---|---|---|---|
| | NK3 | 465 | Neurokinin B | $Gq/G_{11}$; phospholipase C stimulation | Colon, native human airway smooth muscle cells, placenta, platelets | Reproduction | Normal phenotype | Schizophrenia, alcohol dependence |
| Thyrotropin -releasing hormone receptor | TRH1 | 398 | TRH | $Gq/G_{11}$; phospholipase C stimulation | Pituitary gland, lymphocytes, CNS, Testis | Stimulation of thyrotropin Thyroid stimulating hormone (TSH) release and corticosterone release, regulation of blood pressure, lean body mass variations | Reduced growth rate | Hypothyroidism, osteoporotic fracture |
| Urotensin receptor | UT | 389 | Urotensin-related peptide, Urotensin-II (rat and mouse) | $Gq/G_{11}$; phospholipase C stimulation | Heart, arteries, CNS, skeletal muscles, kidney | Vasoconstriction, cardiac contractility | Attenuation of vasoconstrictor to Urotensin-II (UT-II) | Cardiovascular |
| Vasoactive intestinal peptide and pituitary adenylate cyclase-activating polypeptide receptors | PAC1 | 368 | PACAP-27, PACAP-38, VIP | Gs; adenylate cyclase stimulation | Adrenal medulla, pancreatic acini, glands of uterus, myenteric plexus, CNS | Stimulation of catecholamine secretion, proliferation, apoptosis and migration of immature cerebellar granule cells | Reduced glucose tolerance and impaired glucagon response to insulin-induced hypoglycaemia | Colon and prostate cancer |

(*continued*)

**TABLE 3.1 (*Continued*)**

| Name | Receptor Subtypes | Amino Acid | Agonist (endogenous) | 1° Signal Transduction | Tissue Expression | Tissue Function | Knockout Phenotype | Disease Relevance |
|---|---|---|---|---|---|---|---|---|
| | VPAC1 | 457 | Growth hormone-releasing factor (GRF), PACAP-27, PACAP-38, Peptide histidine isoleucine (PHI), Peptide histidine methionine (PHM), Peptide histidine valine (PHV), secretin, VIP | Gs; adenylate cyclase stimulation | Cerebral cortex, hippocampus, amygdala, lung, prostate, kidney | Smooth muscle relaxation, exocrine and endocrine secretion, water and ion flux in lung | | Prostate cancer, HIV1, inflammatory bowel disease, rheumatoid arthritis |
| | VPAC2 | 438 | PACAP-27, PHI, PHV, VIP | Gs; adenylate cyclase stimulation | Stroma of uterus and prostate, smooth muscles in GI tract, seminal vesicles and skin, blood vessels, thymus | Growth, basal energy expenditure, and male reproductive functions. | Loss of normal circadian rhythms of rest/activity behavior | Rheumatoid arthritis, formation of gallbladder stones and gallbladder polyps |
| Vasopressin and Oxytocin receptors | V1A | 418 | Arginine vasopressin | Phosphatidylinositol -calcium stimulation | Brain | Cell contraction and proliferation, platelet aggregation, release of coagulation factor and glycogenolysis | Diabetes, behavorial disorders | Autism, hypertension |

| | | | | | | | |
|---|---|---|---|---|---|---|---|
| V1B | 424 | Vasopressin V3 | Phosphatidylinositol -calcium stimulation | Anterior pituitary | Stimulates ACTH release | Mood disorders | Depression |
| V2 | 371 | Arginine vasopressin | Gs; adenylate cyclase stimulation | Kidney tubule, predominantly in the distal convoluted tubule and collecting ducts, lung tissues | Concentrate the urine and maintain water homeostasis | | Nephrogenic Diabetes Insipidus, lung cancer |
| OT | 389 | Oxytocin | Phosphatidylinositol -calcium stimulation | Uterus, smooth muscle cells and epithelial cells of peritoneal endometriotic lesions and ovarian endometriotic cysts Gastrointestinal (GI) tracts | Parturition, bone metabolism | | Related to pregnancy uterus contraction |

*Source*: Data taken mainly from Reference [11].

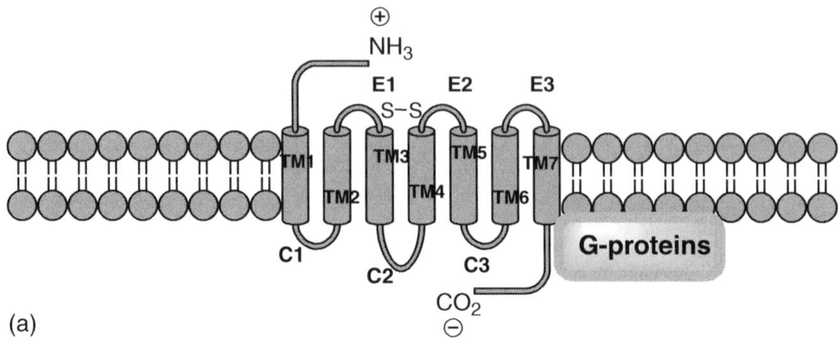

**Figure 3.1** (a) General structure of GPCR; E = exoloop; C = Cytoloop. (b) Schematic presentation of peptide hormone-receptor interaction for peptides of ≤40 amino acids.

structures have been refined at resolutions as high as 2.2 Å [19–21]. The structure of an inactive form confirms the anticlockwise bundle of 7 TM α-helices, connected by extracellular loops of varying lengths [18]. Also, the amino terminal ligand binding segment of the Follicle-stimulating hormone (FSH) receptor was crystallized in complex with its ligand to 2.9 Å, which shed light on the receptor–ligand interactions and receptor activation [22].

Remarkable progress in the analysis of GPCR structures was published in 2007–2008. Crystal structures of a cephalopod rhodopsin showed structural differences, suggesting its coupling to the G-protein Gq rather than for transducin [23, 24]. Crystal structures of another *class A* GPCR, β-adrenergic receptor, which was bound to the inverse agonist, represent the first structures of GPCRs bound to diffusible ligands [25–29]. In addition, the crystal structure of bovine opsin has provided interesting information about the ligand binding and activation pathway [30, 31]. In another important study, native bovine opsin, an inactive form of rhodopsin, was crystallized [30] by optimizing the selective extraction of rhodopsin from rod cell disc membranes. This methodology enabled crystallization without any modification of the protein that might cause structural distortions. Structural examination of this opsin revealed only slight changes relative to rhodopsin for TM helices 1–4. The most obvious differences were found in the region of TM helices 5–7 and were especially prominent at the cytoplasmic ends of these helices and cause rearrangement of the C-2 and C-3 loops [32].

Despite these remarkable findings, there are several obstacles to obtain decent crystals for other GPCRs. For example, the lack of stability of purified GPCRs in detergents compatible with crystallography can be a major limitation. Most of the GPCRs are not stable in the detergents used to obtain rhodopsin crystals; instead, they show stability in nonionic detergents with relatively long alkyl chains, which are capable to form larger micelles that prevent the formation of crystals [33]. Also, posttranslational modifications such as glycosylation, phosphorylation, palmitoylation, and conformational flexibility of the receptors generate structural heterogeneity. Therefore, structural analysis of other GPCRs has largely been limited to the indirect methods such as, the use of site-directed mutagenesis and cysteine scanning mutagenesis [34] to detect receptor–ligand interactions and the use of engineered metal ion binding sites to probe intramolecular interactions [35]. While these approaches provide low resolution structural information, this information can be used to support and improve the accuracy of homology models based on rhodopsin.

Initial rhodopsin structure and subsequent improvements in resolution have provided a template for the creation of homology models [36, 37]. There are various similarities in the structure of the GPCRs but their primary sequences are significantly different. Therefore, it is always been a good strategy to compare the amino acid sequences with other members of the family in attempts to identify specific residues that may be important for molecular recognition. With the year 2007 publications of the crystal structures of the $\beta_1$- and $\beta_2$-adrenergic receptors, it is now possible to utilize both these alternative templates for the creation of homology models as well as to validate the previous rhodopsin-based homology models. Some homology modeling studies suggest that in some cases the adrenergic receptor may better serve as a basis for homology model generation [38, 39]. Structure–function studies, mutagenesis studies, and affinity labeling studies have been used to validate and revise the proposed models. Using this combined approach, the integration of a template and experimental data, molecular models of GPCRs, and receptor–ligand complexes have been generated, which have been reviewed [2, 40].

## 3.5 GPCR ACTIVATION

Activation of a GPCR by an extracellular agonist initiates G-protein-mediated signal transduction that results in the subsequent cascade of intracellular electrochemical signals and release of second messengers. In response to ligand binding, the ligand–receptor complex and cytoplasmic portion of the receptor undergoes conformational change(s), allowing interaction with the G-proteins (which are localized in the cytoplasmic side of the membrane), thereby transmitting the signal across the membrane. Significant advances in our understanding of the structure and function of GPCRs have resulted from the identification of particular residues critical to cell signaling and ligand binding. This has most often resulted from the analysis of *in vitro* mutated forms of the GPCRs. Generally, the event of receptor activation (in case of peptidergic GPCRCs (pGPCRs)) may be divided into at least four steps: ligand binding and signal generation, ligand induced GPCR structural changes, signal transduction, and signal transfer to cytoplasmic signal molecules. Many issues regarding the activation of GPCRs are under rigorous debate and the literature is rapidly expanding in this area. Mechanisms of GPCR/G-protein interaction are reviewed extensively elsewhere [41–45].

### 3.5.1 Ligand (Peptide) Binding and Receptor Activation

Numerous studies have been carried out to identify domains involved in ligand binding to various subclasses of GPCRs. Receptors targeted by bulky ligand molecules, such as large peptides and hormones, have resulted in the generation of experimental data supporting the binding of ligands at the N-terminus, extracellular loops, as well as at the TM segments, depending upon the specific receptors examined. For example, peptides ≤40 amino acids have been reported to bind to both the GPCR core and exoloops, whereas, polypeptides ≤90 amino acids bind to exoloops and N-terminal segment (Figure 3.1b). For the majority of family A peptide receptors, ligands have been postulated to interact with the receptor at the amino terminus and extracellular loop regions. This includes the receptors for angiotensin [46], neuropeptide Y [47], chemokines (interleukin-8, IL-8) [48], vasopressin/oxytocin [49], Gonadotrophin-releasing hormone (GnRH) [50], formyl-Leu-Met-Phe [51], somatostatin [52], bradykinins [53], and cholecystokinin (CCK)/gastrin [54, 55]. Significance of extracellular loop region for peptide binding was demonstrated using affinity cross-linking techniques in the GnRH receptor [50], the bradykinin B2 receptor [53], and the CCK-A receptor [54, 55].

Some examples of peptide ligand binding to their receptors are briefly discussed. Using spectroscopic and mutational studies, a small peptide ligand such as tripeptide *N*-formyl-Met-Leu-Phe binds in the TM core around TMs 2 and 3, whereas the C-terminal region of the ligands associates with the N-terminal segment and exoloops 1 and 2 [51]. In another example, the hydrophobic C-terminal region of angiotensin II (DRVYIHPF) appears to enter the TM core of the angiotensin receptor, and the C-terminal carboxyl group pairs with TM 5 Lys199 [56] 7–14 Å from the extracellular surface. On the contrary, the Asp-Arg of the DRVYIHPF sequence seems to ion pair with exoloop 2 His183 and exoloop 3 Asp281 of the receptor, respectively [46]. Ligand-binding of other receptors are extensively reviewed elsewhere [42].

These receptors have poorly defined binding pockets that can accommodate ligands in many orientations and at alternative binding domains. In addition, many receptors have been found to assume different conformations with distinct signaling functions. This is further complicated by the fact that single receptors may impinge on multiple signaling pathways, whereas groups of receptors may all act on a single intracellular signaling cascade [44]. In many cases the un-liganded receptor has some basal (constitutive) activity. The term *efficacy* is used to describe the effect of a ligand on the functional properties of the receptor (i.e., maximal stimulation or partial agonist pharmacology). Agonists are defined as ligands that fully activate the receptor. Partial agonists induce a submaximal response at saturating concentrations. Inverse agonists decrease the inherent basal activity in a dose-dependent manner. Antagonists have no effect on basal activity, but competitively block access of other ligands that can distinguish between ligand binding and receptor activation by competitively inhibiting agonist binding [57].

A number of kinetic models have been developed to explain GPCR activation using indirect methods such as ligand binding affinity and the activation of G-proteins or effector enzymes [58, 59]. The two-state model is the simplest of all proposals, in which a receptor exists primarily in two states: the inactive state (R) and the active state (R*). In the absence of ligands, the level of basal receptor activity is determined by the equilibrium between R and R*. Full agonists bind to and stabilize R*, while antagonists bind to and stabilize R. Partial agonists have some affinity for both R and R* and are, therefore, less effective in shifting the equilibrium toward R*. The two-state model is very straightforward and describes the systems consisting one receptor and one G-protein. However, there is experimental data to support the concept that multiple conformational states of GPCRs exist in equilibrium and are important for different physiological processes [57]. Within this framework, each ligand may induce or stabilize a unique conformational state that can be distinguished by the activity of that state toward different signaling molecules (e.g., G-proteins, kinases, and arrestins) [1].

### 3.5.2 Common Structural Changes among GPCRs

Agonists bind to a GPCR and induce a conformational change that leads to G-protein activation. Regardless of the mode of ligand binding, the common result is a conformational change in the GPCR. Site-directed spin-labeling experiments of bovine rhodopsin have shown that activation of this receptor primarily results in an outward movement of helix 6, thereby opening a crevice within the intracellular surface of the receptor [15]. This conformational change appears to be essential for transducin (Gt) activation because cross-linking helices 3 and 6 of rhodopsin with artificial disulfide or metal-ion bonds prevents Gt activation [15, 60]. Biophysical and biochemical studies such as fluorescence and NMR have also supported the hypothesis that the GPCRs undergo conformational changes within TM segments and cytoplasmic domains. Spectroscopic studies of $\beta_2$AR labeled with fluorescent probes demonstrate movement in both TM3 and TM6 on activation [61, 62]. Another study of $\beta_2$AR labeled with fluorescent probes at the cytoplasmic end of TM6 provide data supporting the hypothesis that agonists induce a rotation or tilting movement of the cytoplasmic end of TM6 similar to that observed in rhodopsin [63]. Additional support for movement of TM3 and TM6 in the $\beta_2$AR comes from zinc cross-linking studies [64]. Cysteine cross-linking studies on the M3 muscarinic receptor provide evidence for the movement of the cytoplasmic ends of TM5 and TM6 toward each other on agonist activation [65, 66]. Similar findings have been reported for the thyrotropin-releasing hormone receptor [67].

### 3.5.3 G-Protein Coupled Intracellular Signaling Pathways

The majority of peptide-GPCRs transduce a signal through a series of membrane-localized Guanosine diphosphate/Guanosine triphosphate (GDP/GTP) binding proteins known as G-proteins. Classic interactions between receptors, G-protein, and membrane-localized adenylate cyclase are illustrated using the pancreatic hormone glucagon as an example (Figure 3.2). When G-proteins bind to receptors, GTP exchanges with GDP bound to the $\alpha$ subunit of the G-protein. The $G_\alpha$-GTP complex binds adenylate cyclase, activating the enzyme. Effectors are related to secondary messengers to produce metabolic responses. Activation of adenylate cyclase leads to cyclic adenosine monophosphate (cAMP) production in the cytosol and to the activation of protein kinase A (PKA), followed by regulatory phosphorylation of

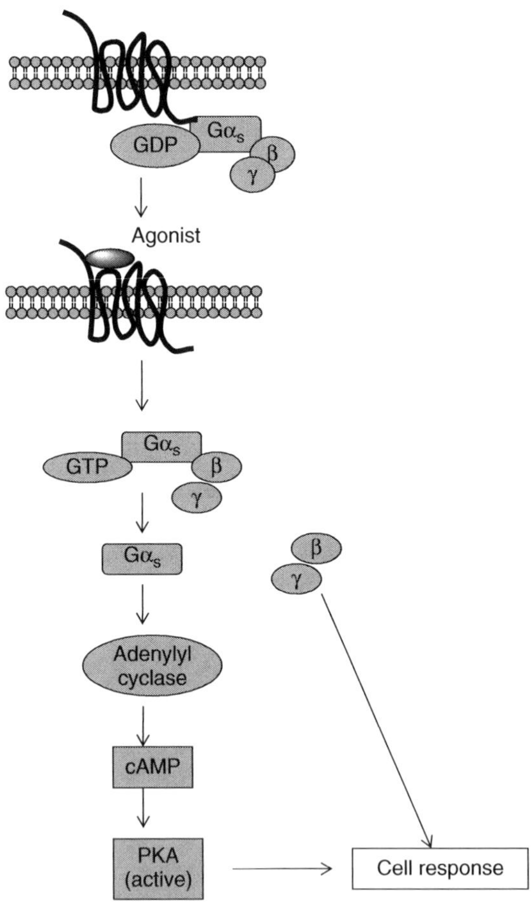


**Figure 3.2** Classical example of signal transduction in seven transmembrane receptor. (*See insert for color representation of this figure.*)

**TABLE 3.2 Heterotrimeric G-Proteins and Their Effectors. Modified from ref. [3].**

| G-Protein Subunits | Effectors | References |
|---|---|---|
| $G\alpha_s$ | ↑Adenylyl cyclase | [68] |
| $G\alpha_{olf}$ | RGS-PX1 (GAP, sorting nexin) | [69] |
| | Calcium channels | |
| | c-Src tyrosine kinases | [70] |
| $G\alpha_T$ (transducin) | ↑cGMP phosphodiesterase | [71] |
| $G\alpha_{gust}$ (gustducin) | Phosphodiesterase (bitter, sweet taste) | [72] |
| $G\alpha_{i\,1,2,3}$ | ↓ Adenylyl cyclase, ↑c-Src tyrosinase kinases | [70, 73–75] |
| $G\alpha_o$ | | |
| $G\alpha_z$ | Rap1Gap1 | |
| $G\alpha_q$, $G_{11}$, $G_{14,\,15,\,16}$ | ↑Phospholipase C | [76–78] |
| | Leukemia associated RhoGEF (LARG) RhoGEF | |
| $G_{12}$, $G_{13}$ | p115 RhoGEF, PDZ-RhoGEF, LARG RhoGEF | [76, 77, 79–81] |
| | (Rho activation, stress-fiber formation) | |
| | E-candherin (β-catenin release) | |
| $G\beta\gamma$ | KIR3. 1–3.4 (GIRK $K^+$ channels) | [82–87] |
| | GRKs | |
| | ↑Adenylyl cyclases (ACII, ACIV) | |
| | ↑Phospholipase (PLC $\beta$1, $\beta$2, $\beta$3) | |
| | PI3K$\gamma$ | |

cGMP, cyclic Guanosine monophosphate (GMP); GAP, GTPase-activating protein; GEF, guanine-nucleotide exchange factor; GIRK, G-Protein-regulated inwardly rectifying potassium channel; PI3K, phosphatidylinositol 3-kinase; RGS, regulator of G-protein signaling.

numerous enzymes. Hydrolysis of GTP to GDP leads to the reassociation of the heterotrimer and the termination of the activation cycle [3].

Generally, GPCR's signal through the major G-proteins (Table 3.2); however, alternative and/or multiple signal transduction pathways are emerging as well as homo and heterodimerization of GPCRs, which may modify a ligands signal transduction pathways. These research areas are still in their infancy, primarily due to technological limitations, and will provide an area of active research for years to come. G-protein is heterotrimeric, comprising $\alpha$, $\beta$, and $\gamma$ subunits. The α-subunit is responsible for GTP and GDP binding and for GTP hydrolysis, whereas the $\beta$ and γ-subunits are associated in a tightly linked $\beta\gamma$ complex. G-proteins are generally referred to by their α-subunits. Therefore, the $G_s$ heterotrimeric complex contains $G\alpha_s$; $G_q$ contains $G\alpha_q$; $G_i$ contains $G\alpha_i$, and so on. Four distinct α-subunits are recognized: $G_s$ (stimulatory protein) proteins couple to stimulate adenylyl cyclase; $G_i$ (inhibitory protein) proteins couple to inhibition of adenylyl cyclase as well as to activation of G-protein coupled inwardly rectifying potassium (GIRK) channels; $G_q$ proteins couple to activate phosholipase C$\beta$; and G12 proteins. Both $\alpha$ subunit and $\beta\gamma$ dimer signal through the activation, or inhibition, of various effectors (Table 3.2) [3].

A second class of peptide hormones induces the transduction of two second messengers, namely, diacylglycerol (DAG) and $IP_3$ (explained below for α-adrenergic stimulation by epinephrine). Hormone binding is followed by an interaction with a stimulatory G-protein, which in turn is followed by G-protein activation of membrane-localized phospholipase C-$\gamma$, (PLC-$\gamma$). PLC-$\gamma$ hydrolyzes phosphatidylinositol 4,5-bisphosphate ($PIP_2$) to produce the second messengers: $IP_3$, which is soluble in the cytosol, and DAG, which remains in the membrane phase. Cytosolic $IP_3$ binds to sites on the endoplasmic reticulum, opening $Ca^{2+}$ channels and allowing stored $Ca^{2+}$ to flood the cytosol. There, it activates numerous enzymes, many by activating their calmodulin or calmodulin-like subunits. DAG has two roles: (i) it binds and activates protein kinase C (PKC) and (ii) it opens $Ca^{2+}$ channels in the plasma membrane, reinforcing the effect of $IP_3$. Like PKA, PKC phosphorylates serine and threonine residues of many proteins, thus modulating their catalytic activity.

## 3.6 STRUCTURE AND FUNCTION OF PEPTIDE HORMONES

Peptide hormones play an essential role in many physiological systems. The noncovalent interaction with their membrane bound GPCRs is highly specific and results in signal transduction from the external milieu to within the cell. Hormones, neuorotransmitters, antigens, cytokines, and growth factors represent key classes of such peptide ligands. Endogenous peptide hormone synthesis originates at the genomic DNA level by a unique nucleotide sequence. Three nucleotides make up a codon, which then is translated into a specific amino acid residue. The nucleotide sequence is contained on a gene that consists of a promoter domain that initiates and determines when the DNA is transcribed to mRNA and subsequently translated into an amino acid sequence. This nascent peptide/protein chain is then transported into the cisternae of the rough endoplasmic reticulum and then to the Golgi elements. Peptides are then pinched off into secretory vesicles within the cellular cytoplasm for further distribution depending upon the type of cell and function of the hormone. If a peptide functions at a neurohormone, generally, then these vesicles are transported (sometime up to relatively long distances) to the neuronal axon terminals awaiting release. Some prohormone peptides are posttranslationally modified by endopeptidases, resulting in one or more distinct peptide hormones. An example of this is the posttranslational processing of the pro-opiomelanocortin (POMC) gene transcript into several peptide hormones with distinct amino acid sequences (Figure 3.3) [88–90]. Once released, hormones can reach their target cells by one or more mechanisms. Generally, peptide hormones have a short half-life (2–60 min), depending on the presence of peptidases (enzymes that cleave peptides), pH, and/or metabolic clearance. Peptidases fall into two general classes. Exopeptidases (carboxy- and aminopeptidases) cleave the peptide from the C- or N-terminal, respectively. Endopeptidases cleave the amide bonds within the peptide as specific recognition sites. Common sites include dibasic amino acids such as Lys and Arg.

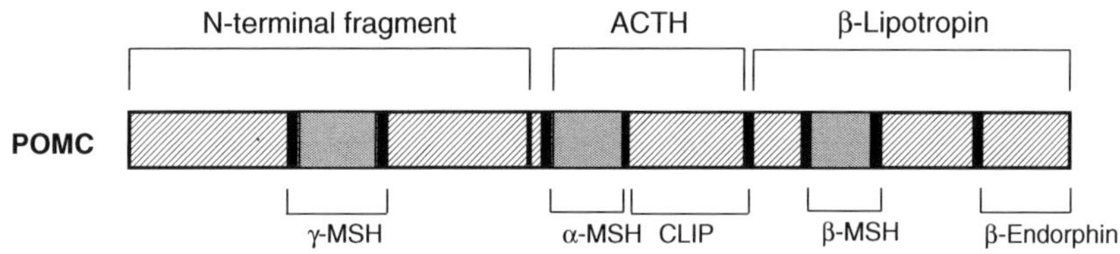


| | |
|---|---|
| **ACTH (1-24)** | SYSME**HFRW**GKPVGKKRRPVKVYP |
| **α-MSH** | Ac-SYSME**HFRW**GKPV-$NH_2$ |
| **β-MSH** | AEKKDEGPYRME**HFRW**GSPPKD-OH |
| **γ-MSH** | YVMG**HFRW**DRFG-OH |

**Figure 3.3** POMC processing of the melanocortins and primary sequence of the melanocortin peptides with core "His-Phe-Arg-Trp."

## 3.7 DESIGN APPROACHES FOR GPCR SELECTIVE PEPTIDE LIGANDS

Peptide hormones interact with their membrane bound receptors in highly specific manner to transduce a cellular signal and play important role(s) in regulating various physiological mechanisms. Structural determination of peptide hormone–receptor complexes is still a challenge in structural biology, therefore, not much is known about these complexes. From structural determination of isolated peptide hormones in conjunction with biochemical and biophysical data, indirect information about the ligand–receptor complex can be postulated. It is well established that the side chain moieties of peptides are involved in the receptor molecular recognition process. The peptide side chain topography and stereoelectronic properties provide the critical information important for specific interactions and receptor stimulation. A general strategy for attaining peptide-based information important for these receptor mediated events is outlined in Figure 3.4. Structure and activity studies of peptide hormones are an important tool to analyze ligand–receptor interaction; other methods are photo-affinity labeling, site-directed mutagenesis, the construction of receptor chimeras, and molecular modeling. We will discuss some of the methods in the following sections by taking examples from peptidergic GPCR family, primarily the melanocortin system, as that is the focus of our research.

### 3.7.1 Structure–Activity Relationship (SAR) Studies

With the ability to synthesize peptides, the concept of exploring the importance of specific peptide amino acids as well as peptide structure became viable. Typical strategies for understanding the importance of a particular amino acid at a particular position in the peptide hormone include: Alanine (Ala) scans, N- and C-terminal truncation peptides, and D-amino acid scans as the most common approaches. Replacement of a desired peptide amino acid with Ala is utilized because this

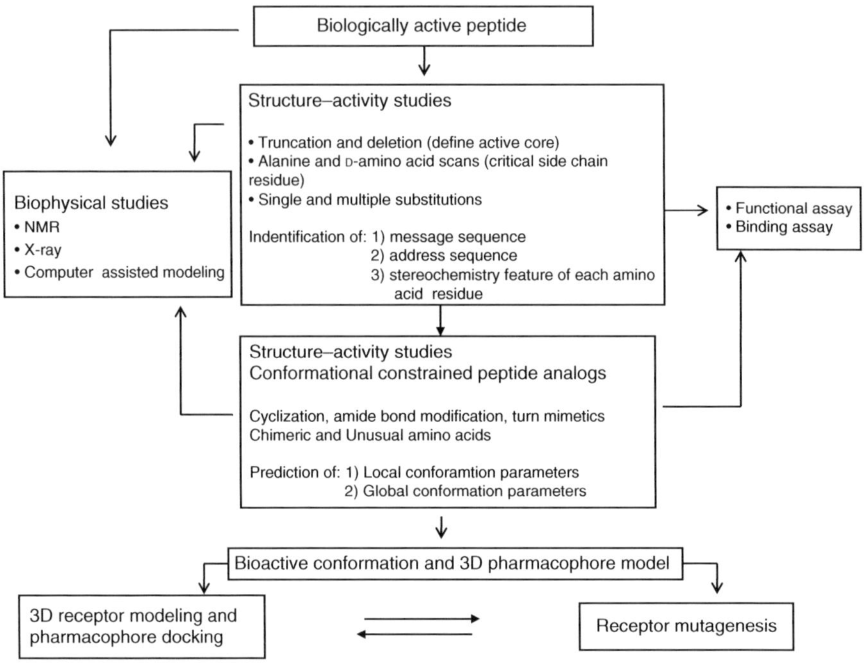


**Figure 3.4** Strategy to design receptor-selective peptide ligands.

residue contains a methyl group at the C$\alpha$ side chain position. This residue possesses the smallest C$\alpha$ side chain of the 20 naturally occurring eukaryotic amino acids besides glycine (Gly) residue. Gly residue contains a proton, which is smaller than the methyl group; however, the amino acid Gly does not possess chirality at the C$\alpha$ carbon, which can be important for the structure of the peptide. Thus, by substituting the peptide side chain with a small relatively neutral amino acid such as Ala, the importance of a particular amino acid side chain moiety interaction with its corresponding target protein or receptor can be examined. If a particular side chain is important for peptide structure or function, then on replacement with Ala, decreased ligand affinity and/or potency is anticipated to result. If the residue is not important for a particular hormone, then a very subtle, or no change, in ligand affinity and/or potency might be observed. These data can allow for the identification of the development of a peptide hormone *pharmacophore* model. This pharmacophore portion of a peptide hormone is considered to be the positioning of key atoms in 3D space important for the peptide to selectively recognize its cognate receptor. This information is highly desirable as it can be used in the design process to generate peptidomimetics and potential small molecule therapeutic ligands.

In the truncation approach, the peptide amino acid residues are deleted singly (or in combination) from either the N- or C-terminal domain of a peptide. Information obtained from these types of studies can narrow down a particular amino acid sequence that might retain acceptable potency while decreasing the number of amino acids needed for molecular recognition, selectivity, and functionality at its corresponding receptor protein. For example, if a 13 amino acid peptide can be truncated to three or four amino acids while retaining nM potency and efficacy, then that would be easier and more cost-effective to perform further structure–activity relationship (SAR) studies. Additionally, for some peptide hormones, both the N- and C-terminal residues are important for the secondary structure important for the physiological activity of the peptide. Thus, the removal of these residues would result in decreased or no activity of the peptide and help determine the amino acids important for the structure of the peptide as well as its function.

A third common approach has been the substitution of the naturally occurring L-configured amino acid with the D-configuration, known as a D-amino acid scan. Because peptides are recognized as being degraded by enzymes within a cell or the body, the incorporation of a D-amino acid might increase enzymatic resistance and increase peptide stability. Additionally, it has been found in several peptide hormones that the incorporation of D-amino acids can increase peptide potency.

It is apparent that systematic structure–function studies provide information about the specific amino acid residues and functional groups in a peptide that are important to biological activity. For example, structure–function studies of α-MSH eventually resulted in a more potent and enzyme-resistant analog [Nle4, DPhe7]-α-MSH (NDP-MSH) that contains the *active core* fragment of melanocortin peptides with Nle substitution in position 4 instead of Met, which is prone to oxidation and DPhe at position 7 in place of Phe [91]. Radiolabeled derivatives of NDP-MSH are extensively used for melanocortin receptor studies. Truncation studies of α-MSH, which involves selective removal of N- and/or C-terminal residues, followed by the evaluation of the truncated analogs for binding and/or functional activity revealed that residues 4, 10, and 12 contributes to the potency of the peptide. The minimum sequence identified for the biological activity was Ac-His-Phe-Arg-Trp-$NH_2$ (core message sequence) for α-MSH in classical frog and lizard bioassay [92–94]. Positional Ala scan of α-MSH revealed that positions 4, 6, 7, 8 and 9 are important for the receptor binding at MC1R [95].

Once a peptide template has been identified that meets the investigators criteria for potency and minimal length, a variety of other types of peptide SAR studies can be performed in attempts to optimize receptor potency, selectivity, ligand structure, and stability. Table 3.3 summarizes some of the standard types of SAR studies that can be performed. Since the majority of peptide hormones are linear and highly flexible in solution, they can adopt a plethora of different structural conformations and global structures depending upon a local environment. For example, a linear peptide in solution that contains a large component of hydrophilic amino acid side chains may possess an extended conformation in an aqueous environment. However, this same peptide when exposed to a hydrophobic environment, such as a lipid bilayer or interior binding domain of its receptor protein, may form a constrained conformation where the hydrophilic side chains are facing the interior of the peptide and the hydrophobic portion of the peptide is interacting within the hydrophobic local environment. Thus, the concept of identifying the *biologically active* peptide conformation reasons that by restraining a flexible peptide into this biologically active

**TABLE 3.3 General SAR Studies to Design Receptor Selective and Potent Ligands.**

| | Study | Feature |
|---|---|---|
| 1 | Substitution by D-amino acids | Stereochemical requirement; secondary structures (β-turns, a-helix, etc.) |
| 2 | Substitution of side chain moieties by a methyl group | Stereoelectronic properties of the side chain and its importance in interaction |
| 3 | Substitution of peptide bonds | Importance of specific amide bonds for ligand–receptor interactions |
| 4 | Cyclization approaches | Define topography of the amino acid residues; secondary structure |
| 5 | Reduction or increase in ring size | The optimum ring size for biological activity |
| 6 | Backbone $N^{\alpha}$-alkylation | Conformational constraint; less prone to enzymatic hydrolysis |
| 7 | Backbone $C^{\alpha}$-alkylation | Conformational constraint, generally to α-helix |

peptide, the energy associated with ligand structural conversion can be minimized and, therefore, the ligand potency can be enhanced by decreasing the overall energy required by the system. With that rationale, a common approach to restricting conformational freedom of a peptide is the incorporation of cyclization strategies. These cyclization approaches can include side chain to side chain, backbone to backbone, and side chain to backbone. A common cyclization strategy used by nature is the disulfide bridge, but the synthetic opportunities to create different types of cyclizations in peptides are only limited by the creativity of the investigator and the available orthogonal synthetic strategies. One of the most common synthetic cyclization that has historically been incorporated into peptides is the lactam bridge. This has been primarily due to the same chemistry as the typical amide bond formation of a growing peptide chain. Backbone cyclization (BC) is one of the approaches that utilizes atoms in the backbone (N and/or C) of a target linear peptide through a linker to form a ring and shown to dramatically enhance the metabolic stability and pharmacological stability of peptides [96]. Advantage of BC over other peptide cyclization methods is that they use the backbone atoms leaving the side chain intact, which are essential for biological activity at the receptor. Utilizing this approach, Hess et al. [96] have synthesized a library of backbone cyclic analogs where the bridge was formed connecting the N-terminus to the N$\alpha$ of the C-terminal Gly building unit by a dicarboxylic acid spacer. All the peptides in the library consist of the same parent sequence, but differ in ring size. From this study, they found that the compound BL3020-1, which was selective for the MC4R, had favorable metabolic and pharmacokinetic properties. In another example of cyclization approach, Ahn et al. [97] have used *positional cyclization scanning* approach to identify the bioactive conformation of glucagon. Once hypotheses regarding a particular peptide pharmacophore

**TABLE 3.4 Amide Bond Replacements and Their Applications for Peptide Ligands [98].**

| Amide Bond Replacement | Application | References |
|---|---|---|
| $\Psi[CH_2NH]$ | Neurokinin antagonist | [99] |
| $\Psi[CH_2O]$ | Gastrin releasing peptide antagonist | [100] |
| $\Psi[CH_2S]$ and $\Psi[CH_2SO]$ | Reverse turn stabilizers | [101] |
| $\Psi[COCH_2]$ | Neurotensin analog | [102] |
| $\Psi$[(E)-CH=CH] | Determination of bioactive conformation of cholecystokinin terminal hexapeptide | [103] |
| $\Psi$[(E)-CF=CH] | Opioid agonist | [104] |
| $\Psi[CN_4]$ | Somatostatin and bradykinin analogs | [105] |
| $\Psi$[CH(CN)NH] | Neurotensin analog | [102] |

and/or biologically active conformations are generated, a common approach is then to incorporate unusual amino acids in attempts to increase ligand-binding interactions as well as enhance desired receptor selectivity profiles by restricting the conformational flexibility of the peptide backbone. Constrained amino acids strategy has led to the discovery of peptides that show increased binding affinity, potency, and selectivity toward one or more of the receptors.

Peptide backbones consist of amide bonds that are most commonly found in a trans configuration under *normal* conditions and are very susceptible to the biodegradation, which limit the ability of peptides to act as therapeutic agents. However, the modification of the amide backbone can help stabilize a postulated pharmacophore model, add increased enzymatic and biological stability. Modifications of the peptide amide bond with a bioisosteric group that resembles an amide without the drawbacks listed above, result in the somewhat rigid or locked conformation of the ligand that may have enhance binding affinity to specific target. Amide bond surrogates range from simple olefinic groups to more sophisticated heterocycles. Table 3.4 lists some of the common amide bond isosteres that have been reported to be applicable in the case of peptide ligands [106].

### 3.7.2 Chimeric Peptide Analogs

Another approach is to place key structural moieties into novel templates or link them together on alternate templates to produce chimeric analogs to examine selectivity and/or potency. Recent examples in this context are the novel chimeric melanotropin–deltorphin analogs by Han et al. [107]. Chimeric melanocortin-AGRP (agouti-related protein) peptides were synthesized to test the hypothesis that the Arg-Phe-Phe motif human agouti-related protein(hAGRP)(111–113) mimics the DPhe-Arg-Trp of the melanocortin agonists in interactions with melanocortin receptors [108–110].

### 3.7.3 Combinatorial Libraries

Combinatorial chemistry was developed with peptide synthetic strategies in combination with high throughput bioassays. These approaches resulted in the emergence of the privileged scaffold concept, which was proposed by Evans et al. [111]. The reason behind choosing a common (privileged) structure was related to the G-proteins that have evolved from the same ancestral gene and have similar structural features, especially in the TM region, where ligands bind in most cases. Therefore, GPCRs that act through G-proteins should respond to the certain privileged structures. Therefore, the basic principle is that if molecular scaffold is able to bind and stimulate multiple GPCR systems, it can be used to generate a library that can be successfully screened against other GPCRs, particularly orphaned receptors. Benzodiazepine scaffold utilized by Evans et al. [111], thought to mimic a reverse turn, has resulted in lead *hits* against multiple peptide receptors such as neuropeptide Y receptor, [112] bradykinin receptor [113], and melanocortin receptor [114]. Combinatorial chemistry has shown that different chemical structures can interact with a given GPCR and yield therapeutically useful analogs with nanomolar affinities on optimization. Parallel synthesis and testing of multiple analogs with different conformational constraints, D-amino acids, dehydroamino acids, amide and disulide cyclic constraint, and reverse turn mimetics offer a rapid approach to the determination of the receptor-bound conformation [115].

### 3.7.4 Three-Dimensional (3D) GPCR Homology Molecular Modeling

The overall goal of 3D homology molecular modeling is to aid structure-based design strategy and provide further insight for rational drug design. Identification of putative receptor residues important for ligand binding is a key component to homology modeling goals. A desirable approach to identify peptide–receptor interactions is to obtain an X-ray structure based on the formation of a protein complex crystal. Since GPCRs are functional in the cellular lipid bilayer, the generation of the desirable high resolution structures is lacking. As mentioned earlier, one approach that has become a viable strategy is using homology molecular modeling of GPCRs based on originally bacteriorhodopsin [116] and bovine rhodopsin [117] templates. Subsequently, in 2000, the first high resolution GPCR crystal structure of the rhodopsin and more recently in 2007, key studies of crystal structures of the human $\beta_2$-adrenergic receptor GPCR were reported [25, 26]. Thus, using these structures of GPCRs in their various forms and atomic resolution, scientists have been developing the concept of homology molecular modeling in attempts to identify putative ligand–receptor interactions since the 1990s [118, 119]. In the absence of 3D structures of the other GPCRs, one approach is to construct model for the GPCR and to refine the model based on experimental results [40]. Homology molecular modeling has been traditionally utilized for decades by X-ray crystallographers, and thus is a widely accepted and validated computational experimental approach. Computer-generated models for many GPCRs have been constructed and available at http://cssb.biology.gatech.edu/skolnick/files/gpcr/gpcr.html. There are a variety of tools and approaches currently available to align a desired GPCR of interest using the available crystal structures (Protein data bank (PDB) Id: 1U19, 1GZM, 2I37, 3CAP, 2Z73, 2VT4, 2RH1, and 3D4S). These approaches range from manual alignment to computer-assisted programs on a variety of different platforms. It is important to examine the specific GPCR amino acid residues to identify residues that are common to the majority of GPCRs, as they are most likely involved in structural integrity important for general GPCR structure or function versus amino acids that are unique for a particular peptide receptor class. These unique residues might be postulated to be important for molecular recognition and peptide specificity for a given GPCR. Computational modeling approaches for GPCRs were extensively reviewed by Fanelli and Bandetti [120] in 2005.

The next common approach is to dock the peptide hormone into the refined GPCR model and perform energy-minimization studies to ensure the steric and overlapping side chains interactions are removed. Determining the putative structure of the ligand and the conformation that will be used to dock into the receptor can occur using a variety of approaches. Generally, peptide SAR (as described earlier) is performed, so that key ligand atoms have been identified as important for pharmacological activity. Biophysical studies such as NMR and computer-assisted molecular modeling are also a strategy that can be used. Typically, the actual docking process is now performed by a computer program, but oversight and chemical *intuition* by scientists is still an important aspect for performing and monitoring these types of studies.

Once a 3D model of the peptide–receptor complex has been generated, putative key ligand–receptor interactions between side chains as well as backbones are postulated. The side chain interactions can be experimentally tested by receptor mutagenesis, peptide SAR, or both in a complementary fashion. In these experiments, it is advantageous to have both binding as well as functional data, so that changes in pharmacology can be associated as important for receptor structure–function, peptide molecular recognition, and/or receptor activation.

## 3.8 CONCLUSIONS

Although classical drug-screening programs have been successful, more structural knowledge is needed for rational-based drug design. Paucity of structural knowledge about GPCRs has severely limited the application of structure-based drug design. Even though >30% of all marketed therapeutics act on GPCRs, these drugs target only ~30 members of this class so there is enormous potential to exploit the remaining family members, including the >100 orphan receptors for which no existing ligands have yet been identified.

## ACKNOWLEDGMENTS

We would like to acknowledge financial support from NIH Grants RO1DK57080, RO1DK64250, RO1DK063974, and an American Diabetes Research Award.

## REFERENCES


1. Ji TH, Grossmann M, Ji I. G protein-coupled receptors. J Bio Chem 1998;273: 17299–17302.
2. Klabunde T, Hessler G. Drug design strategies for targeting G-protein-coupled receptors. ChemBioChem 2002;3:928–944.
3. Pierce KL, Permont RT, Lefkowitz RJ. Seven-transmembrane receptors. Nat Rev 2002;3:639–650.
4. Wilson S, Bergsma D. Orphan G-protein coupled receptors: novel drug targets for the pharmaceutical industry. Drug Des Discovery 2000;17:105–114.
5. Imming P, Sinning C, Meyer A. Drugs, their targets and the nature and number of drug targets. Nat Rev Drug Discovery 2006;5:821–834.
6. Lundstrom K. Latest development in drug discovery on G protein-coupled receptors. Curr Protein Pept Sci 2006;7:465–470.
7. Baldwin JM. Structure and function of receptors coupled to G proteins. Curr Opin Cell Biol 1994;6:180–190.
8. Tyndall JDA, Pfeiffer B, Abbenante G, Fairlie DP. Over one hundred peptide-activated G protein-coupled receptors recognize ligands with turn structure. Chem Rev 2005;105:793–826.
9. Conn PJ, Pin JP. Pharmacology and functions of metabotropic glutamate receptors. Annu Rev Pharmacol Toxicol 1997;37:205–237.
10. Fredriksson R, Lagerstrom MC, Lundin LG, Schioth HB. The G-protein-coupled receptors in the human genome from five main families.phylogenetic analysis, paralogon groups and fingerprints. Mol Pharmacol 2003;63:1256–1272.
11. Pawson AJ, Sharman JL, Benson HE, et al. The IUPHAR/BPS guide to PHARMACOLOGY: an expert-driven knowledgebase of drug targets and their ligands. Nucl Acids Res 2014;42 (Database Issue):D1098–D1106.
12. Anastasiadis-Pool A, Schwalbe H. Websites about G protein-coupled receptors. ChemBioChem 2002;3:1029–1030.
13. Liu J, Schoneberg T, Rhee MV, Wess J. Mutational analysis of the relative orientation of transmembrane helices I and VII in G Protein-coupled receptors. J Biol Chem 1995;270:19532–19539.
14. Elling CE, Nielsen SM, Schwartz TW. Conversion of antagonist-binding site to metal-ion site in the tachykinin NK-1 receptor. Nature 1995;374:74–77.
15. Farrens DL, Altenbach C, Yang K, et al. Requirement of rigid-body motion of transmembrane helices for light activation of rhodopsin. Science 1996;274:768–770.
16. Farrens DL, Khorana HG. Structure and function in rhodopsin. Measurement of the rate of metarhodopsin II decay by fluorescence spectroscopy. J Bio Chem 1995;270:5073–5076.
17. Krebs A, Villa C, Edwards PC, Schertler GF. Characterisation of an improved two-dimensional p22121 crystal from bovine rhodopsin. J Mol Biol 1998;282:991–1003.
18. Palczewski K, Kumasaka T, Hori T, et al. Crystal structure of rhodopsin: a G protein-coupled receptor. Science 2000;289:739–745.
19. Okada T, Sugihara M, Bondar AN, et al. The retinal conformation and its environment in rhodopsin in light of a new 2.2 A ° crystal structure. J Mol Biol 2004;342:571–583.

20. Teller DC, Okada T, Behnke CA, et al. Advances in determination of a high-resolution three dimensional structure of rhodopsin, a model of G-protein-coupled receptors (GPCRs). Biochemistry 2001;40:7761–7772.
21. Li J, Edwards PC, Burghammer M, et al. Structure of bovine rhodopsin in a trigonal crystal form. J Mol Biol 2004;343:1409–1438.
22. Fan QR, Hendrickson WA. Structure of human follicle-stimulating hormone in complex with its receptor. Nature 2005;433:269–277.
23. Murakami M, Kouyama T. Crystal structure of squid rhodopsin. Nature 2008;453: 333–363.
24. Shimamura T, Hiraki K, Takahashi N, et al. Crystal structure of squid rhodopsin with intracellularly extended cytoplasmic region. J Bio Chem 2008;283:17753–17756.
25. Rasmussen SGF, Choi HJ, Rosenbaum DM, et al. Crystal structure of the human beta(2) adrenergic G-protein-coupled receptor. Nature 2007;450:383–384.
26. Rosenbaum DM, Cherezov V, Hanson MA, et al. GPCR engineering yields high-resolution structural insights into beta(2)-adrenergic receptor function. Science 2007;318:1266–1273.
27. Cherezov V, Rosenbaum DM, Hanson MA, et al. High-resolution crystal structure of an engineered human beta(2)-adrenergic G protein-coupled receptor. Science 2007;318:1258–1265.
28. Warne T, Serrano-Vega MJ, Baker JG, et al. Structure of a beta(1)-adrenergic G-protein-coupled receptor. Nature 2008;454:482–486.
29. Hanson MA, Cherezov V, Griffith MT, et al. A specific cholesterol binding site is established by the 2.8 angstrom structure of the human beta(2)-adrenergic receptor. Structure 2008;16:897–905.
30. Park JH, Scheerer P, Hofmann KP, et al. Crystal structure of the ligand-free G-protein-coupled receptor opsin. Nature 2008;454:133–183.
31. Weis WI, Kobilka BA. Structural insights into G-protein-coupled receptor activation. Curr Opin Struct Biol 2008;18:7734–7740.
32. Scheerer P, Park JH, Hildebrand PW, et al. Crystal structure of opsin in its G-protein-interacting conformation. Nature 2008;455:497–503.
33. Ostermeier C, Michel H. Crystallization of membrane proteins. Curr Opin Struct Biol 1997;7:697–701.
34. Shi L, Javitch JA. The binding site of aminergic G protein-coupled receptors: the transmembrane segments and second extracellular loop. Annu Rev Pharmacol Toxicol 2002;42:437–467.
35. Elling CE, Thirstrup K, Nielsen SM, et al. Metal-ion sites as structural and functional probes of helix–helix interactions in 7TM receptors. Ann N Y Acad Sci 1997;814:142–151.
36. Zhang Y, Devries ME, Skolnick J. Structure modeling of all identified G protein-coupled receptors in the human genome. PLoS Comput Biol 2006;2:88–99.
37. Filipek S, Teller DC, Palczewski K, Stenkamp R. The crystallographic model of rhodopsin and its use in studies of other G protein-coupled receptors. Annu Rev Biophys Biomol Struct 2003;32:375–397.
38. Costanzi S. On the applicability of GPCR homology models to computer-aided drug discovery: a comparison between in silico and crystal structures of the beta2-adrenergic receptor. J Med Chem 2008;51:2907–2914.

39. Lodowski DT, Angel TE, Palczewski K. Comparative analysis of GPCR crystal structures. Photochem Photobiol 2009;85:425–430.
40. Gershengorn MC, Osman R. Insight into G-Protein-coupled receptor functioning using molecular models. Endocrinology 2001;142:2–10.
41. Gether U, Kobilka BK. G protein-coupled receptors. Mechanism of agonist activation. J Biol Chem 1998;273:17979–17982.
42. Gether U. Uncovering molecular mechanisms involved in activation of G protein-coupled receptors. Endocr Rev 2000;21:90–113.
43. Kobilka BA. G protein coupled receptor structure and activation. Biochim Biophy Acta 2007;1768:794–807.
44. Rashid AJ, O'Dowd BF, George SR. Diversity and complexity of signaling through peptidergic G protein-coupled receptors. Endocrinology 2004;145:2645–2652.
45. Oldham WM, Hamm HE. Heterotrimeric G protein activation by G-protein-coupled receptors. Nat Rev Mol Cell Biol 2008;9:60–71.
46. Feng Y, Noda K, Saad Y, et al. The docking of Arg2 of angiotensin II with Asp281 of AT1 receptor is essential for full agonism. J Bio Chem 1995;270:12846–12850.
47. Walker P, Munoz M, Martinez R, Peitsch M. Acidic residues in extracellular loops of the human-y1 neuropeptide-y receptor are essential for ligand binding. J Bio Chem 1994;269:2863–2869.
48. Leong S, Kabakoff R, Hebert C. Complete mutagenesis of the extracellular domain of interleukin-8 (IL-8) type A receptor identifies charged residues mediating IL-8 binding and signal transduction. J Bio Chem 1994;269:19343–19348.
49. Cotte N, Balestre M, Phalipou S, et al. Identification of residues responsible for the selective binding of peptide antagonists and agonists in the V2 vasopressin receptor. J Bio Chem 1998;273:29462–29468.
50. Davidson J, Assefa D, Pawson A, et al. Irreversible activation of the gonadotropin-releasing hormone receptor by photoaffinity cross-linking: localization of attachment site to Cys residue in Nterminal segment. Biochemistry 1997;36:12881–12889.
51. Perez H, Vilander L, Andrews W, et al. Human formyl peptide receptor ligand binding domain(s). Studies using an improved mutagenesis/expression vector reveal a novel mechanism for the regulation of receptor occupancy. J Bio Chem 1994;269:22485–22487.
52. Liapakis G, Fitzpatrick D, Hoeger C, et al. Identification of ligand binding determinants in the somatostatin receptor subtypes 1 and 2. J Bio Chem 1996;271:20331–20339.
53. AbdAlla S, Jarnagin K, Muller-Esterl W, Quitterer U. The N-terminal amino group of [Tyr8]bradykinin is bound adjacent to analogous amino acids of the human and rat B2 receptor. J Bio Chem 1996;271:27382–27387.
54. Ji Z, Hadac E, Henne R, et al. Direct identification of a distinct site of interaction between the carboxyl-terminal residue of cholecystokinin and the type A cholecystokinin receptor using photoaffinity labeling. J Biol Chem 1997;272:24393–24401.
55. Hadac E, Pinon D, Ji Z, et al. Direct identification of a second distinct site of contact between cholecystokinin and its receptor. J Biol Chem 1998;273:12988–12993.
56. Noda K, Saad Y, Kinoshita A, et al. Tetrazole and carboxylate groups of angiotensis receptor antagonists bind to the same subsite by different mechanisms. J Bio Chem 1995;270:2284–2289.

57. Kenakin T. Ligand-selective receptor conformations revisited: the promise and the problem. Trends Pharmacol Sci 2003;24:346–354.
58. Leff P. The two-state model of receptor activation. Trends Pharmacol Sci 1995;16:89–97.
59. Weiss JM, Morgan PH, Lutz MW, Kenakin TP. The cubic ternary complex receptor-occupancy model. III. Resurrecting efficacy. J Theor Biol 1996;181:381–397.
60. Sheikh SP, Zvyaga TA, Lichtarge O, Sakmar TP. Rhodopsin activation blocked by metal-ion-binding sites linking transmembrane helices C and F. Nature 1996;383:347–350.
61. Gether U, Lin S, Ghanouni P, et al. Agonists induce conformational changes in transmembrane domains III and VI of the beta2 adrenoceptor. EMBO 1997;16:6737–6747.
62. Jensen AD, Guarnieri F, Rasmussen SG, Asmar F, Ballesteros JA, Gether U. Agonist-induced conformational changes at the cytoplasmic side of transmembrane segment 6 in the beta2 adrenergic receptor mapped by site-selective fluorescent labeling. J Biol Chem 2001;276:9279–9290.
63. Ghanouni P, Steenhuis JJ, Farrens DL, Kobilka BK. Agonist-induced conformational changes in the G-protein-coupling domain of the beta 2 adrenergic receptor. Proc Natl Acad Sci U S A 2001;98:5997–6002.
64. Elling CE, Frimurer TM, Gerlach LO, et al. Metal ion site engineering indicates a global toggle switch model for seven-transmembrane receptor activation. J Bio Chem 2006;281:17337–17346.
65. Han SJ, Hamdan FF, Kim SK, et al. Identification of an agonist-induced conformational change occurring adjacent to the ligand-binding pocket of the M(3) muscarinic acetylcholine receptor. J Bio Chem 2005;280:34849–34858.
66. Ward SD, Hamdan FF, Bloodworth LM, Siddiqui NA, Li JH, Wess J. Use of an in situ disulfide crosslinking strategy to study the dynamic properties of the cytoplasmic end of transmembrane domain VI of the M3 muscarinic acetylcholine receptor. Biochemistry 2006;45:675–685.
67. Huang W, Osman R, Gershengorn MC. Agonistinduced conformational changes in thyrotropin releasing hormone receptor type I: disulfide crosslinking and molecular modeling approaches. Biochemistry 2005;44:2419–2433.
68. Rodbell M, Birnbaumer L, Pohl S, Krans H. The glucagon-sensitive adenyl cyclase system in plasma membranes of rat liver. V. An obligatory role of guanylnucleotides in glucagon action. J Bio Chem 1971;246:1877–1882.
69. Zheng B, Ma Y, Ostrom R, et al. RGS-PX1, a GAP for GalphaS and sorting nexin in vesicular trafficking. Science 2001;294:1939–1942.
70. Ma Y, Huang J, Ali S, et al. Src tyrosine kinase is a novel direct effector of G proteins. Cell 2000;102:635–646.
71. Kwok-Keung Fung B, Stryer L. Photolyzed rhodopsin catalyzes the exchange of GTP for bound GDP in retinal rod outer segments. Proc Natl Acad Sci U S A 1980;77:2500–2504.
72. McLaughlin S, McKinnon P, Margolskee R. Gustducin is a taste-cell-specific G protein closely related to the transducins. Nature 1992;357:563–569.
73. Meng J, Glick J, Polakis P, Casey P. Functional interaction between Galpha(z) and Rap1GAP suggests a novel form of cellular cross-talk. J Bio Chem 1999;274:36663–36669.
74. Katada T, Bokoch G, Northup J, et al. The inhibitory guanine nucleotide-binding regulatory component of adenylate cyclase. Properties and function of the purified protein. J Bio Chem 1984;259:3568–3577.

75. Bokoch G, Katada T, Northup J, et al. The inhibitory guanine nucleotide-binding regulatory component of adenylate cyclase. Properties and function of the purified protein. J Bio Chem 1984;259:3560–3567.

76. Chikumi H, Vázquez-Prado J, Servitja J, et al. Potent activation of RhoA by Galpha q and Gq-coupled receptors. J Bio Chem 2002;277:27130–27134.

77. Booden M, Siderovski D, Der C. Leukemia-associated Rho guanine nucleotide exchange factor promotes G alpha q-coupled activation of RhoA. Mol Cell Biol 2002;22:4053–4061.

78. Smrcka A, Hepler J, Brown K, Sternweis P. Regulation of polyphosphoinositide-specific phospholipase C activity by purified Gq. Science 2002;251:804–807.

79. Dixon R, Kobilka B, Strader D, et al. Cloning of the gene and cDNA for mammalian beta-adrenergic receptor and homology with rhodopsin. Nature 1986;321:75–79.

80. Meigs TE FT, McKee DD, Casey PJ. Interaction of Galpha 12 and Galpha 13 with the cytoplasmic domain of cadherin provides a mechanism for beta -catenin release. Proc Natl Acad Sci U S A 2001;98:519–524.

81. Kozasa T, Jiang X, Hart M, et al. p115 RhoGEF, a GTPase activating protein for Galpha12 and Galpha13. Science 1998;280:2109–2111.

82. Logothetis D, Kurachi Y, Galper J, et al. The beta gamma subunits of GTP-binding proteins activate the muscarinic K+ channel in heart. Nature 1987;325:321–326.

83. Boyer J, Waldo G, Harden T. Beta gamma-subunit activation of G-protein-regulated phospholipase C. J Bio Chem 1992;267:25451–25456.

84. Camps M, Carozzi A, Schnabel P, et al. Isozyme-selective stimulation of phospholipase C-beta 2 by G protein beta gamma-subunits. Nature 1992;360:684–686.

85. Pitcher J, Inglese J, Higgins J, et al. Role of beta gamma subunits of G proteins in targeting the beta-adrenergic receptor kinase to membrane-bound receptors. Science 1992;257:1264–1267.

86. Tang W, Gilman A. Type-specific regulation of adenylyl cyclase by G protein beta gamma subunits. Science 1991;254:1500–1503.

87. Stephens L, Smrcka A, Cooke F, et al. A novel phosphoinositide 3 kinase activity in myeloid-derived cells is activated by G protein beta gamma subunits. Cell 1994;77:83–93.

88. Eipper BA, Mains RE. Structure and biosynthesis of pro-ACTH/endorphin and related peptides. Endocr Rev 1980;1:1–26.

89. Smith AI, Funder JW. Proopiomelanocortin processing in the pituitary, central nervous system and peripheral tissues. Endocrin Rev 1988;9:159–179.

90. Pritchard LE, White A. Neuropeptide processing and its impact on melanocortin pathways. Endocrinology 2007;148:4201–4207.

91. Sawyer TK, Sanfillippo PJ, Hruby VJ, et al. 4-Norleucine, 7-D-phenylalanine-$\alpha$-melanocyte-stimulating hormone: a highly potent $\alpha$-melanotropin with ultra long biological activity. Proc Natl Acad Sci U S A 1980;77:5754–5758.

92. Haskell-Luevano C, Sawyer TK, Hendrata S, et al. Truncation studies of alpha-melanotropin peptides identify tripeptide analogues exhibiting prolonged agonist bioactivity. Peptides 1996;17:995–1002.

93. Hruby VJ, Wilkes BC, Hadley ME, et al. alpha-Melanotropin: the minimal active sequence in the frog skin bioassay. J Med Chem 1987;30:2126–2130.

94. Castrucci A, Hadley M, Sawyer T, et al. Alpha-melanotropin—the minimal active sequence in the lizard skin bioassay. Gen Comp Endocrinol 1989;73:157–163.

95. Sahm U, Olivier G, Branch S, et al. Synthesis and biological evaluation of a-MSH analogs substituted with alanine. Peptides 1994;15:1297–1302.

96. Hess S, Linde Y, Ovadia O, et al. Backbone cyclic peptidomimetic melanocortin-4 receptor agonist as a novel orally administrated drug lead for treating obesity. J Med Chem 2008;51:1026–1034.

97. Ahn JM, Gitu PM, Medeiros M, et al. A new approach to search for the bioactive conformation of glucagon: positional cyclization scanning. J Med Chem 2001;44:3109–3116.

98. Ahn J-M, Boyle NA, MacDonald MT, Janda KD. Peptidomimetics and peptide backbone modifications. Mini Rev Med Chem 2002;2:463–473.

99. Harbeson SL, Shatzer SA, Le TB, Buck SH. A new class of high affinity ligands for the neurokinin A NK2 receptor: psi (CH2NR) reduced peptide bond analogues of neurokinin A4-10. J Med Chem 1992;35:3949–3955.

100. Saari WS, Heimbrood DC, Friedman A, et al. A gastrin-releasing peptide antagonist containing A psi (CH2O) amide bond surrogate. Biochem Biophys Res Commun 1989;165:114–117.

101. Ma S, Richarson JF, Spatola AF. Solution and solid-state structures of a cyclic psi[CH2SO]. Pseudopentapeptide. J Am Chem Soc 1991;113:8529–8530.

102. González-Muñiz R, García-López MT, Gómez-Monterrey I, et al. Ketomethylene and (cyanomethylene)amino pseudopeptide analogues of the C-terminal hexapeptide of neurotensin. J Med Chem 1995;38:1015–1021.

103. Shue Y-KT, Tufano MD, Carrera GMJ, et al. Double bond isosteres of the peptide bond: synthesis and biological activity of cholecystokinin (CCK) C-terminal haxapeptide analogs. Bioorg Med Chem 1993;1:161–171.

104. Chandrakumar NS, Yonan PK, Stapelfeld A, et al. Preparation and opioid activity of analogues of the analgesic dipdeptide 2,6-dimethyl-L-tyrosyl-N-(3-phenylpropyl)-D-alaninamide. J Med Chem 1992;35:223–233.

105. Beusen DD, Zabrocki J, Slomczynska U, et al. Conformational mimicry: synthesis and solution conformation of a cyclic somatostatin hexapeptide containing a tetrazole cis amide bond surrogate. Biopolymers 1995;36:181–200.

106. Spatola AF. Peptide backbone modifications: a structure-activity analysis of peptides containing amide bond surogates. In: *Chemistry and Biochemistry of Amino Acids, Peptides and Proteins*. Basel: Marcel Dekker, Inc.; 1983. p 267–357.

107. Han G, Quillan J, Carlson K, et al. Design of novel chimeric melanotropin-deltorphin analogues. discovery of the first potent human melanocortin 1 receptor antagonist. J Med Chem 2003;46:810–819.

108. Wilczynski A, Wilson K, Wilson KR, Scott JW, Edison AS, Haskell-Luevano C. Structure-activity relationships of the unique and potent agouti-related protein (AGRP)-melanocortin chimeric Tyr-c[beta-Asp-His-DPhe-Arg-Trp-Asn-Ala-Phe-Dpr]-Tyr-NH2 peptide template. J Med Chem 2005;48:3060–3075.

109. Joseph CG, Wilczynski A, Holder JR, Xiang Z, Bauzo RM, Scott JW, Haskell-Luevano C. Chimeric NDP-MSH and MTII melanocortin peptides with agouti-related protein (AGRP) Arg-Phe-Phe amino acids possess agonist melanocortin receptor activity. Peptides 2003;24:1899–1908.

110. Jackson PJ, Yu B, Hunrichs B, et al. Chimeras of the agouti-related protein: insights into agonist and antagonist selectivity of melanocortin receptors. Peptides 2005;26:1978–1987.

111. Evans BE, Rittle KE, Bock MG, et al. Methods for drug discovery: development of potent, selective, orally effective cholecystokinin antagonists. J Med Chem 1988;31:2235–2246.

112. Shigeri Y, Ishikawa M, Ishihara Y, Fujimoto M. A potent nonpeptide neuropeptide Y Y1 receptor antagonist, a benzodiazepine derivative. Life Sci 1998;63:151–160.

113. Dziadulewicz E, Brown M, Dunstan A, et al. The design of non-peptide human bradykinin B2 receptor antagonists employing the benzodiazepine peptidomimetic scaffold. Biorg Med Chem Lett 1999;9:463–468.

114. Joseph CG, Wilson KR, Wood MS, et al. The 1,4-benzodiazepine-2,5-dione small molecule template results in melanocortin receptor agonists with nanomolar potencies. J Med Chem 2008;51:1423–1431.

115. Marshall GR. Peptide interactions with G protein coupled receptors. Biopolymers 2001;60:246–277.

116. Henderson R, Baldwin JM, Ceska TA, et al. Model for the structure of bacteriorhodopsin based on high-resolution electron cryo-microscopy. J Mol Biol 1990;213:899–929.

117. Schertler GFX, Villa C, Henderson R. Projection structure of rhodopsin. Nature 1993;362:770–772.

118. Trumpp-Kallmeyer S, Hoflack J, Bruinvels A, Hibert M. Modeling of G-protein-coupled receptors: application to dopamine, adrenaline, serotonin, acetylcholine, and mammalian opsin receptors. J Med Chem 1992;35:3448–3462.

119. Takeuchi Y, Shands EF, Beusen DD, Marshall GR. Derivation of a three-dimensional pharmacophore model of substance P antagonists bound to the neurokinin-1 receptor. J Med Chem 1998;41:3609–3623.

120. Fanelli F, Benedetti GD. Computational modeling approaches to structure-function analysis of G protein-coupled receptors. Chem Rev 2005;105:3297–3351.

# 4

# PEPTIDE-BASED INHIBITORS OF ENZYMES

ANNA KNAPINSKA AND SABRINA AMAR

*Departments of Chemistry and Biology, Torrey Pines Institute for Molecular Studies, Port St. Lucie, FL, USA*

TRISTA K. ROBICHAUD

*Departments of Periodontics and Biochemistry, University of Texas Health Science Center, San Antonio, TX, USA*

GREGG B. FIELDS

*Departments of Chemistry and Biology, Torrey Pines Institute for Molecular Studies, Port St. Lucie, FL, USA; Departments of Biochemistry, University of Texas Health Science Center, San Antonio, TX, USA*

## 4.1 INTRODUCTION

Enzymes are responsible for catalyzing a great variety of critical biological reactions. In turn, disease initiation and progression are often marked by aberrant enzyme activity. Peptide substrates have been utilized to study the mechanisms of action of many enzymes of various classifications. Concurrently, information derived from peptide substrate studies has been used to develop peptide-based inhibitors of enzymes. In this chapter, we describe peptide-based inhibitors of enzymes representative of several classifications.

*Peptide Chemistry and Drug Design*, First Edition. Edited by Ben M. Dunn.

Enzymes are divided into six different classifications: according to their enzyme commission (EC) number EC 1, oxidoreductases; EC 2, transferases; EC 3, hydrolases; EC 4, lyases; EC 5, isomerases; and EC 6, ligases. The majority of peptide-based studies have been performed with enzymes of the hydrolase classification (EC 3), primarily peptidases/proteases (EC 3.4), deacetylases (EC 3.5), and phosphatases (EC 3.1.3). Proteolytic enzymes represent a significant portion of the human genome (~2%) and have been shown to be viable targets for drug development [1–3]. Extensive reviews of clinical applications of protease inhibitors are available elsewhere [2, 3], so this chapter focuses on a few select examples of peptide-based inhibitors of proteases (angiotensin-converting enzyme (ACE), human acquired immunodeficiency virus (HIV) protease, matrix metalloproteinases (MMPs), and anthrax lethal factor (LF)).

Significant peptide-based studies have been performed with transferase enzymes (EC 2), namely, kinases (EC 2.7). Kinases account for approximately 2% of the human genome. Their strong implication in numerous diseases, particularly cancer, led to the development of peptide and peptidomimetic inhibitors. Other transferases targeted by peptide-based inhibitors are glycosyltransferases (EC 2.4), telomerase (EC 2.7.7.49), and histone methyltransferase (EC 2.1.1.43). Peptide-based inhibitors have also been developed for members of the oxidoreductase classification (EC 1), such as tyrosinase (EC 1.14.18.1), and the isomerase classification (EC 5), such as peptidyl-prolyl isomerase (EC 5.2).

## 4.2 ANGIOTENSIN-CONVERTING ENZYME AND NEPRILYSIN/NEUTRAL ENDOPEPTIDASE

Active octapeptide angiotensin II was first isolated in 1934, and 20 years later ACE was shown to be the enzyme responsible for cleaving the C-terminal dipeptide from the inactive decapeptide angiotensin I [4].

If a popularity contest for peptide inhibitors of enzyme action were held today, the inhibitors of ACE would be a clear front-runner. These peptide-derived drugs are among the first doctors turn to in cases of congestive heart failure and hypertensive disease [4]. In healthy systems, the hydrolase ACE cleaves angiotensin I into active angiotensin II, a potent vasorepressor active in all tissues. This step is required for angiotensin receptor activation [5]. Constriction of blood vessels results in a net increase in blood pressure as the heart increases effort to transport blood throughout the body.

Inhibitors for ACE were sought to artificially lower blood pressure in hypertensive patients, and in the late 1960s Bristol Myers-Squibb took up the search. At that time, the rennin-angiotensin system (RAS) for control of blood pressure was poorly understood with no extant quantitative assays for ACE activity. Kineticist David Cushman pioneered the structure-function approach by purifying rabbit lung ACE, then developing a spectrophotometric assay to quantify its activity [4]. After chelation and ion deprivation analysis, ACE was determined to be a zinc metallopeptidase, and Cushman along with collaborator Miguel Ondretti theorized its catalytic center might be similar to one of the few known protein crystal structures, carboxypeptidase A [4].

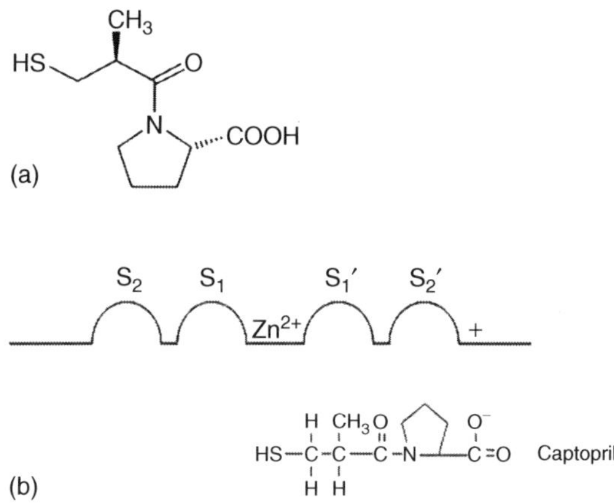


**Figure 4.1** Structure of captopril (a). Schematic of captopril bound to the active site of tACE (testis ACE) (b). *Source*: Reprinted with permission from Reference 10.

Concurrently, Sergio Ferreia found peptides from pit viper venom that inhibited ACE and receptors of the nonapeptide bradykinin [6, 7]. Tepretide, the venom-derived peptide with the most ACE activity was tested in patients and showed significant blood-pressure lowering effects [8]. Unfortunately, tepretide was costly to make, and had low oral bioactivity and low solubility. In 1972, Byers and Wolfrenden [9] published work on a *biproduct analog* inhibitor of carboxypeptidase A; its high affinity was ascribed to its aromatic structure, which mimicked both products of the enzyme. Cushman and Ondretti had, by this point, developed reliable guinea pig ileum assays for ACE inhibition. Knowledge obtained from venom peptides that ACE preferred Phe-Ala-Pro at the C-terminus, a model that ACE released a dipeptide rather than a single amino acid, and the success of the guinea pig model system guided them to the eventual synthesis of captopril, a dipeptide analog with high oral bioavailability (Figure 4.1). More ACE inhibitors have been developed to avoid the side effects of captopril, but the overall Phe-Ala-Pro analog structure remains. ACE inhibitors also inhibit the inactivation of bradykinin and substance P. These peptides mediate some of the side effects of ACE inhibitors, such as cough and angioedema [10].

In the year 2002, ACE inhibitors were the most commonly prescribed drugs for the treatment of hypertension in the United States and are definitively the major protease inhibitor success story [3]. Current-generation ACE inhibitors are widely used for cardiovascular diseases, including high blood pressure, heart failure, heart attack, and kidney failure, and have combined annual sales in excess of US $6 billion [10]. Thirteen ACE inhibitors are currently approved for clinical use and several others are in clinical trials. ACE inhibitors developed in the late 1970s and early 1980s possess detrimental side effects, prompting further exploration. These side effects could be explained by the fact that the first-generation ACE inhibitors were designed based on the structure of carboxypeptidase A, which we now know is considerably different from ACE. Once the crystal structure of ACE was solved, it revealed that

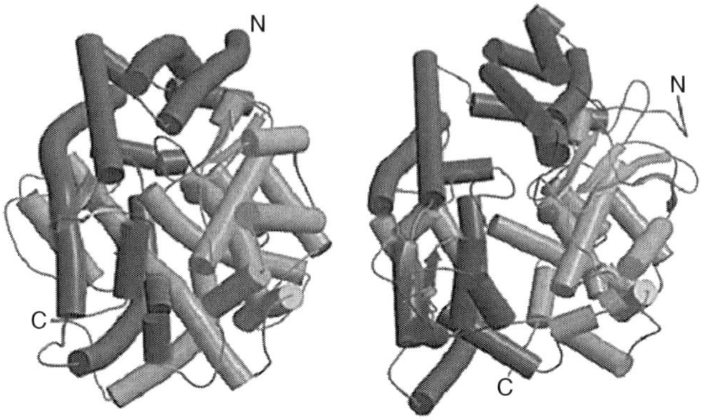


**Figure 4.2** Comparison of the X-ray crystallographic structures of ACE and neurolysin folds. Active site zinc molecules are shown in green [11].

ACE consists of N- and C-domains that have different functions and specificities (Figure 4.2) [11]. The C-terminal domain seems to be primarily responsible for the conversion of angiotensin I to angiotensin II, with the major effect on blood-pressure regulation. Early ACE inhibitors were relatively nonselective and inhibited both domains with similar activities. Therefore, the design of domain-selective ACE inhibitors is expected to produce safer and more effective drugs.

Based on large body of experimental evidence, it has become apparent that the RAS, the kallikrein-kinin pathway, and the natriuretic peptides are important modulators of cardiovascular homeostasis. These findings have provided the impetus to develop inhibitors that simultaneously block angiotensin II and increase atrial natriuretic peptide (ANP), which are regulated by endothelial, membrane-bound ACE, and neural endopeptidase (NEP) [12].

NEP belongs to the family of zinc-dependent endopeptidases. Catalytic properties of NEP resemble thermolysin, a zinc-dependent bacterial endopeptidase. It is located at the cell surface with the bulk of the protein, including the active site, facing the extracellular space, and therefore functions as an ectoenzyme, catalyzing peptide hydrolysis at the surface of the plasma membrane [13, 14]. NEP has been implicated in the regulation of opioid peptide action through the degradation of endogenously released enkephalins [15]. NEP is involved in the physiological degradation of the peptides modulating blood pressure, such as the cardiac hormone ANP, bradykinin, and endothelin [16]. More recently, NEP has been implicated in the degradation of amyloid $\beta$ peptide (A$\beta$1-42) [17, 18], the primary pathogenic agent in Alzheimer's disease, and has been shown to play a role in the degradation of the incretin hormone glucagon-like peptide (GLP)-1, which is a potent stimulator of insulin secretion. Potent inhibitors of NEP produce a pharmacological response through an increase in opioid or vasoactive peptide levels, indicating their therapeutic potential as novel analgesics or antihypertensive agents [19–21].

The concept of dual inhibition of the two enzymes by a single molecule has shown major benefits and potential superiority versus other agents in various experimental models of hypertension, heart failure, and renal diseases. The underlying presumed rationale for the combined inhibition of ACE and NEP is to block the vasoconstrictor angiotensin II and simultaneously increase the vasodilator ANP by decreasing its enzymatic degradation. However, it remains controversial as to whether dual ACE/NEP inhibitors confer superior cardiovascular effects when compared to ACE inhibition alone. A major impediment for routine use of these agents remains the potentially life threatening side effect of angioedema, or excessive, painful swelling beneath the skin. Omapatrilat (Bristol-Myers Squibb) [22], the most advanced dual ACE/NEP inhibitor, although shown to be superior over existing agents in reducing hypertension, was halted by the US Food and Drug Administration (FDA) in phase III clinical trials because of increased side effects, such as severe angioedema [10].

## 4.3 PEPTIDE INHIBITORS OF THE HIV-1 VIRAL LIFE CYCLE

HIV-1 is the causative agent for acquired immune deficiency syndrome (AIDS). AIDS was a great plague of the last century, and remains a significant public health problem worldwide. Heroic public health and research efforts have been made to counteract this disease, with research targeted at generating therapeutics at every stage in the viral life cycle. Drug resistance is a great challenge in HIV-1 treatment therapies, resulting in the development of combination therapies [23].

The major hurdle in HIV-1 immune response and drug resistance is its high mutation rate. Since HIV-1 reverse transcriptase does not possess editing functionality, HIV-1 has highly error-prone replication [23, 24]. A patient infected with HIV-1 undergoes constant reinfection by self-generated mutant virus, presenting an ever-changing challenge to the immune system. Over time, the viral load increases as immune reserves become depleted, and normally trivial secondary infections by rhinovirus or fungus may prove lethal to the patient's overloaded immune system [24].

HIV-1 protease is an aspartic protease and an attractive secondary target for peptide-based therapeutic intervention. The 17-protein viral mRNA is read out as one long polypeptide [24]. HIV-1 protease autolyses itself from this polypeptide and subsequently cleaves other viral proteins into active forms (Figure 4.3) [24]. The first FDA-approved antiretroviral protease inhibitor was saquinivir (Table 4.1), developed by Roche and approved in 1995 [25]. This 680 Da peptidomimetic compound had excellent binding affinity but poor oral viability [24]. In the clinic, single protease inhibitor treatments quickly resulted in drug resistance, so other protease inhibitors were rushed to market (Table 4.1).

Because viral resistance is a complex, moving target, a recurring theme in HIV-1 treatment is combinatorial therapy. As a first example, a mixture of saquinavir and ritonavir therapy in clinics preserved saquinavir serum levels, and ritonavir was later shown to inhibit not only HIV-1 but also the breakdown of saquinavir by cytochrome P450 peroxidase [24]. However, inhibiting HIV-1 protease alone can result in side effects such as diabetes, cardiovascular events, and lipidystrophy [26]. Simultaneously, utilizing nonscissile substrate mimics as inhibitors of HIV-1

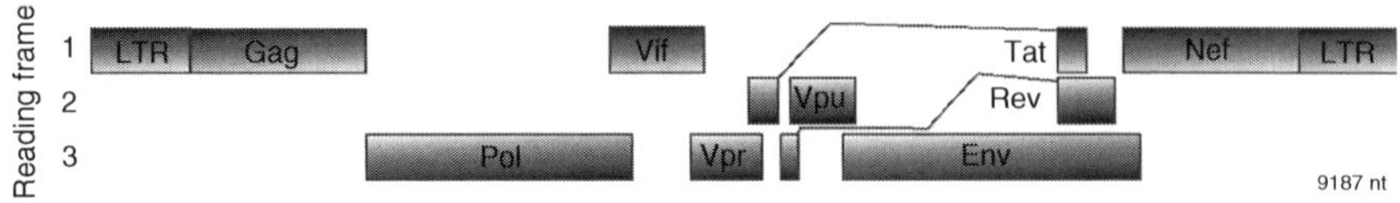


**Figure 4.3** HIV-1 Viral Genome. Gag: group specific antigen codes for the Gag polyprotein, processed by HIV-1 protease into structural proteins such as matrix protein (MA), capsid protein (CA), nucleocapsid Protein (NC) and spacer peptides. The Pol polyprotein is most often targeted for drug therapies; it codes for the viral enzymes, including reverse transcriptase, viral integrase, and HIV-1 protease. Env or "envelope" codes for gp160, which is a precursor to the fusion proteins gp120 and gp41. Regulator of virion (Rev), negative regulatory factor (Nef), and transcriptional AcTivator (Tat) are viral regulatory genes to augment and time expression. Viral protein U (Vpu) is involved in virion budding, while viral protein R is critical to nuclear import of the RNA genome.

protease typically produced resistance mutations to several drugs at once, caused by heightened similarities of catalytic site and peptide inhibitor interaction [27–29]. *In vivo* cases have reported that as few as four amino-acid mutations in HIV-1 protease (M46I/L63P/V82T/I84V) were sufficient to confer viral resistance to early drugs in clinical trials [30]. Current research strategies on inhibitor design for HIV-1 protease have moved from blocking the catalytic site to binding inhibitory peptides to conserved regions on the protein backbone [31, 32]. Furthermore, the treatment of HIV infection has moved to a vertical system of inhibition known as highly active antiretroviral therapy (HAART). HAART became standard care following the Vancouver-based International AIDS Conference of 1996 [26]. This strategy converts HIV from a bed-ridden death sentence to a treatable, chronic health condition, provided the patient follows a strict dosing regimen. HAART is a daily combination therapy; it comprises of a nucleoside analog reverse transcriptase inhibitors, fusion inhibitors, and HIV protease inhibitors. This *cocktail* approach is advantageous because it may be tailored to a given patient's drug resistances, resulting in viral loads low enough to significantly suppress (95%) disease transfer between infected mother to prenatal child [26].

The successes of HAART and further advances in HIV-1 protease inhibition secure peptide-based inhibitors as a predominant success story for computational inhibitor design, from the initial development of saquinavir [25] to raltegravir [33] and beyond.

## 4.4 MATRIX METALLOPROTEINASES

MMPs are a large family of zinc-dependent neutral endopeptidases involved in the degradation of extracellular matrix (ECM) components, and thus MMPs play a crucial role in the homeostasis of normal tissue remodeling [34, 35]. MMP activity is controlled and finely balanced at many levels: RNA transcription, protein translation, secretion, localization, activation of zymogen, inhibition by endogenous proteins,

TABLE 4.1 **Peptide and Peptidomimetic Inhibitors of HIV-1 Protease.**

| Name | Developed by | Approval Date, Name | Target | Structure/Sequence | References |
|---|---|---|---|---|---|
| Saquinavir | Hoffman-La Roche | 1995, Invirase<br>1997, Fortovase | HIV-1 protease | 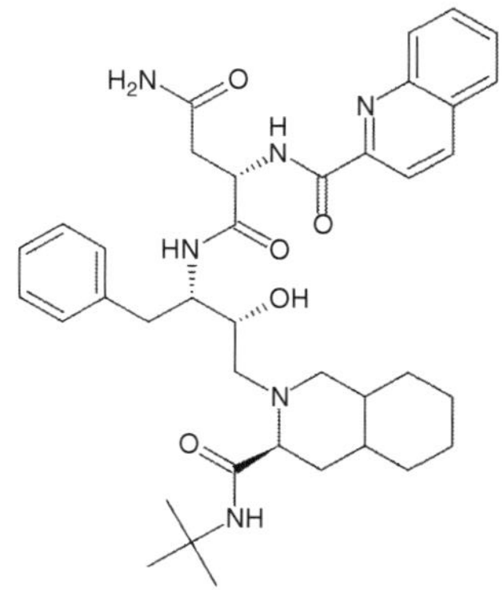 | [25] |
| Indinavir | Merck | 1996, Crixivan | HIV-1 protease | 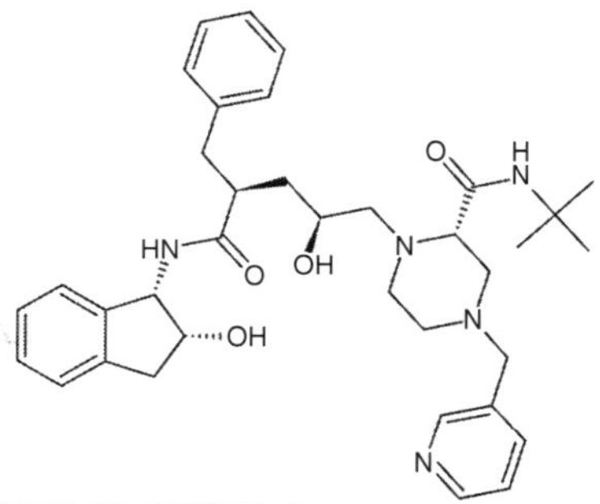 | [25, 149] |

(*continued*)

**TABLE 4.1** (*Continued*)

| Name | Developed by | Approval Date, Name | Target | Structure/Sequence | References |
|---|---|---|---|---|---|
| Ritonavir | Abott | 1996, Norvir | HIV-1 protease | 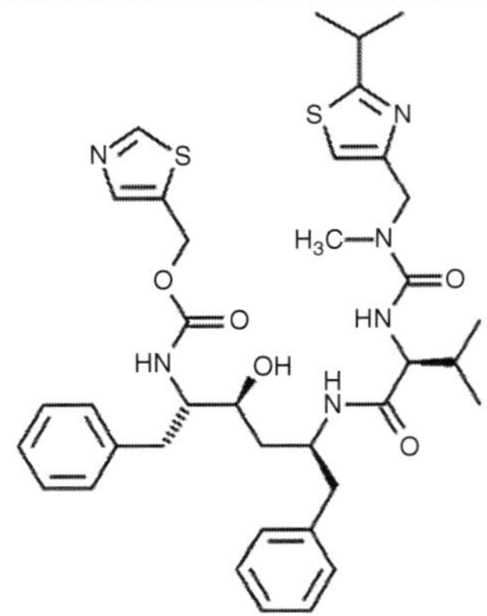 | [150] |
| Nelfinavir | Agouron Pharm. | 1997, Viracept | HIV-1 protease | 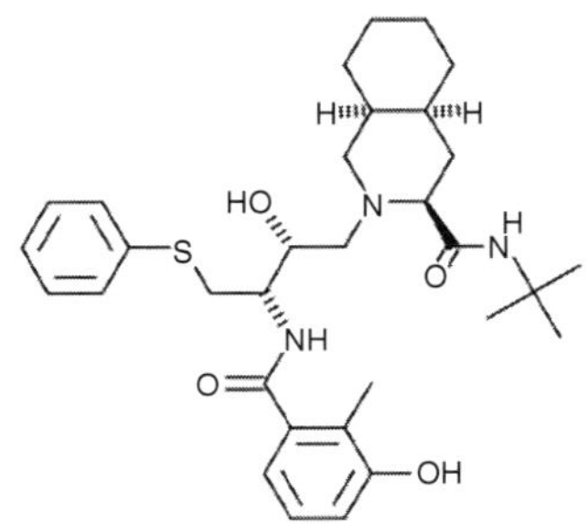 | [151] |

| | | | | | |
|---|---|---|---|---|---|
| FosAmprenavir | ViiV Healthcare | 2003, Lexiva | HIV-1 protease | 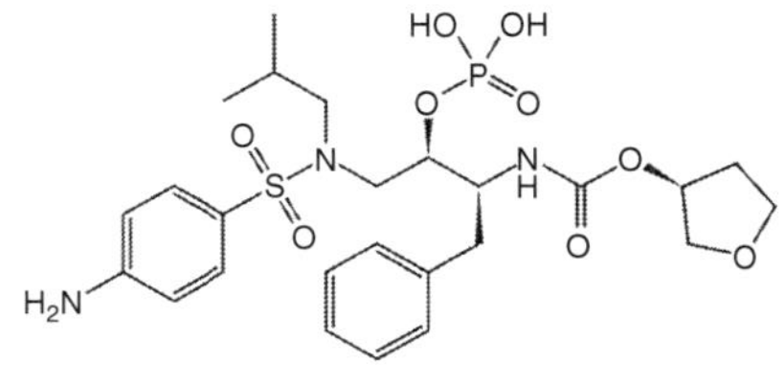 | [152] |
| Amprenavir | Glaxo-SmithKline | 1999, Agenerase | HIV-1 protease | 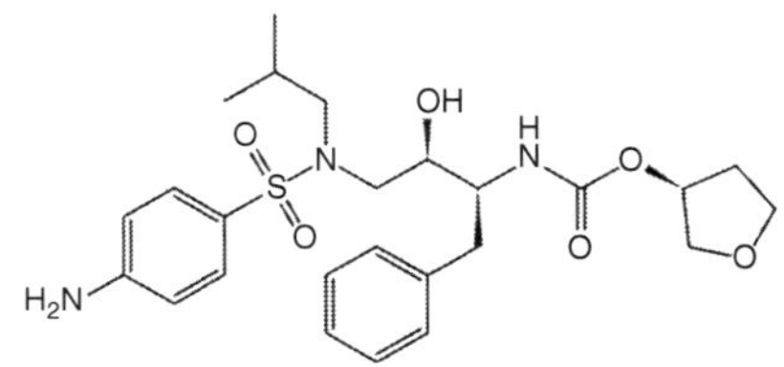 | [153] |
| Atazanavir | Bristol-Myers | 2003, Reyataz | HIV-1 protease | 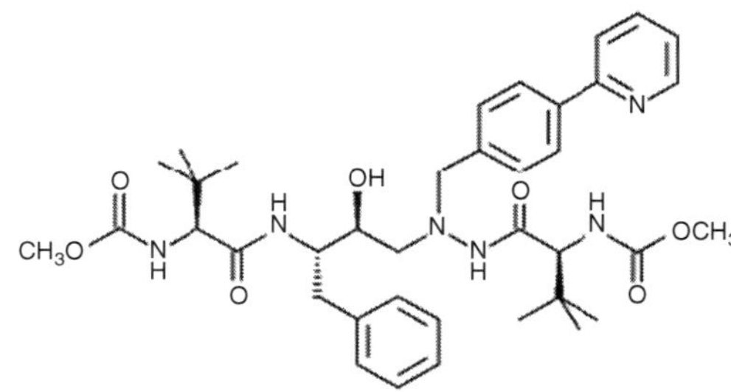 | [154] |

*(continued)*

**TABLE 4.1** *(Continued)*

| Name | Developed by | Approval Date, Name | Target | Structure/Sequence | References |
|---|---|---|---|---|---|
| Darunavir | Tibotec | 2006, Prezista | HIV-1 protease | 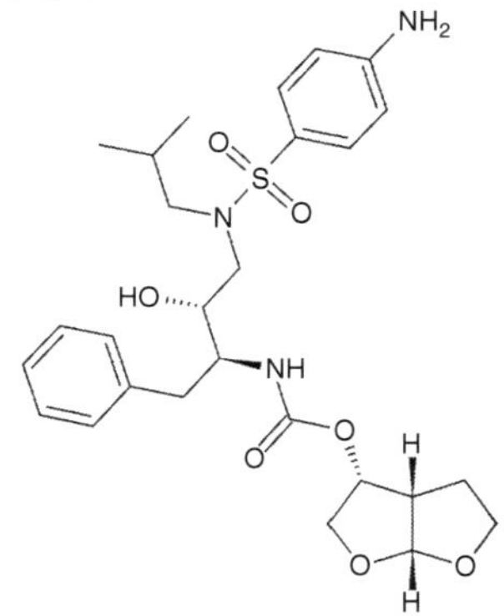 | [155] |
| Raltegravir | Merck | 2007, Isentress | HIV-1 integrase | 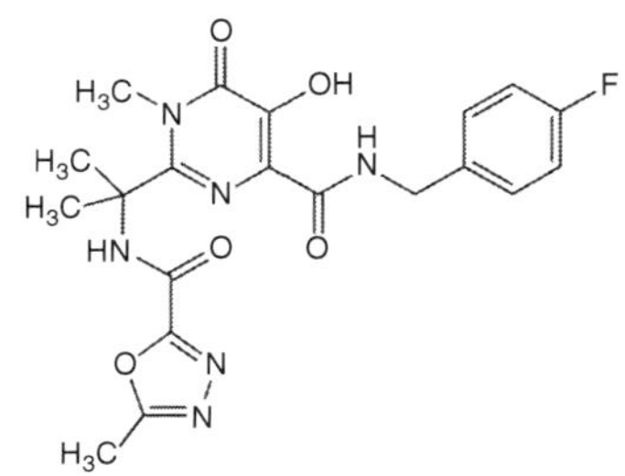 | [33] |
| Enfuvirtide | Roche | 2003, Fuzeon | Fusion inhibitor | Ac-Tyr-Thr-Ser-Leu-Ile-His-Ser-Leu-Ile-Glu-Glu-Ser-Gln-Asn-Gln-Gln-Glu-Lys-Asn-Glu-Gln-Glu-Leu-Leu-Glu- Leu-Asp-Lys-Trp-Ala-Ser-Leu-Trp-Asn- Trp-Phe-$NH_2$ | [156] |

and degradation. MMPs are usually minimally expressed in normal adult physiological conditions. Overexpression of MMPs results in an imbalance between the activity of MMPs and endogenous inhibitors of MMPs, tissue inhibitors of metalloproteinases (TIMPs), leading to tissue degradation and consequently facilitating a variety of pathological disorders, including arthritis and cancer [36–38]. Accordingly, MMPs became important pharmaceutical targets for treatment of these diseases [39, 40]. MMPs were the first proteases seriously considered as targets to combat cancer because of their role in ECM degradation. The compelling results of preclinical studies on MMP inhibition in tumor models raised the idea that the development of strategies to inhibit MMPs may prove to be a powerful tool to fight cancer. Unfortunately, the results of MMP inhibitor clinical trials have been disappointing. MMP inhibitors such as the hydroxamates batimastat (British Biotech), marimastat (British Biotech), and prinomastat (Aguron) and the nonhydroxamates neovastat (Aeterna), rebimastat (Bristol-Myers Squibb), and tanomastat (Beyer) have failed clinical trials because of severe side effects and/or offering no significant therapeutic advantage. The tetracycline analog periostat (CollaGenex), which inhibits both the activity and synthesis of MMPs, is the only MMP inhibitor on the market. Periostat is prescribed for treatment of adult periodontitis.

The failure of MMP inhibitors can be attributed to numerous factors, but a major one is the lack of enzyme selectivity [41–43]. The critical examination of previous results has prompted serious reevaluation of MMP inhibition strategies focusing attention on the identification of specific MMP targets at different stages of tumor progression, both in order to improve efficacy and to reduce side effects. The most attractive design strategies include the development of novel zinc-binding ligands and to exploit alternative ways to increase inhibitory potency by exploring the differences between the various enzyme subtypes [44, 45].

Identification of protease secondary binding sites (exosites), that is, nonactive site regions that facilitate or modulate protease activity, could be utilized for the design of selective inhibitors within protease families. Exosite identification has allowed for the design of selective inhibitors for coagulation Factors VIIa, IXa, and Xa [46–48], caspases [49, 50], pregnancy-associated plasma protein-A (PAPP-A) [51], and cathepsin K [52]. Exosite-binding, small molecule inhibitors have been described that are highly selective for MMP-13 [53–59].

Exosite binding may also be combined with active site binding to create selective MMP inhibitors. One such approach has taken advantage of the unique ability of several MMPs to catalyze the hydrolysis of collagen triple-helical structure. Metallo(zinc)-proteases use the nucleophilic attack of a water molecule as one of the steps of amide bond hydrolysis [60]. The tetrahedral intermediate that results from water addition to the amide carbonyl has been the focus of many protease inhibitor designs. Phosphinic peptides/phosphinates have been shown to behave as transition state analog inhibitors of MMPs [61]. Subsequently, phosphinate triple-helical MMP inhibitors have several potential advantages over other inhibitor constructs. These analogs allow the incorporation of specificity elements for both the S and S′ subsites of the enzyme. Although binding to the nonprimed region of the active site is generally weaker than the primed site to prevent product inhibition [60], it does add

**TABLE 4.2 Inhibition of MMP-2, MMP-9, and MMP-1 by Triple-Helical Peptide Inhibitors (THPIs).**

| Enzyme | Inhibitor | Temperature (°C) | $K_i^{(app)}$ (nM) |
|---|---|---|---|
| MMP-2 | $\alpha$1(V)GlyΨ{$PO_2H$-$CH_2$}Val THPI | 10 | 4.14 ± 0.47 |
| | | 37 | 19.2 ± 0.6 |
| | MMP inhibitor III | 10 | 3.17 ± 0.23 |
| | | 37 | 0.83 ± 0.03 |
| | $\alpha$1(I-III)GlyΨ{$PO_2H$-$CH_2$}Leu THPI | 10 | 0.18 ± 0.00 |
| | | 37 | 0.08 ± 0.01 |
| MMP-9 | $\alpha$1(V)GlyΨ{$PO_2H$-$CH_2$}Val THPI | 10 | 1.76 ± 0.05 |
| | | 37 | 1.29 ± 0.00 |
| | $\alpha$1(I-III)GlyΨ{$PO_2H$-$CH_2$}Leu THPI | 10 | 0.02 ± 0.01 |
| | | 37 | 0.09 ± 0.00 |
| MMP-1 | $\alpha$1(I-III)GlyΨ{$PO_2H$-$CH_2$}Leu THPI | 10 | 7.83 ± 1.03 |
| | | 37 | 26.7 ± 5.2 |
| | MMP inhibitor III | 10 | 2.48 ± 0.35 |
| | | 37 | 4.72 ± 0.38 |

sequence diversity and potential selectivity. The triple-helical structure allows for interaction with both the active site and exosites [62]. Triple-helical conformation is also less susceptible to general proteolysis than peptides and other folded proteins [63, 64].

In order to create the desired phosphinate transition state analogs, our laboratory prepared protected Fmoc-phosphinodipeptides [57, 65–67]. An Fmoc-phosphinodipeptide was utilized to create $C_6$-(Gly-Pro-Hyp)$_4$-Gly-Pro-Pro -GlyΨ{$PO_2H$-$CH_2$}(*R,S*)Val-Val-Gly-Glu-Gln-Gly-Glu-Gln-Gly-Pro-Pro-(Gly-Pro-Hyp)$_4$-$NH_2$ [designated $\alpha$1(V)GlyΨ{$PO_2H$-$CH_2$}Val THPI] [66], based on the cleavage site in type V collagen by MMP-9 [68].

The $\alpha$1(V)GlyΨ{$PO_2H$-$CH_2$}Val THPI (which contains the S configuration in the $P_1'$ position, equivalent to an L-amino acid) was initially tested against MMP-2 and MMP-9 (Table 4.2). Due to the low melting temperature of the potential inhibitor ($T_m \sim 25\,°C$), $K_i$ values were first determined at 10 °C. $\alpha$1(V)GlyΨ{$PO_2H$-$CH_2$}Val THPI was found to be a very effective inhibitor of MMP-2 and MMP-9, with $K_i$ values of 4 and 2 nM, respectively. When inhibition assays were repeated at 37 °C, the $K_i$ value increased for MMP-2 but not for MMP-9 (Table 4.2). Thus, triple-helical structure modulated inhibition of MMP-2 but not MMP-9.

To determine if an increase in $K_i$ as a function of temperature was a general trend for inhibition of MMP-2, inhibition of MMP-2 by MMP inhibitor III (a hydroxamic acid-Leu-homoPhe dipeptide) was examined. At 10 °C, the $K_i$ value for MMP-2 inhibition was 3 nM (Table 4.2). Increasing the temperature to 37 °C decreased the $K_i$ to 0.8 nM (Table 4.2). Thus, for a small molecule inhibitor, an increase in temperature slightly increased the affinity toward MMP-2, most likely due to enhanced hydrophobic interactions. This further suggested that the decreased inhibition of MMP-2 by

$\alpha$1(V)Gly$\Psi${$PO_2H$-$CH_2$}Val THPI as a function of increasing temperature is due to unfolding of the inhibitor triple-helical structure.

MMP-1, MMP-3, MMP-8, MMP-13, and MT1-MMP were tested for inhibition by $\alpha$1(V)Gly$\Psi${$PO_2H$-$CH_2$}Val THPI. No inhibition of MMP-1, MMP-3, or MT1-MMP was observed up to an $\alpha$1(V)Gly$\Psi${$PO_2H$-$CH_2$}Val THPI concentration of 25 μM. MMP-8 and MMP-13 were inhibited weakly, with $K_i$ values in the range of 50 and 10 μM, respectively. Thus, this study utilized a Gly$\Psi${$PO_2H$-$CH_2$}Val transition state analog to bind selectively at the $S_1$-$S_1'$ site of MMP-2 and MMP-9. Selective inhibition of these MMPs is desirable, as MMP-2 has been validated as an anticancer drug target, whereas MMP-9 inhibition may be useful in treating early-stage cancers [69].

Our second inhibitor design utilized a Triple helical peptide substrate mimicking $\alpha$1(II)769-783, which is hydrolyzed by MMP-1, MMP-2, MMP-8, MMP-9, MMP-13, and MT1-MMP [70, 71]. The $P_1$-$P_1'$ subsites of the triple-helical peptide, which incorporate Gly-Leu in the substrate, were substituted by a Gly$\Psi${$PO_2H$-$CH_2$}Leu transition state analog. Because the $T_m$ value for $\alpha$1(V)Gly$\Psi${$PO_2H$-$CH_2$}Val THPI was low (see above) [72], the $\alpha$1(I-III)Gly$\Psi${$PO_2$ H-$CH_2$}Leu THPI incorporated (4*R*)-Flp to enhance triple-helicity [73–75]. Thus, the sequence of this inhibitor was $C_6$-Gly-Pro-Flp-(Gly-Pro-Hyp)$_4$-Gly-Pro-Gln-Gly$\Psi${$PO_2H$-$CH_2$}(*R*,*S*)Leu-Ala-Gly-Gln-Arg-Gly-Ile-Arg-(Gly-Pro-Hyp)$_4$ -Gly-Pro-Flp-$NH_2$, and it exhibited a $T_m$ value of 30 °C [76]. Studies revealed low nanomolar $K_i$ values for inhibition of MMP-1, MMP-2, and MMP-9 (Table 4.2). Our second transition state analog inhibitor appears to be effective against a broader range of collagenolytic MMPs than the first inhibitor. Interestingly, MMP-1 was sensitive to the triple-helical structure of the inhibitor ($K_i$ increased approximately four times when the inhibitor was thermally unwound), but neither MMP-2 nor MMP-9 was. This contrasts with the sensitivity of MMP-2 to the triple-helical structure of $\alpha$1(V)Gly$\Psi${$PO_2H$-$CH_2$}Val THPI (Table 4.2), and indicates that there is *a sequence-dependent sensitivity to triple-helical structure* for some MMPs.

A single-stranded peptide model of the $\alpha$1(I)715-721 collagen sequence has been identified as a ligand for the MMP-2 fibronectin type II (FN II) insert and inhibited MMP-2 gelatinolysis [77]. Our laboratory assembled a triple-helical version of this ligand [$\alpha$1(I)715-721 THP], and evaluated it for the ability to inhibit MMP-2 and MMP-9 triple-helical peptidase and gelatinase activities [76]. $\alpha$1(I)715-721 THP inhibited type V collagen-model triple-helical peptidase activity but not interstitial collagen-model triple-helical peptidase activity. To our knowledge, this demonstrated the first use of an exosite binder to selectively inhibit one collagen-based MMP activity but not another.

## 4.5 ANTRAX LETHAL FACTOR INHIBITION BY DEFENSINS

Defensins are small peptides that display broad antibacterial, antifungal, and antiviral properties and are believed to be the first responders to microbial and viral attacks in primates and humans [78, 79]. The θ-defensins are the only animal-derived

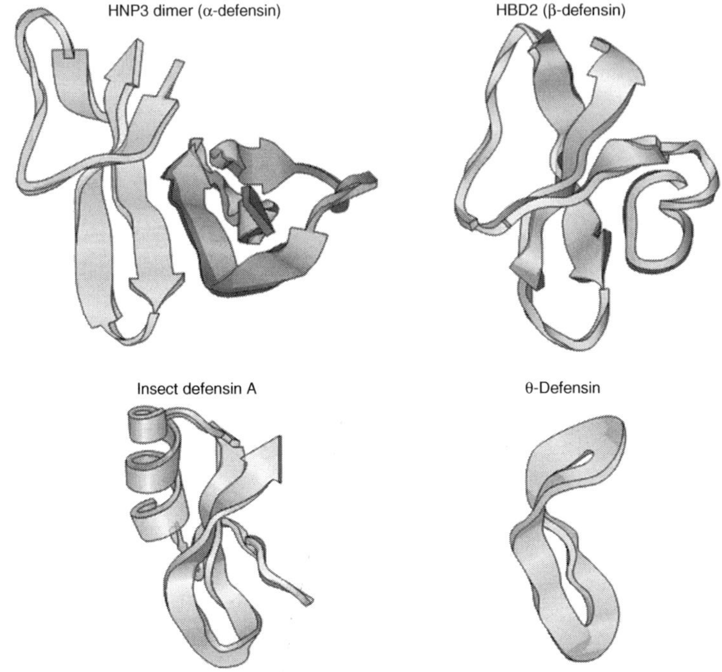


**Figure 4.4** Cartoon structures of representative mammalian defensins and an insect defensin. β-Sheet structures are indicated by flat ribbons and arrows. Human neutrophil peptide 3 (HNP3, α-defensin) forms a β-sheet rich dimer. Human β-defensin 2 (HBD2) in solution is a monomer with the same general shape (defensin fold) despite the change of the disulfide-bond pattern. In addition, there is a short α-helical segment at the N-terminus. The conformation of insect defensins (132 from Ganz) is distinct, with a prominent α-helical segment that is linked by two disulfide bonds to the C-terminal β-sheet. The θ-defensin structure is cyclic, forming a simple β-sheet. *Source*: Adapted with permission from Reference 142.

head-to-tail cyclic octadecapeptides possessing three disulfide bonds [80–82]. These cyclic peptides have been purified from nonhuman primates, and are encoded by mutated α-defensin genes (Figure 4.4).

Retrocyclins are synthetic θ-defensin peptides designed based on the human genome, lacking the premature stop codon present in their natural counterparts. Human α-defensins, as well as θ-defensin peptides, have been found to be non-competitive/uncompetitive inhibitors of *Bacillus anthracis* LF, a zinc-dependent metalloprotease [83, 84]. Both defensins were recently shown to be active against *B. anthracis* bacilli, their spores, and in the inactivation of the LF *in vitro*. Wang et al. [84] demonstrated that low concentrations of θ-defensins not only killed vegetative *B. anthracis* bacilli and rendered the germinating spores nonviable, but also inactivated the enzymatic activity of the LF and protected murine macrophage cells (RAW-264.7) from lethal toxin, which consists of LF and protective antigen. Insight into the three-dimensional interactions revealed that the cyclic backbone, intramolecular tri-disulfide ladder, and Arg residues of θ-defensins are responsible for those protective effects. Retrocyclin derivatives studied by the group also displayed high affinity and rapid binding to the anthrax LF. Interestingly, preincubation of the LF with retrocyclins increased inhibition of the enzyme and resulted in a slowly progressive extent of enzyme inhibition. These findings suggest that postbinding effects, such as *in situ* oligomerization, may contribute to the antitoxic properties of retrocyclins [84].

In addition to the direct LF inhibition, θ-defensins were also shown to induce orientational disorder of the anionic bacterial membrane as compared to a zwitterionic mammalian membrane. Moreover, *B. anthracis* peptidoglycan cell wall contains a myriad of pores called *tesserae* with the diameter of 41 Å, which are large enough for a passage of a 25 kDa globular protein. The above effects improve the likelihood of the θ-defensin delivery to bacterial intracellular destinations and are supported by radial diffusion assays, in which θ-defensins were shown to be effective against *B. anthracis* spores [84].

## 4.6 KINASES

Signal transduction regulates many cellular processes, such as metabolism, survival, growth, division, and death. Phosphorylation is one of the most important signals within the cell. Protein kinases catalyze the transfer of the terminal phosphoryl group from Adenosine triphosphate to a protein or peptide substrate (Figure 4.5). There are at least 500 kinases, and they account for approximately 2% of the human genome [85]. Kinases play an important role in regulating many aspects of cellular function and their dysregulation has been implicated in numerous diseases. Kinases are presently the second most popular drug target, behind G-protein-coupled receptors, comprising 25% of the drug development effort in pharmaceutical companies [86].

Protein kinases share a high degree of structure similarity within conserved regions that include a catalytic domain, an ATP-binding site, and an activation loop. Due to this conservation, it has been difficult to develop small molecule inhibitors exhibiting high selectivity. Although all kinases share these important domains, their substrate specificity is highly diverse due to varying regulatory sequences outside of these domains [85]. Most of the kinase inhibitors to date have exploited the ATP-binding site. However, the drawback of this approach is that potential inhibitors have to compete with very high intracellular ATP concentrations, as well as greater than 200 other proteins that utilize ATP [86].

Protein–protein interactions are the major determinants of kinase specificity and they have been intensely studied in order to develop effective inhibitors. Kinases preferentially bind to specific recognition motifs within the substrate they phosphorylate. Peptides mimicking these binding sites should have better specificity than small molecules due to the larger binding surface, thus offering greater opportunity for the development of effective inhibitors (Table 4.3 and Figure 4.6).

One of the first successful kinase inhibitors developed was a 20 amino-acid peptide derived from the sequence of naturally occurring protein kinase A (PKA)

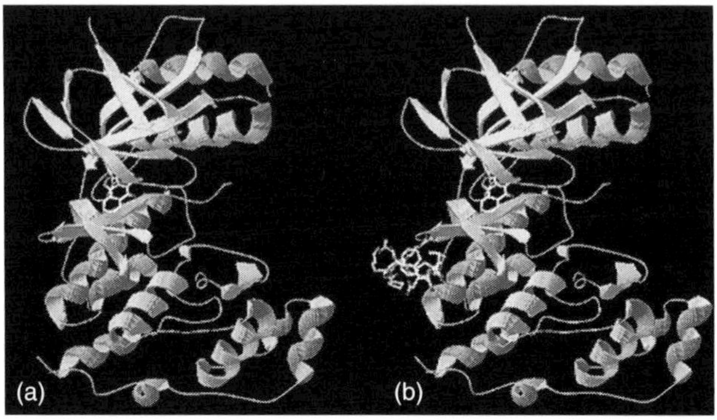


**Figure 4.5** Protein kinase interactions with ATP-competitive and ATP-noncompetitive inhibitors: c-Jun N-terminal kinase. (a) Structural analysis has shown that the ATP-binding site of the protein kinase C-Jun N-terminal Kinase (JNK) is occupied by an ATP-competitive inhibitor of JNKs, SP600125. (b) The structure of the complex between JNK1 and the peptide inhibitor derived from the JNK pathway scaffold protein JNK-Interacting-Protein-1 showed the interaction of the JNK1 protein with the peptide inhibitor at a site remote from the ATP-binding pocket. (*See insert for color representation of this figure.*) *Source*: Reprinted with permission from Reference 86.

inhibitor [87]. PKA is a Tyr kinase, catalyzing phosphorylation of the Tyr residue in the substrate. PKA recognizes a RRNYL motif on proteins targeted for preferential phosphorylation. The PKA-derived peptide contains an Ala substitution within the regular PKA phosphorylation site and displays a high potency of inhibition [88]. Aside from binding of the active site of PKA, the peptide also contributes binding of amino acids located further upstream from the recognition motif [89, 90]. This underscores the importance of extensive protein–protein interactions for the inhibition specificity.

Other peptides derived from the recognition motif of glycogen synthase kinase-3 (GSK-3) have also proven to be effective modulators of the kinase's activity [85]. GSK-3 is a serine kinase, recognizing the already phosphorylated motif SXXXS(p), where X denotes any amino acid. Mutated peptides in which Ala replaced the pre-phosphorylated Ser have been shown to be effective inhibitors of GSK-3 activity toward substrates such as cAMP Response Element-Binding protein and heat shock factor-1 (HSF-1). In addition, the sequence upstream from the GSK-3 active site was shown to exhibit a regulatory role, as the substitution of a key Glu residue improved the inhibitory potency of the peptide [91].

In order to improve specificity of inhibitory peptides, inhibitors can be derived from an autoinhibitory domain. This pseudosubstrate often contains an exact copy of the kinase specific recognition motif. Protein kinase C (PKC) isoforms contain this type of pseudosubstrate within their $NH_2$-terminal region and the derived peptides displayed high selectivity toward each isoform [92, 93]. Other kinases containing this type of pseudosubstrate are $Ca^{2+}$/CaM-dependent protein kinase II (CaMK-II), myosin light chain kinase (MLCK), and phosphorylase kinase [85].

**TABLE 4.3 Synthetic Peptide Inhibitors of Kinases.**

| Peptide sequence ($NH_2 \rightarrow COOH$) | Peptide Target (Protein Kinase) | Derived from | References |
|---|---|---|---|
| Substrate | | | |
| **TYADFIASGRTG**RRNAI *PKI-(6-22)-amide* | PKA | PS site of PKA | [87, 88] |
| **KEAPP**APPQS**(p)P** *L803-mts* | GSK-3 | GSK-3 substrate recognition site of HSF-1 | [91, 157, 158] |
| **A: FA**RKG**ALRQ** **B: RFA**RKG**ALRQ KNV** | PKC$\alpha/\beta$ | PS site of PKC$\alpha/\beta$ | [159, 160] |
| **SIYRRGA**RRWR**KL** | PKC$\zeta$ | PS site of PKC$\zeta$ | [161] |
| **LKKFNARR**KLKGA**ILTTMLA** | CAMII kinase | PS site of CaMK II | [162, 163] |
| Docking sites | | | |
| 1. D-site | | | |
| **MP**KKK**PTPIQLNPAPDG** | ERK2 | D-site of MEK1 | [95] |
| **MQ**GKRK**ALKLNFANPP** | JNK1/2 | D-site of MEK4 | [94] |
| RPKR**PTT**LNL**FPQVPRSQDT** *(L)-JNKI 1 (Stress-Activated Protein Kinase Inhibitor I)* | JNK | JNK-binding domain (JIP-1) | [97–99] |
| 2. FXFP docking motif | | | |
| **RRPRSPAKLSF**QFPS | ERK | FQFP docking site Elk1 | [164] |
| 3. HJ and $\alpha$D regions | | | |
| **GG**Y**NQN**H**QKL**F**QL** *KRX-014$_{H151}$* | PKB | HJ-$\alpha$G region of PKB | [101] |
| **GGRAGNQYL** *KRX-702$_{H105}$* | PDK1 | HJ-$\alpha$G region of PDK1 | [101] |
| 4. Other protein docking sites | | | |
| **HAKRRLIF** *p21$^{WAF1}$* | CDK2 | COOH-terminal cyclin-binding domain of p21$^{WAF1}$ | [165] |
| **SQPETRTGDDDPHRLL QQLVLS-GNLIKEAV RRLHSRRLQ** *FRATide* | GSK-3 | COOH-terminus of FRAT1 | [166] |
| **DIHVDPEKFAAELISR LEGVLRDR** *GID* | GSK-3 | GSK-3β-interacting domain of axin | [167, 168] |
| Cellular targeting | | | |
| **EAVSLKPT** *$\varepsilon$V1-2* | PKC$\varepsilon$ | V1 domain of PKC$\varepsilon$ | [169, 170] |
| **SLNPEWNET** *$\beta$C2-4* | PKC$\beta$ | C2 domain of PKC$\beta$ | [171] |
| **DLIEEAASRIVD AVIEQVKAAGAY** *S-Ht31* | PKA | AKAP | [102, 104, 172, 173] |

PS, pseudosubstrate. Recognition or docking motives are marked in bold.

*Source*: Table adapted with permission from Reference 85.

Another viable target for interruption of kinase activity is the docking interaction site. Since protein kinases are usually a part of larger complexes involved in signaling cascades, protein–protein interactions play a major role in substrate recognition and processing. Many signaling pathways, such as the mitogen-activated protein kinase (MAPK) family, involve docking and/or scaffold proteins, which bring together the upstream regulators and downstream substrates. Peptides derived from the docking sites of upstream MAPK kinases (MEKs) have been used to inhibit phosphorylation of their dependent kinases such as ERK1/2, c-Jun N-terminal kinase (JNK1), JNK2, as well as terminal substrates, such as S6 ribosomal protein kinase, p90 Ribosomal S6 Kinase, and MAPK dual specific phosphatases [94–96].

Disruption of docking interactions has proven to be successful at the cellular and animal model levels. Peptides derived from a JNK scaffold protein termed *JIP1* displayed a high potency inhibition toward JNKs, leading to neuroprotection [97], blocking of pancreatic B-cell death [98], as well as antidiabetic effects in a diabetic animal model [99]. Docking peptides derived from PDK-1 (3-phosphoinositide-dependent protein kinase-1) and protein kinase B (PKB) were shown to disrupt the phosphorylation of their downstream targets, leading to prostate cancer cell growth inhibition [100]. Other peptides of this type displayed favorable outcomes in diabetes and angiogenesis [101].

Another way a docking site could be exploited is the inhibition of sites on proteins that transport kinases between cellular compartments. Translocation of kinases is isoform specific, thus designing peptides that mimic those sites should provide very specific inhibitory effects. An example of successful application of this type of inhibitory peptide can be found in the PKA pathway. A peptide derived from A-kinase

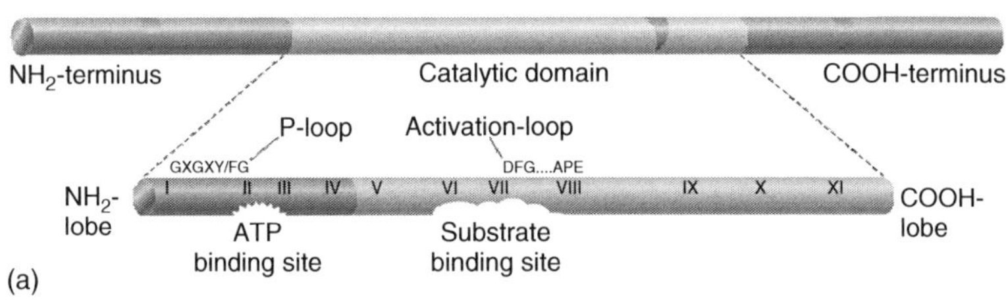


**Figure 4.6** Inhibitory mechanisms of peptides targeting protein kinases. (a) Schematic presentation of domain structure of protein kinase with conserved bilobal catalytic domain composed of 11 conserved subdomains. Conserved domains including the P-loop involved in ATP binding and the activation loop with the DFG … APE motif are marked. Red triangles mark docking site position with proteins such as substrates, upstream regulators, and scaffold proteins. (b) Inhibition of protein kinase by peptide (1) competition with substrate, (2) competition with docking sites derived from substrates, (3) competition with docking sites derived from the kinase, or (4) competition with cellular targeting anchor proteins. Peptides marked in yellow indicate the position of derivation; peptides marked in pink indicate the interaction with the target site; X describes disruption location; red chain represents the peptide inhibitor. *Source*: Reprinted with permission from Reference 85. (*See insert for color representation of this figure.*)

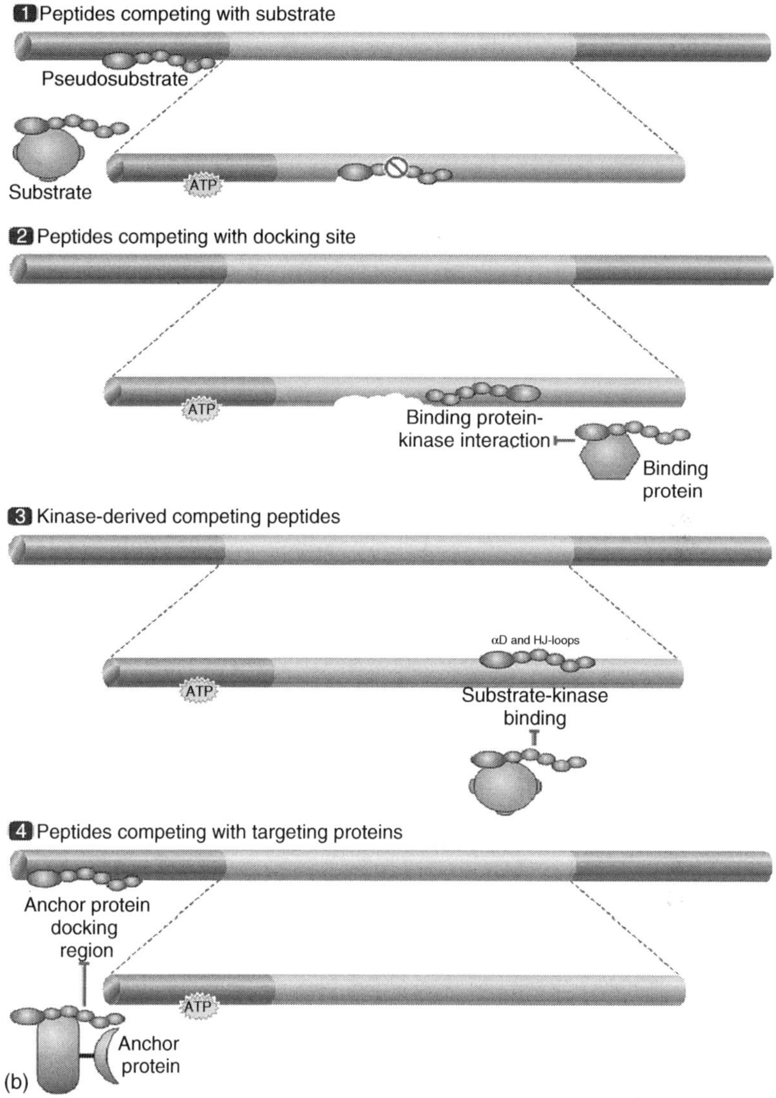


**Figure 4.6** *(Continued)*

anchor protein (AKAP) mimiced the role of PKA in synaptic transmission [102], oocyte maturation, and sperm mobility [103, 104].

## 4.7 GLYCOSYLTRANSFERASES (OLIGOSACCHARYLTRANSFERASES)

Aspargine-linked glycosylation (N-glycosylation) is a common co- and posttranslational modification of membrane and secretory proteins. This process occurs in eukaryotes, archaea, and some bacteria and is catalyzed in the lumen of the endoplasmic reticulum (ER) by the oligosaccharyltransferase (OSTase) complex. OSTase is a membrane bound multimeric protein complex that binds to ribosomes during active protein translation [105]. As a ternary complex, OSTase accommodates simultaneously the polypeptide and oligosaccharide substrates [106]. OSTase typically transfers the oligosaccharide ($Glucose_3Mannose_9N$-$acetylglucosamine_2$) from the dolichol-pyrophosphate donor (Dol-*PP*-oligosaccharide) to designated Asn side-chains present in glycosylation *sequons* (Asn-Xaa-Thr/Ser, where Xaa≠Pro) within the nascent polypeptide chains. N-glycans enhance proper protein folding by providing a hydrophilic environment and stabilizing β-turns [107] as well as by indirectly recruiting molecular chaperones. As new proteins are transported through the Golgi, their N-glycan residues can be modified, resulting in highly diverse structures that can dramatically alter their function [105]. Dysregulation of glycosylation patterns can lead to aberrant immune, developmental, and oncogenic events.

Since the early 1980s, it has been recognized that the presence of a hydroxyamino acid in the Asn-Xaa-Thr(Ser) cognitive sequence is required for the catalysis of transglycosylation and later it was shown that this hydroxyamino acid participates actively in transglycosylation through hydrogen-bond interactions [108]. One of the first inhibitors synthesized toward the OSTase enzyme complex was designed as a consensus sequence hexapeptide motif Arg-Asn-Gly-Yaa-Ala-Val. The side chain of the amino acid Yaa was replaced with epoxyethyl-, epoxypropyl-, allyl-, and vinyl-functional groups and the hexapeptide was tested against N-glycosyltransferase activity. Substitution of the Yaa amino acid with epoxyethylglycine produced irreversible inhibition of the membrane-bound enzyme complex, presumably through a suicide mechanism, in which OSTase catalyzes its own inactivation [109, 110].

Owing to the challenges in the expression and purification of the OSTase complex, little is known about how the complex coordinates binding to large substrates and catalyzes transglycosylation [106]. In an effort to elucidate the mechanism of the OSTase inhibition by the epoxy-inhibitor, Bause et al. undertook double labeling studies. They incubated the OSTase with labeled Dol-*PP*-[$^{14}$C]Oligosaccharides and *N*-dinitrobenzoylated epoxy-inhibitor peptide after which they resolved the complex by SDS-PAGE. Even though the gel displayed four OSTase subunits (40, 48, 63, and 66 kDa), only the 48 and 66 kDa polypeptides were radioactively labeled. The presence of other glycosyl receptors or an epoxy-inhibitor analog, in which the Asp was substituted by Gln abolished this labeling (Figure 4.7). These findings imply that the oligosaccharide transfer was not able to proceed with the inhibitor in place due to the direct inactivation of the active site of the OSTase complex (Figure 4.8) [108].

To date, substrate and product-based probes remain the best tools for the mechanistic studies of OSTase. Development of a potent peptide-based OSTase inhibitor, c[Hex-Dab-Cys]Thr-Val-Thr-Nph-$NH_2$ with $K_i = 0.037\ \mu M$, led Peluso et al. to hypothesize that the naphthyl group might be involved in π-stacking interactions within the saccharide binding pocket. As aromatic amino acids are often involved in carbohydrate–protein interactions, this hypothesis led them to investigate *neopeptides* (glycopeptide mimetics) in an effort to identify more potent OSTase inhibitors as well as to provide tools for structural studies. The

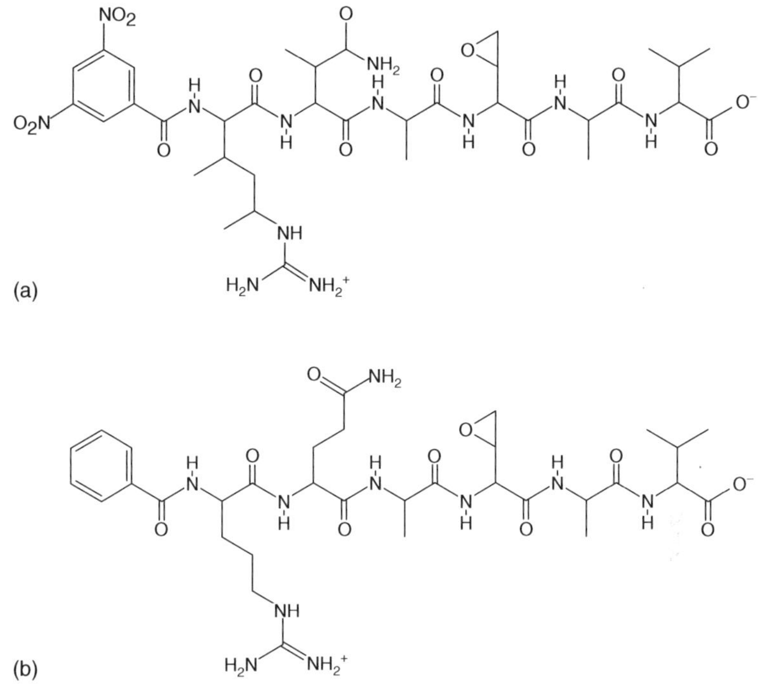


**Figure 4.7** Structure of the epoxy-inhibitor (a) and its glutamine analog (b). *Source*: Adapted with permission from Reference 108.

strategy utilized alanine-β-hydroxylamine (A*β*x), alanine-β-hydrazide (A*β*z), and 1,3-diaminobutanoic acid (Dab) substitution of the Asn normally present in the Xaa position within the Bz-Xaa-Ala-Thr-Val-Thr-Nph-$NH_2$ glycosylation sequence. *N*-acetylglucosamine conjugates of the above peptides were also prepared to test the product inhibition potential. Interestingly, the unnatural glycopeptides displayed similar binding affinity to those of the parent peptides, while the natural glycopeptide had significantly diminished binding compared to its parent peptide (Figure 4.9).

It was speculated that the replacement of the naturally occurring glycosyl-amide bond with glycosyl-hydrazide, glycosyl-oxime, glycosyl-hydroxylamine, or glycosyl-amine contributes to low nanomolar to micromolar inhibition due the improved flexibility of the newly formed linkage (Figure 4.10). This observation suggested that conformational changes within the OSTase complex dictated substrate binding and release. Most likely neopeptides are accommodated within the active site only when the nitrogen carrying the glycosyl moiety is not locked into a transorientation relative to the Asn. Based on these suppositions, it was proposed that the OSTase active site may be represented with cis or twisted amide geometry. The energetically favorable equilibration to the *trans* amide species may participate in the release of the *N*-glycosylated peptide from the OSTase active site and in this manner eliminate product inhibition through the reduced binding affinity for the

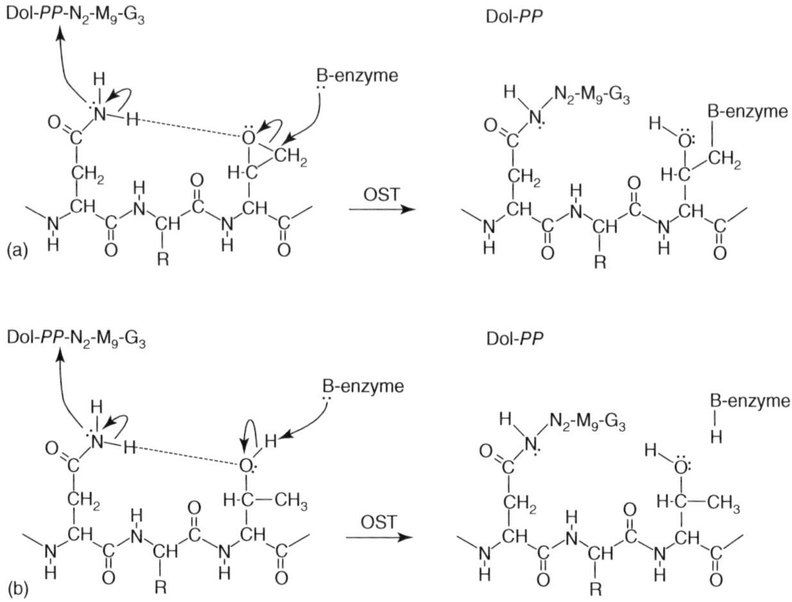


**Figure 4.8** Models describing (a) the suicide inactivation of OST leading to subunit double-labeling, and (b) the catalytic mechanism of OST. *N,N*-acetylglucosamine; M, mannose; G, glucose. *Source*: Reprinted with permission from Reference 110.

*trans* amide isomer. This explanation reveals that inhibition of the OSTase active site requires both the geometry and electronic interactions of the peptide–saccharide linkage (Figure 4.11) [106].

## 4.8 TELOMERASE INHIBITORS

The number of cell divisions a cell can undertake is limited due to the loss of telomeres. Telomeres are noncoding DNA located at the end of the chromosomes and are on average 10 kilobases (kb) in normal cells. They are responsible for maintaining the integrity of the chromosomes and preventing the replication of defective genes occurring during cell divisions [111]. On average 30–150 base-pairs (bp) are lost during each division [112, 113] and senescence is observed after 50 divisions [114].

Telomerases are holoenzymes with a molecular weight of ~650 kDa and are composed of three main components (Figure 4.12): a well-conserved telomerase reverse transcriptase (TERT), a telomerase RNA (TR), which constitutes the template to synthesize new telomeric DNA at the end of the chromosomes, and dyskerin, a protein that binds to TR or proteins that ensure the stabilization of the protein/RNA complex. Human telomerase appears to form dimers, in contrast to other organisms such as *T. thermophilla*. Ability of telomerases to extend the 3′ ends of linear chromosomes preserves the length of telomeres.

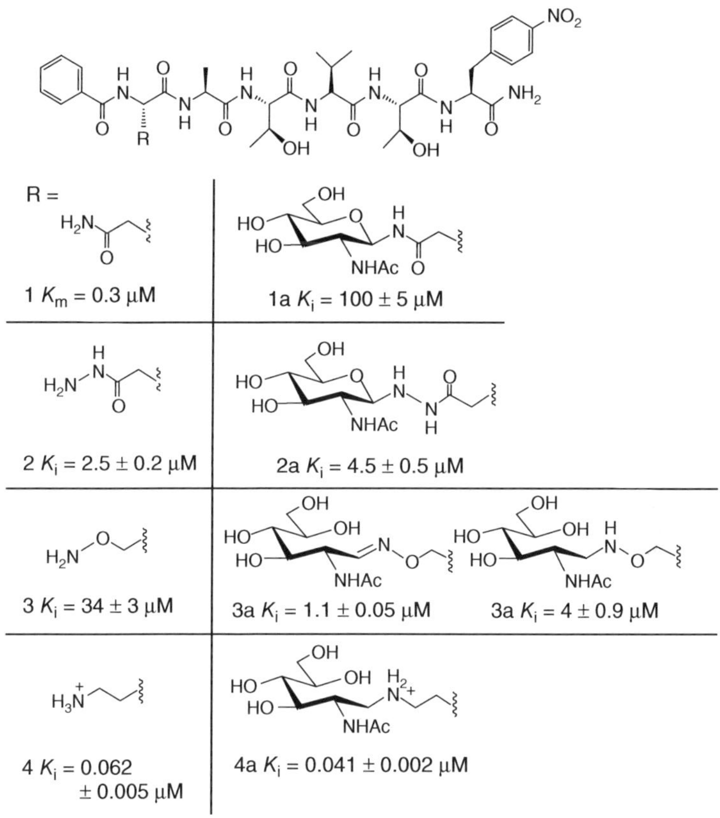


**Figure 4.9** Peptides and glycopeptide conjugates for OSTase. *Source*: Reprinted with permission from Reference 106.

The length of the telomere is an important factor in processes such as tumorigenesis, cell proliferation, and aging. Cancer cells possess shorter telomeres than healthy cells (on average 5 kb). Their DNA shortens during the early phase of tumorigenesis but then the length is maintained causing the immortalization of proliferating cancer cells. Maintenance of telomere length in cancer cells is due to the expression of telomerase, allowing cells to divide indefinitely. There is no evidence of telomerase expression in normal tissues (with the exception of stem cells) [115], whereas telomerases were detected in 85% of studied cancerous tissues [116–118]. The length of telomeres is already used for the prognosis of metastasis in breast and prostate cancer [119]. Telomerase could therefore be used as potential target for selectively inducing cancer cell death. However, the loss of DNA on the telomeres is estimated at 30–150 bp per cell division, which implies that all inhibition requires a lag time to cause toxicity. This depends on the initial length of the telomere before inhibition (on average 555 cell divisions for a 5 kb telomere, for example). Limitation of a quick

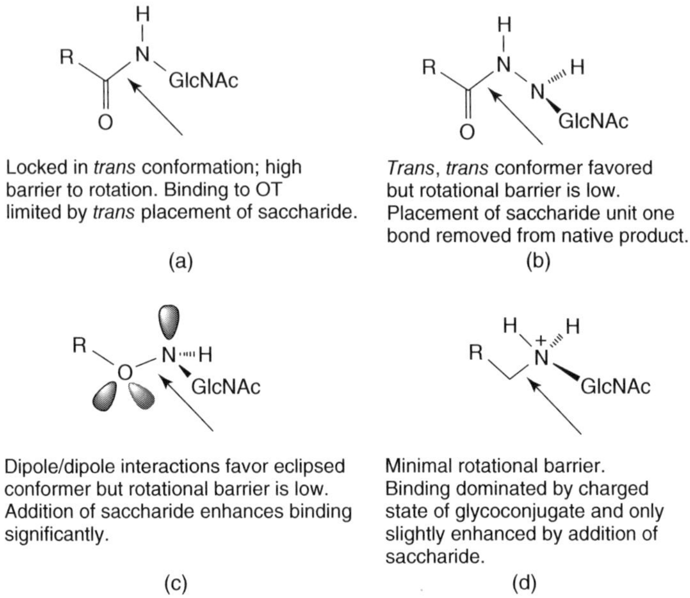


**Figure 4.10** Conformational preferences of (a) glycosyl amide, (b) glycosyl hydrazide, (c) glycosyl oxime, and (d) glycosylamine. *Source*: Reprinted with permission from Reference 106.

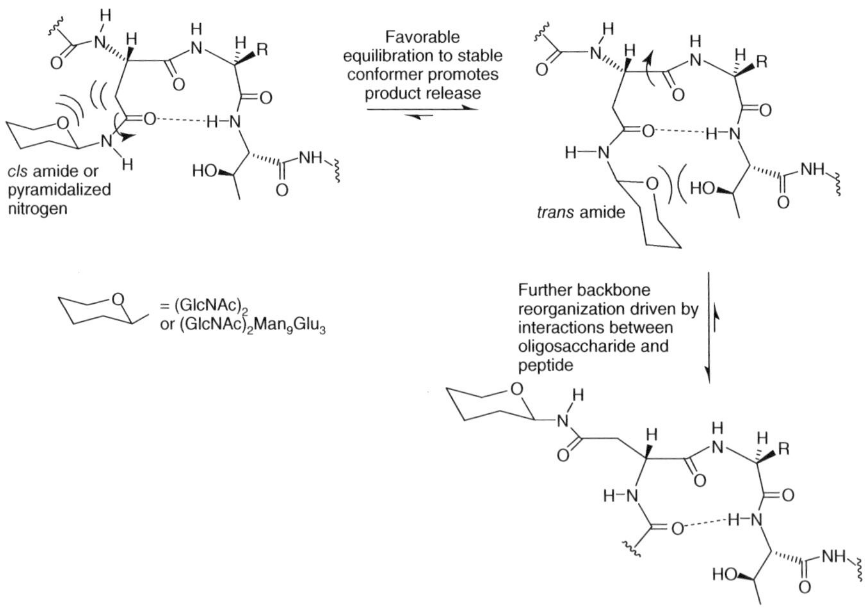


**Figure 4.11** Proposal for the role of isomerization in promoting OSTase product release. *Source*: Reprinted with permission from Reference 106.

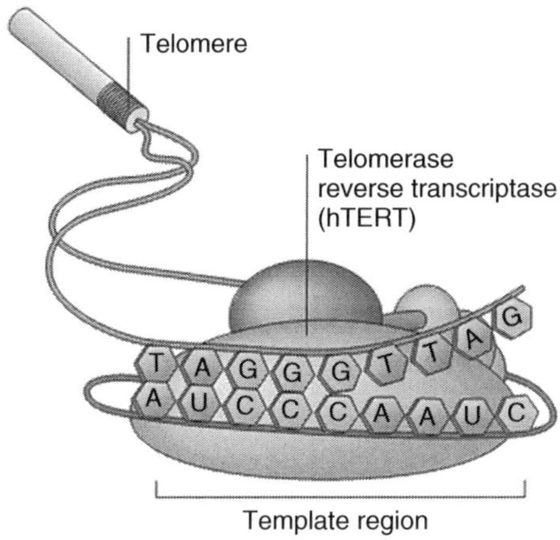


**Figure 4.12** Telomerase components. Human telomerase is a cellular reverse transcriptase with two essential components: telomerase reverse transcriptase catalytic subunit (hTERT) and functional telomerase RNA (hTR), which serves as a template for the addition of telomeric repeats (left side). *Source*: Figure adapted from Reference 143.

response to treatment suggests that telomerase inhibitors cannot be efficient solely and will necessitate the use of complementary therapy.

Telomerase inhibitors are currently designed to target either the TERT catalytic subunit or the TR component. Inhibition of TR involves the use of antisense oligonucleotides complementary to the RNA associated with the telomerase. Inhibition of the TERT can be obtained by targeting different subunits. Recent structural studies [120] have unraveled key domains in order to design inhibitors with higher efficacy and specificity. Inhibitors targeting the transcriptase subunit have already been designed and tested in clinical trials. Along with gene therapy, immunotherapy, and the design of small compounds such as BIBR1532, the development of small peptides in order to inhibit telomerase activity has been initiated. However, peptides encounter specificity problems and have not showed significant outcomes so far. Specificity issues have so far been thought achievable by targeting the nucleic acid unit of the telomerase. However, sequence selectivity remains poor. Design of second-generation inhibitors combining the use of nucleic acid complementary to the telomeric RNA with the use of synthesized peptides has been explored [121].

Peptide nucleic acid (PNA) has the property of mimicking the DNA or RNA backbone (Figure 4.13), achieved by utilizing *N*-(2-aminoethyl)-glycine units exhibiting structural similarities to the sugar/phosphate backbone present in nucleic acids. Unlike the native backbone of DNA or RNA, the internucleotide linkage is uncharged as the phosphate group is replaced by an amine and is less prone to degradation by nucleases, therefore stabilizing the oligomers. PNAs were reported to inhibit telomerases at nanomolar concentration in a cell-free environment. However, they do not cross the cytoplasmic membrane. Addition of a cellular transporter peptide to the

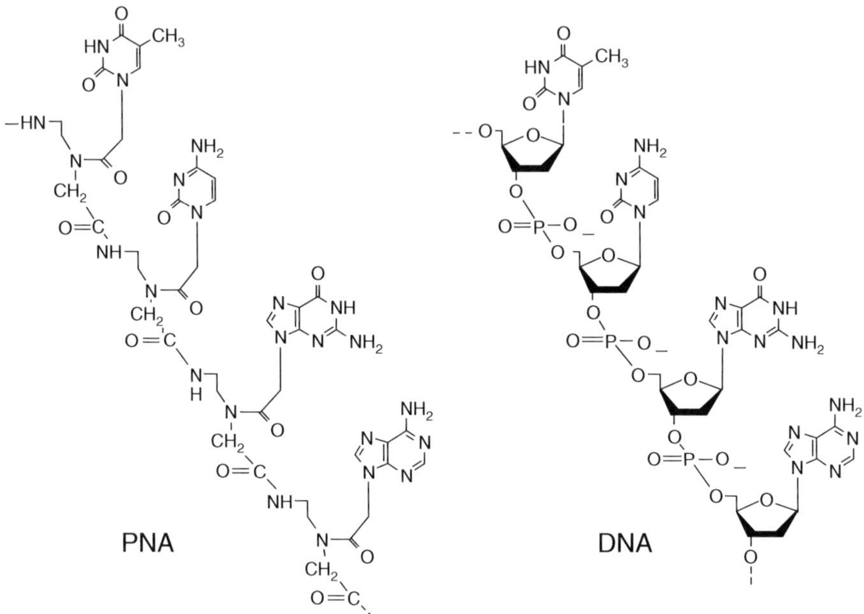


**Figure 4.13** PNA consists of repeating units of *N*-(2-aminoethyl)-glycine linked by amide bonds. The nucleobases are attached to the backbone by ethylene carbonyl linkages. *Source*: Reprinted with permission from Reference 144.

PNAs in order to improve cell internalization failed to induce telomerase shortening in human melanoma cell lines even after 20 days of treatment [121]. Although PNAs were initially reported in the literature as a promising new class of telomerase inhibitors in terms of selectivity and efficacy [122], the development of such inhibitors was abandoned.

## 4.9 TYROSINASE

Tyrosinase is a copper-dependent phenol oxidase responsible for hydroxylation of monophenol (such as Tyr) to *o*-diphenol (such as dihydroxyphenylalanine) and the corresponding oxidation of *o*-diphenol to *o*-quinone [123]. The reactive quinones are then involved in a cascade of oxidative condensation and addition reactions with phenolic compounds, thiols, and primary or secondary amines, leading to the formation of melanin (Figure 4.14).

Production of epidermal melanin is responsible for natural skin and hair color. Melanogenesis can be stimulated by a multitude of factors, such as UV radiation, and its dysregulation can lead to many cosmetic problems [124]. The expression and activation of tyrosinase controls the production of melanins within mammalian melanocytes, as well as causes browning that occurs upon bruising or long-term storage of vegetables, fruits, and mushrooms. Tyrosinase is responsible for normal skin

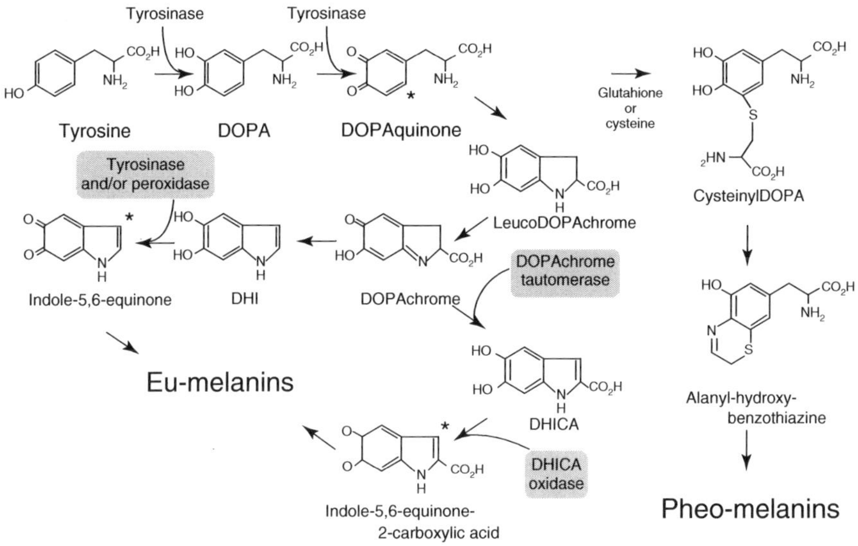


**Figure 4.14** Melanogenic pathway. The series of chemical reactions involved in the production of eumelanins and pheomelanins are shown, as are the regulatory enzymes involved. *Source*: Adapted with permission from Reference 145.

browning in response to UV exposure, but also participates in abnormal skin pigmentation, such as vitiligo and freckles and brown spots. In addition, this enzyme may play a role in some neurodegenerative diseases as well as cancer [125, 126].

Prevention of undesirable pigmentation has led the search for tyrosinase inhibitors. Tyrosinase activation is accomplished through PKC-$\beta$ on a phosphorylation site within a C-terminal sequence of 11 amino acids [127, 128]. An 11-mer tyrosinase mimetic peptide has been shown to be an effective inhibitor of PKC-$\beta$ binding, resulting in decreased tyrosinase phosphorylation [129]. Another effective tyrosinase inhibitor is kojic acid, a fungal metabolite widely used as a cosmetic skin-lightening agent or food additive for antibrowning effect. However, its use in cosmetics has been limited because of short storage life as well as skin irritation [130]. Although kojic acid alone is a potent inhibitor of tyrosinase, its derivatives have been synthesized in an effort to reduce side effects and provide longer shelf life. C-7 carboxyl group of kojic acid has been converted into esters, hydroxyphenyl ethers, glycosides, and amino-acid derivatives [130]. Kojic acid conjugation to the tripeptide FWY has been investigated and studies have shown that the tripeptide decreases the acid's toxicity, increases its storage stability, and most importantly, increases the inhibitory effect by 100-fold [131].

Recently, Schurink et al. undertook two SPOT screens of an octapeptide library from natural sources in an effort to identify novel tyrosinase inhibitors. One screen utilized phenol oxidase inhibitor (POI)-derivatized peptide library consisting of a scan of overlapping peptides with one amino acid shift ranging from 10 to 6 residues in length. The second library consisted of octameric peptides derived from industrial proteins from milk, egg, and wheat. Both libraries were spotted on a polyvinylidene difluoride membrane and binding and activity of fluorescently labeled tyrosinase was detected. Several novel peptide sequences with inhibitory activities were identified, and it was noted that Arg was the most important residue contributing to this inhibition [125, 126].

Another group led by Abu Ubeid has identified two competitive peptide tyrosinase inhibitors from a screen of an internal peptide library. The inhibitors, termed *P3* (RADSRADC) and *P4* (YRSRKYSSWY), displayed $K_i$ values of 123 and 40 μM, respectively. Consistent with the findings of Schurink et al., both of these peptides contain Arg residues, thus corroborating the importance of this residue in the tyrosinase inhibitory mechanism. These findings show promise toward the identification of potential novel antipigmenting agents without deleterious side effects. Unfortunately, further studies with these peptides displayed no effect on melanocyte proliferation or cytotoxicity up to 100 μM.

## 4.10 PEPTIDYL-PROLYL ISOMERASE

Peptidyl-prolyl isomerases (PPIases) are enzymes regulating cis and trans conformations in Xaa-Pro peptides (Figure 4.15) and play multiple roles in several biological systems. Cyclophilin and FK506 Binding Protein were the first identified PPIases due to their ability to bind and regulate immunosuppressant prodrugs cyclosporine and FK506. Other PPIases possess chaperone activities, improving the rate and yield of protein folding [132].

Pin1 belongs to a class of PPIases referred to as parvulins. It is a unique PPIase that exhibits preferential activity toward phosphorylated substrates. Protein phosphorylation plays a critical role in cell cycle progression, as alterations in phosphorylation levels of many proteins drives their turnover and leads to significant structural changes [133]. Pin1 catalyzes isomerization of phosphoSer/phosphoThr-Pro amide bonds in multiple cell cycle proteins (Figures 4.15 and 4.16). On the recognition of such a bond Pin1 efficiently interconverts *cis* and *trans* amide isomers, leading to cell cycle progression, transformation, and cellular proliferation. Pin1 is thought to play an important role in oncogenesis, as its overexpression has been identified in numerous cancers, including breast, prostate, oral squamous cell, lung, cervical, and colon cancers. Furthermore, a study of 580 prostate cancer patients revealed a correlation of Pin1 overexpression and highest levels of recurrence. Although the precise role of Pin1 in oncogenesis is not known, Pin1 has been shown to target the tumor suppressor p53 and the antiapoptotic protein Bcl-2, as well as influence signaling through Ras, Neu, and Wnt pathways. The potential of Pin1 as a prospective anticancer target is underscored by the studies showing that the reduction in Pin1 expression levels result in the inhibition of cancer cell proliferation and reversion of Ras- and Neu-induced transformation [133].

In addition to regulation of cell-cycle progression, Pin1 also controls the processing ratio of the amyloid precursor protein (APP), producing excess of the benign α-APP and reduced amounts of A$\beta$42, which is thought to be a precursor of

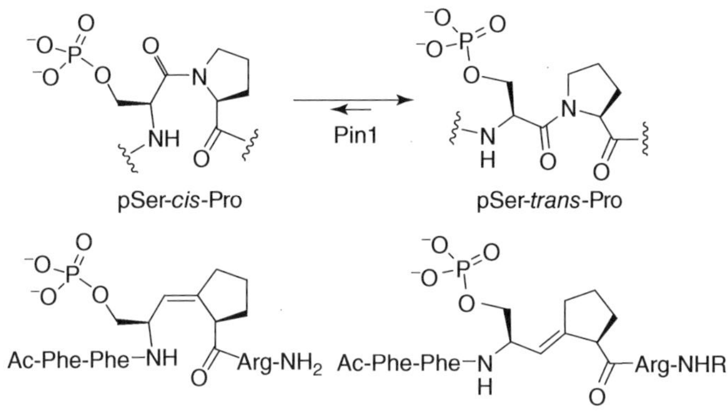


**Figure 4.15** Isomerization of -pSer-Pro- amides by Pin1 (a). Conformationally locked inhibitors, cis isostere **5** and *trans* isosteres **6** and **7** (b). *Source*: Reprinted with permission from Reference 146.

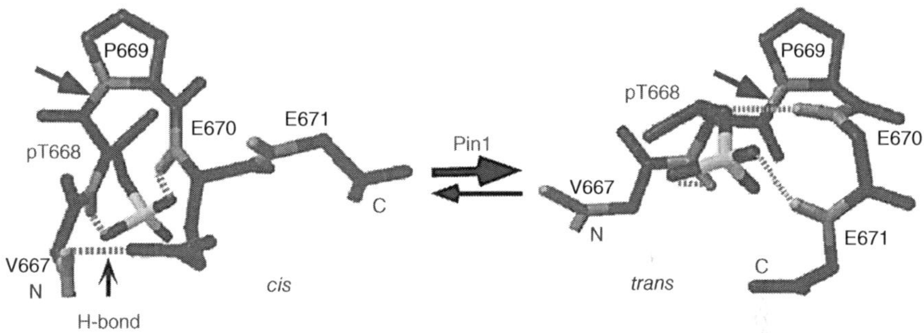


**Figure 4.16** Pin1 catalyzes isomerization of the pThr668-Pro motif in APP as visualized by NMR spectroscopy. Pin1-catalyzed isomerization between cis and trans conformations of the pThr688-Pro peptide bond (red arrows). Structural models display local backbone conformations with associated hydrogen bonds. The size and direction of black arrows represent the catalysis reaction accelerated by Pin1. (*See insert for color representation of this figure.*) *Source*: Adapted with permission from Reference 135.

amyloid plaque formation [134]. Pin1 is thought to decrease the concentration of pThr668-*cis*-Pro APP resulting in decreased levels of amyloidogenic A$\beta$ peptides secreted from cells overexpressing this protein. Conversely, Pin1 knockout mice display increased levels of insoluble toxic peptide A$\beta$42 in age-dependent manner, leading to plaque formation in dorsal medial cortical neurons (Figures 4.16–4.18) [135].

Owing to the potential importance of Pin1 as druggable target in anticancer and Alzheimer's disease therapy, there has been considerable effort placed in the development of the inhibitors to this protein. Etzkorn et al. has demonstrated the design of

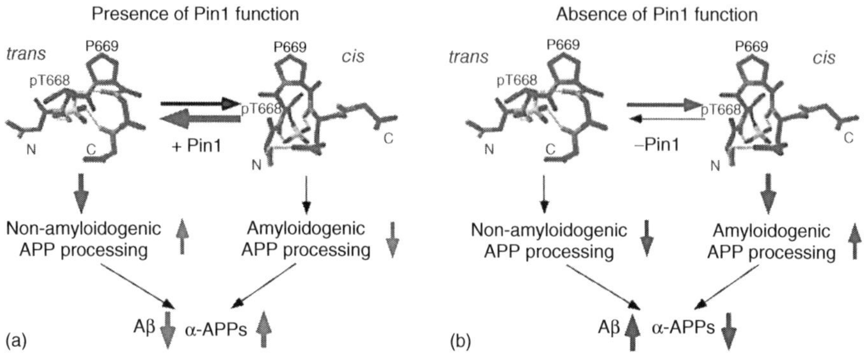


**Figure 4.17** Although the pThr668-Pro motif of APP tends to be in cis after phosphorylation, functional Pin1 would greatly accelerate cis to trans isomerization, which might favor non-amyloidogenic APP processing (a). Without proper Pin1 function, the *cis* pThr688-Pro motif would not be isomerized to trans in a timely manner, which might favor amyloidogenic APP processing (b). *Source*: Adapted with permission from Reference 135. (*See insert for color representation of this figure.*)

a stereo-controlled synthesis of the *cis* and *trans* Ser-Pro dipeptide isosteres, which were subsequently phosphorylated and fitted into a pentapeptide substrate analog. This pentapeptide not only displayed competitive inhibition of Pin1 with $K_i$ values of 1.7 and 40 μM but also demonstrated antiproliferative activity in an ovarian cancer cell line A2780 with $K_i$ values of 8.3 and 140 μM (cis and trans forms, respectively) [132].

The nonphosphorylated cyclic peptide CRYPEVEIC inhibits Pin1 with $K_i$ of 0.5 μM [133]. Cyclic peptides inhibit cis–trans isomerization by placing conformational constraints on the isomerase. A phage-display panning of cyclic peptides provided an unbiased approach to inhibitor discovery and resulted in peptides with a YP(E/D)V motif that interacted specifically with the PPIase domain of Pin1. Upon further panning, the cyclic peptide CRYPEVEIC was found to be not only specific for the active site but also displayed potent inhibition and caused chemical shift changes that are localized to the residues in and immediately surrounding the active site (Figure 4.19). Furthermore, binding analysis revealed a tight interface without the requirement of interaction with the phosphate binding site of the enzyme. In an effort to make the CRPEVEIC peptide cell membrane-permeable, two peptides were synthesized, in which the terminal Cys residues were replaced either with Lys or Glu residues. Unfortunately, these peptides lost their inhibitory properties with regards to Pin1. This effort demonstrated that most likely Pin1 inhibition is based on the sequence specificity and the geometry of the ring. Although cell-based studies with this inhibitor could not be performed, it was demonstrated that the phosphate group was not absolutely required for high affinity binding [133]. These findings could be useful in determining inhibitors for other enzymes known for phosphate-guided binding and possibly open new avenues of peptide-based inhibitor discovery.

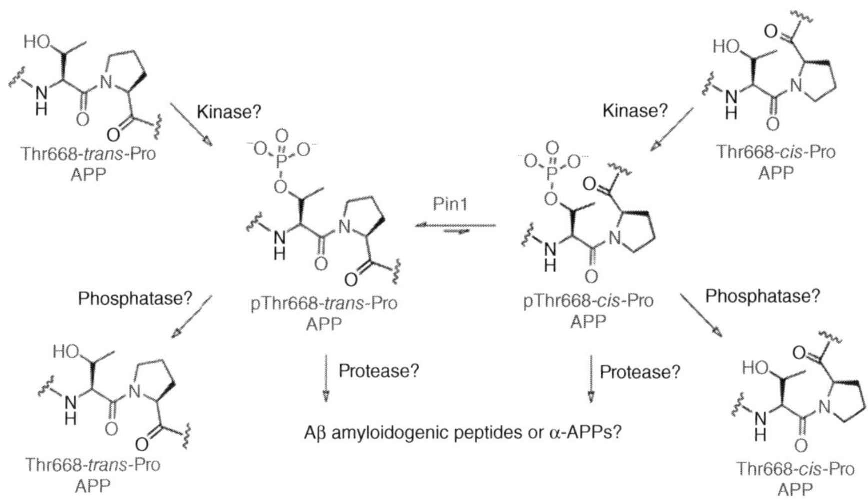


**Figure 4.18** Alternative model to explain the central role of Pin1 in processing APP based on equilibrium upset by a kinase or a phosphatase. APP must rapidly cross the central equilibration catalyzed by Pin1 to avoid processing to amyloidogenic products. This model assumes a conformation-specific kinase or phosphatase to upset the conformational equilibrium. Whether the left (magenta) or the right (blue) side of the scheme leads to amyloidogenic processing remains to be seen [134].

Liu et al. developed membrane-permeable phosphorylated cyclic inhibitors of Pin1 from a peptide library. In general, the cyclization of a peptide renders it resistant to proteolysis and it may improve its target binding affinity due to reduced conformational freedom. The library of cyclic peptides was designed based on the linear Pin1 inhibitor sequences. All of the isolated peptides contained phosphoThr-Pip-Nal (where Pip is L-piperidine-2-carboxylic acid and Nal is L-2-napthylalanine) and were rendered membrane-permeable by incorporating an $Arg_8$ sequence onto a side chain or into the peptide backbone. These cyclic peptides successfully entered cells and slowed down cell proliferation, displaying the first example of macrocyclic Pin1 inhibitors active *in vivo* [136].

## 4.11 HISTONE MODIFYING ENZYMES

Epigenetic studies throughout the human genome revealed that gene sequences are not solely responsible for different phenotypes. Organization of chromatin is also a key player in gene regulation and can be passed on from one generation to another. Structure of chromatin is regulated by factors including DNA methylation and histone posttranslational modifications such as acetylation, methylation, and phosphorylation. Mapping of these features at key genes (involved in stem cell maintenance and differentiation, cancer markers, oncogenes, etc.) in a variety of cells showed a characteristic pattern of epigenetic marks. Alteration of these marks correlates with several

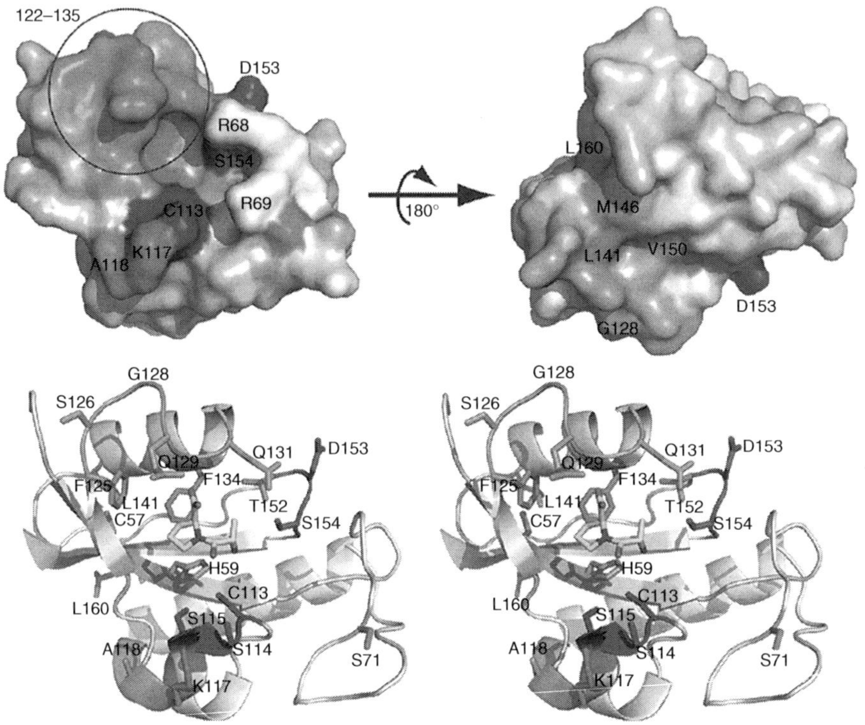


**Figure 4.19** Mapping chemical shift changes to Pin1 structure. The Pin1 PPIase domain is shown with residues colored from white to green according to the magnitude of the chemical shift change observed upon binding cyclic CRYPEVEIC. White indicated no change in chemical shift, while the darkest green represents the largest change observed (a 9 $\sigma$ change for residue L160). Residues colored red are those whose chemical shifts broadened to the point that they could not be assigned in the CRYPEVEIC-bound NMR spectrum. (Top) Surface of Pin1 showing the active site (top left) and the opposite side of the protein (top right). The circled region represents residues 122–135 all of which showed generally large chemical shifts. Residues K63 and R68 implicated in binding the phosphate of substrate peptides, exhibited little or no chemical shift change, consistent with the absence of phosphate in CRYPEVEIC. (Bottom) A stereodiagram of Pin1, shown as a ribbon along with side chains of residues exhibiting a 1.5 $\sigma$ or greater change in their chemical shift upon CRYPEVEIC binding. Also shown is the Ala-Pro dipeptide present in the original structure of Pin1. *Source*: Adapted with permission from Reference 133.

diseases, suggesting that histone-modifying enzymes could be utilized as a new target for anticancer agents.

### 4.11.1 Histone Deacetylase

The imbalance between acetylation and deacetylation at key genes has been correlated to diseases such as cancer. It is assumed that the balance between acetylation and

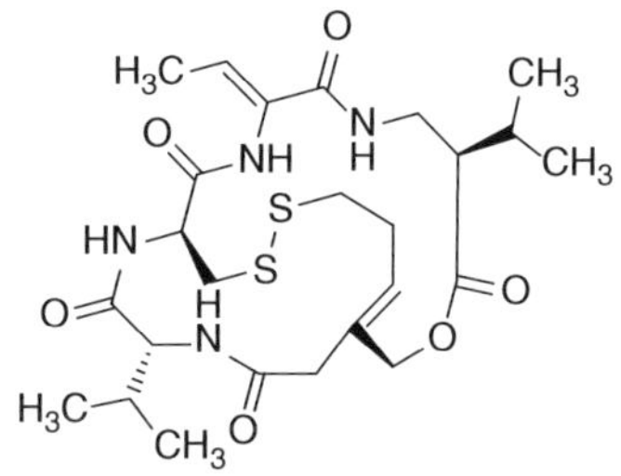


**Figure 4.20** Depsipeptide (FK228/romidepsin/ISTODAX®) is a natural cyclic peptide prodrug type, which inhibits HDAC1 and 2 selectively. *Source*: Figure adapted from Reference 147.

deacetylation regulates the compaction of DNA by neutralizing the negative charges of the phosphates. As histone acetylation loosens up the DNA, it allows the binding of proteins such as transcription factors, which will regulate gene expression. Histone acetylation is regulated by acetyl transferases (HATs) and deacetylase (HDACs).

Efforts have been focused on HDAC inhibitors, which are categorized into four classes. Cyclic peptides represent one of the four classes and include depsipeptides, apicidin, and cyclic hydroxamic acid-containing peptides. Originally isolated from *Chromobaterium violaceum*, depsipeptide FR901228 (Gloucester Pharmaceutical) (Figure 4.20) was approved for commercialization in 2009. After investigation of the toxicity of the drug on cell lines and mice, the compound was subject to a phase I trial. A case study on a patient suffering from T-cell lymphoma reported that complete remission was observed after treatment with depsipeptide FR901228 (also called FK228) [137]. Three patients with cutaneous T-cell lymphoma (CTCL) showed partial remission following treatment. However, as the drug entered phase II clinical trial, adverse cardiac side effects were observed and the trial was suspended. It became clear that a better understanding of mechanism of action of this compound was necessary and its synthesis was published [138, 139].

### 4.11.2 Histone Methyl-Transferase

Like all histone posttranslational modifications, methylation is a reversible and is controlled by histone methyl-transferase (HMT) and histone demethylase (HDM). Methylation occurs on Lys and Arg residues of histone H3 and H4.

One of the most studied HDMs, LSD1 (also called BHC110), is an amine oxidase and regulates the methylation status of histone H3 by removing one or more methyl groups on Lys4. The inhibition of LSD1 may allow maintenance of H3 in a methylated state and therefore reactivate gene expression of silenced genes, whereas the repression of these genes may enhance diseases such as cancer. Owing to its recent discovery and characterization, the development of LSD1 inhibitors remains at early stages. However, Culhane et al. proposed the use of a synthesized propargyl-Lys-derivatized peptide of H3 tail (Figure 4.21) as a promising inhibitor

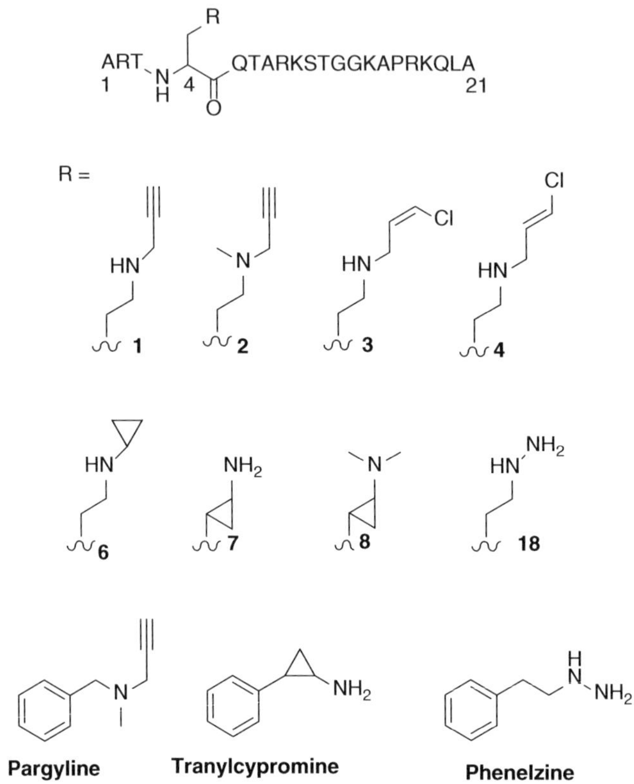


**Figure 4.21** Structures of evaluated LSD1 inhibitors and inactivators. *Source*: Reprinted with permission from Reference 141.

of LSD1. Peptides **8** and **9** (Figure 4.21) were initially synthesized and showed inhibition of LSD1 in the range of micromolar *in vitro* [140]. Since then, a panel of modifications was applied to the synthesized peptide in order to improve its potency as an inhibitor (Figure 4.21) [141]. All inhibitors were tested using LSD1 bound to glutatione S-transferase-tag and were not tested on cell lines as further modifications are needed in order to improve on the cellular internalization of the compound.

## 4.12 PUTTING IT ALL TOGETHER: PEPTIDE INHIBITOR APPLICATIONS IN SKIN CARE

Inhibition of many of the enzymes described in this chapter is being pursued in the clinic. Since multiple enzymes are expressed in any given organ, several inhibitors may be used to achieve a desired effect. Skin conditions are constantly at the forefront of enzyme inhibitor discovery. Figure 4.22 summarizes novel peptide-based therapeutics currently approved for treatment of skin conditions, including several enzyme

| Trade name | Generic name | Target | Indication | Year |
|---|---|---|---|---|
| Forteo | Teriparatide | PTH1R agonist | Osteoarthritis | 2002 |
| Fuzeon | Enfuvirtide | Protein–protein inh. | HIV | 2003 |
| Prialt | Aiconotide | $Ca^{2+}$ channel inh. | Pain | 2004 |
| Byetta | Exenatide | GLP-1 R agonist | T2 diabetes | 2005 |
| Symlin | Pramlintide | Calcitonin agonist | T1/T2 diabetes | 2005 |
| Somatuline | Lanreotide | SST agonist | Acromegaly | 2007 |
| Nplate | Romiplostim | Thrombopoietin agonist | Haematology | 2008 |
| Egrifta | Tesamorelin | GHRF agonist | Lipodystrophy | 2010 |
| Victoza | Liraglutide | GLP-1 R agonist | T2D | 2010 |
| Bydureon | Exenatide LAR | GLP-1 R agonist | T2 diabetes | 2011 |
| Surfaxin | Lucinactant | | IRDS | 2012 |
| Omontys | Peginesatide | Erithropoeitin analog. | Anemia | 2012 |
| Slgnifor | Pasireotide | Somatostatin analog | Cushing's disease | 2012 |
| Kyprolis | Carfilzomib | Proteasome inhibitor | Multiple myeloma | 2012 |
| Linzess | Linaclotide | Guanidyl cyclase 2C agonist | IBS-C and CIC | 2012 |
| Gattex | Teduglutide | Gluc-like peptide analog | SBS | 2012 |

**Figure 1.1** Peptide therapeutics marketed since 2002.

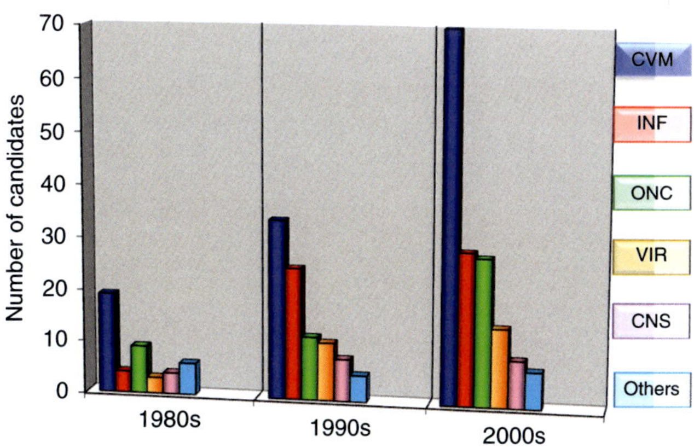


**Figure 1.2** Peptides in development over the last three decades.

*Peptide Chemistry and Drug Design*, First Edition. Edited by Ben M. Dunn.

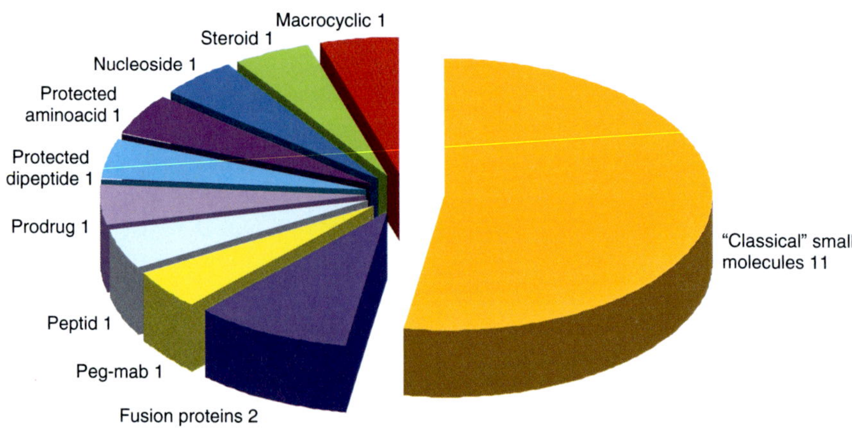


**Figure 2.1** Distribution by chemical structure of the new drugs approved by the FDA in 2008.

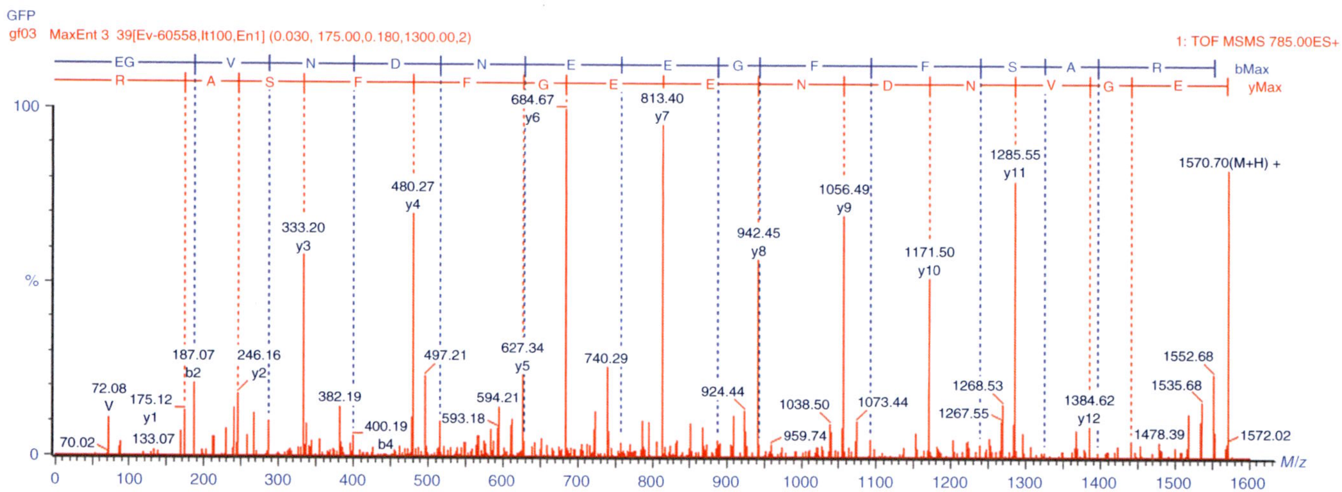


**Figure 2.22** Amino-acid sequence assigned to MSMS spectrum of Glu-fibrinogen peptide (theoretical amino acid sequence: EGVNDNEEGFFSAR; theoretical monoisotopic mass, $M+H^+ = 1570.677$ Da; acquired in a nano-ESI-Q-TOF ultima (waters/micromass).

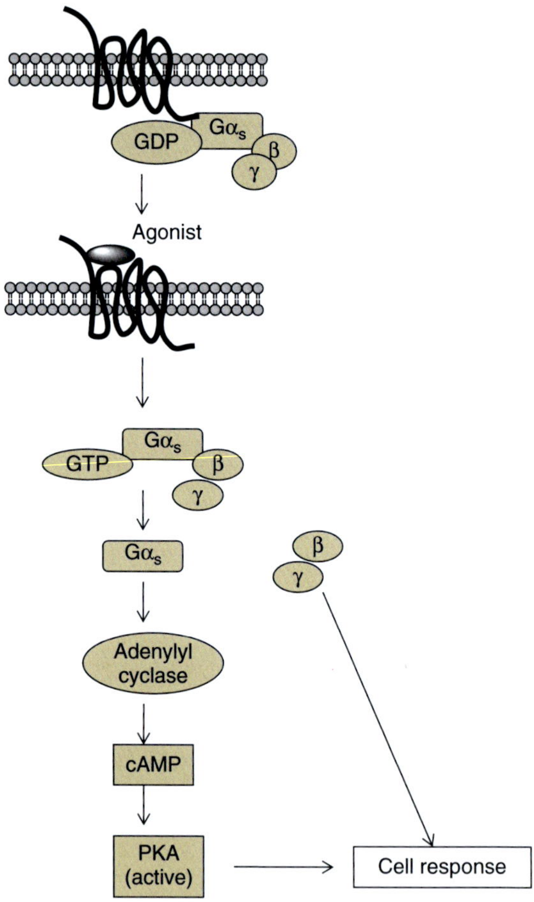


**Figure 3.2** Classical example of signal transduction in seven transmembrane type receptor.

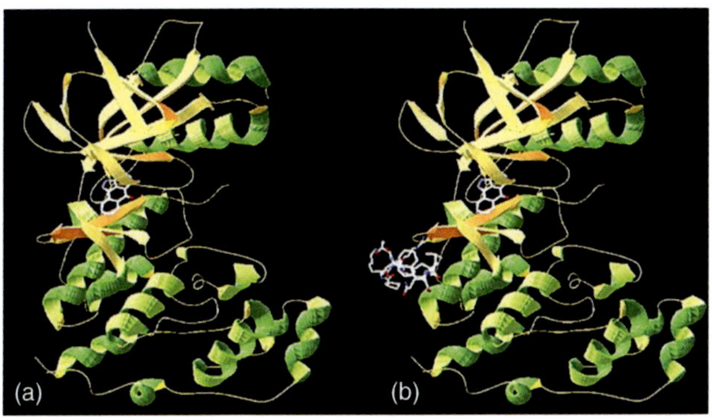


**Figure 4.5** Protein kinase interactions with ATP-competitive and ATP-noncompetitive inhibitors: c-Jun N-terminal kinase. (a) Structural analysis has shown that the ATP-binding site of the protein kinase C-Jun N-terminal Kinase (JNK) is occupied by an ATP-competitive inhibitor of JNKs, SP600125. (b) The structure of the complex between JNK1 and the peptide inhibitor derived from the JNK pathway scaffold protein JIP1 showed the interaction of the JNK1 protein with the peptide inhibitor at a site remote from the ATP-binding pocket. *Source*: Reprinted with permission from Reference 86.

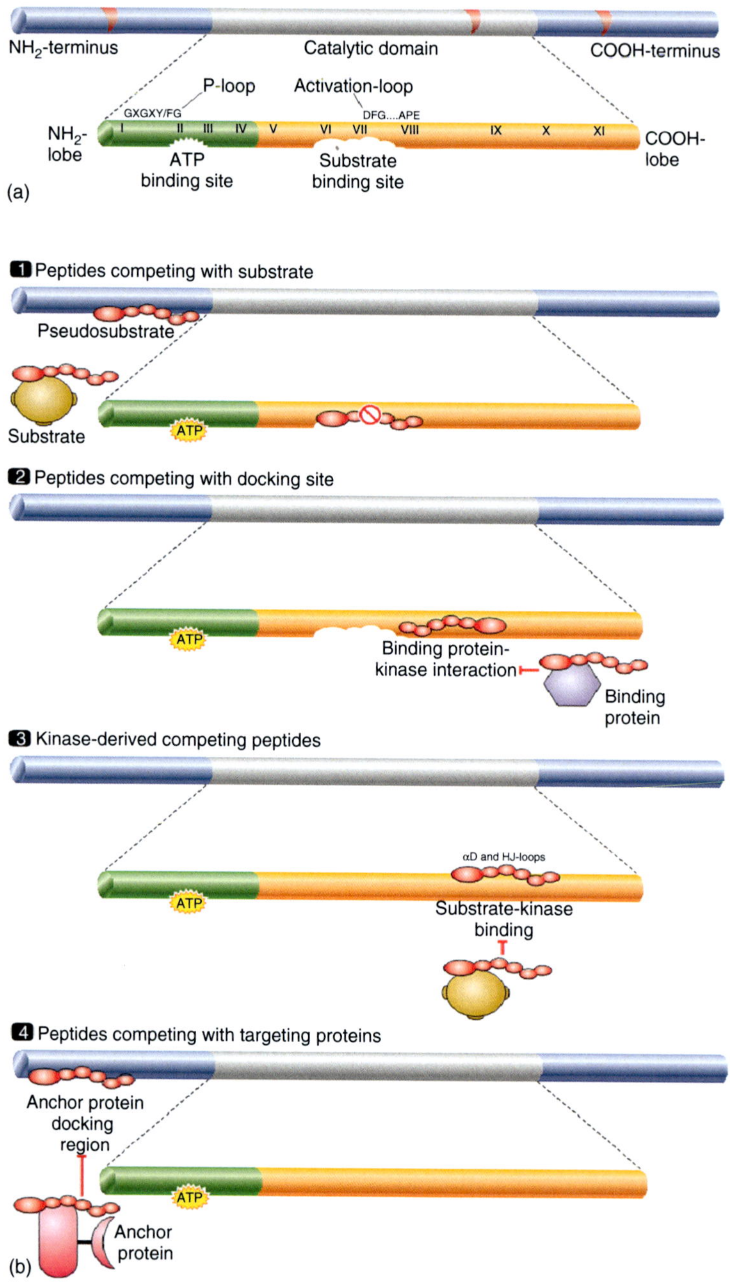


**Figure 4.6** Inhibitory mechanisms of peptides targeting protein kinases. (a) Schematic presentation of domain structure of protein kinase with conserved bilobal catalytic domain composed of 11 conserved subdomains. Conserved domains including the P-loop involved in ATP binding and the activation loop with the DFG...APE motif are marked. Red triangles mark docking site position with proteins such as substrates, upstream regulators, and scaffold proteins. (b) Inhibition of protein kinase by peptide (1) competition with substrate, (2) competition with docking sites derived from substrates, (3) competition with docking sites derived from the kinase, or (4) competition with cellular targeting anchor proteins. Peptides marked in yellow indicate the position of derivation; peptides marked in pink indicate the interaction with the target site; T describes disruption location; red chain represents the peptide inhibitor. *Source*: Reprinted with permission from Reference 85.

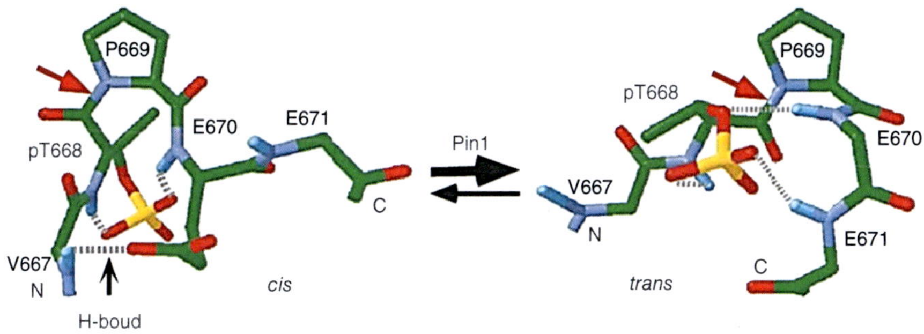


**Figure 4.16** Pin1 catalyzes isomerization of the pThr668-Pro motif in APP as visualized by NMR spectroscopy. Pin1-catalyzed isomerization between cis and trans conformations of the pThr688-Pro peptide bond (red arrows). Structural models display local backbone conformations with associated hydrogen bonds. The size and direction of black arrows represent the catalysis reaction accelerated by Pin1. *Source*: Adapted with permission from Reference 135.

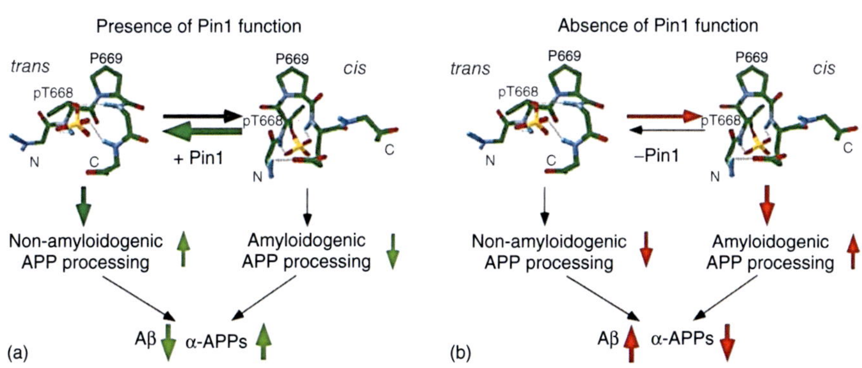


**Figure 4.17** Although the pThr668-Pro motif of APP tends to be in cis after phosphorylation, functional Pin1 would greatly accelerate cis to trans isomerization, which might favor non-amyloidogenic APP processing (a). Without proper Pin1 function, the *cis* pThr688-Pro motif would not be isomerized to trans in a timely manner, which might favor amyloidogenic APP processing (b). *Source*: Adapted with permission from Reference 135.

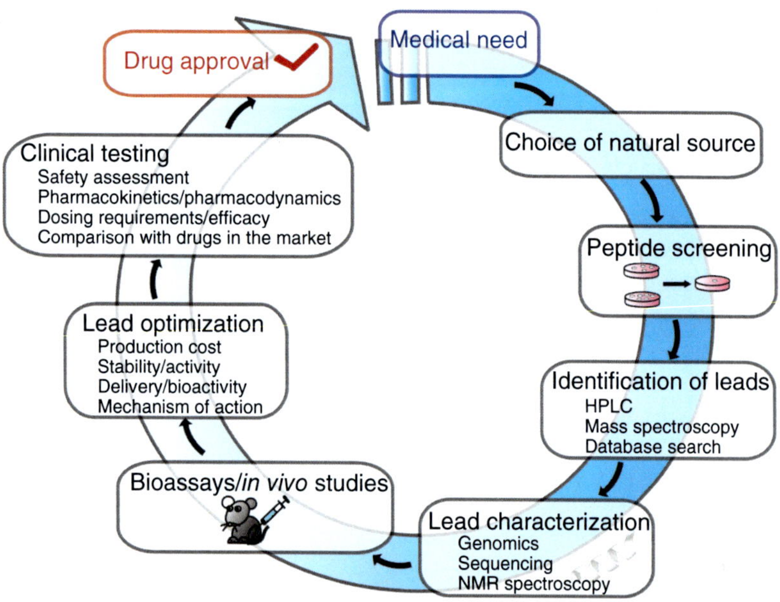


**Figure 6.1** Overview of the drug-development pathway for peptide-based drug design.

| Company | Name | Activity | Premix products | Source |
|---|---|---|---|---|
| Atrium | Tripeptide-2 | ECM stimulation via MMP-1 inhibition | ECM-protect | Undisclosed |
| Atrium | Tripeptide-1 | ECM stimulation via growth factor | Kollaren | HGF |
| Atrium | Acetyl tetrapeptide-2 | Reduce loss of thymic factors | Thymulen 4 | Thymopoieten |
| Atrium | Acetylpeptide-1 | Melanin increase via MSH regulation | Melitene | MSH agonist |
| Atrium | Nonapeptide-1 | Tyrosinase activation inhibition | Melanostatine | MSH antagonist |
| Grant Indust. | Palmitoyl hexapeptide-6 | Dermal repair | Matrix Rebuilder | Innate immunity |
| Grant Indust. | Oligopeptide-10 | Dermal protection | InvisaSkin-64 | Innate immunity |
| Lipotec | Tripeptide-1 | Inhibits collagen glycation | Aldenine, trylagen | Human serum |
| Lipotec | Tripeptide-10 citrulline | Collagen fibrillogenesis | Decorinyl, trylagen | Decorin |
| Lipotec | Acetyl tetrapeptide-5 | Edema reduction by ACE inhibition | Eyeseryl | Undisclosed |
| Lipotec | Pentapeptide-3 | Botox-like via mimicing enkephalins | Leuphasyl | Undisclosed |
| Lipotec | Acetyl hexapeptide-3 (or -8) | Botox-like via SNARE inhibition | Argireline | SNAP-25 |
| Lipotec | Acetyl octatapeptide-1 | Botox-like via SNARE inhibition | SNAP-8 | SNAP-25 |
| Lipotec | Hexapeptide-10 | Increases cell proliferation and laminin V | Serilesine | Laminin |
| Pentapharm | Palmitoyl tripeptide-5 | Collagen synthesis via TGF-β | Syn-coll | Thrombospondin I |
| Pentapharm | Dipeptide diaminobutyroyl benzylamide diacetate | Botox-like via acetycholine receptor | Syn-ake | Waglerin 1 |
| Pentapharm | Oligopeptide-20 | MMP inhibitor via TIMP | Pepha-timp | TIMP-2 |
| Pentapharm | Pentapeptide-3 | Botox-like via acetycholine receptor | Vialox | Undisclosed |
| Procyte | Copper GHK/AHK | Wound healing | Brand example Neova | Human serum |
| Sederma | Dipeptide-2 | Lymph drainage via ACE inhibition | Eyeliss | Rapeseed |
| Sederma | Palmitoyl oligopeptide | Collagen synthesis via signalling | Eyeliss, Matrixyl 3000 | Human serum |
| Sederma | Palmitoyl tetrapeptide-7 (formally -3) | Elasticity via IL6 reduction | Matrixyl 3000, Rigin | IgG/matrikine |
| Sederma | Palmitoyl pentapeptide-3 | Collagen stimulation via signalling | Matrixyl | Procollagen |
| Sederma | Palmitoyl oligopeptide | Retinoic acid-like activity | Biopeptide-CL | Collagen |
| Sederma | Palmitoyl oligopeptide | Increases collagen and HA | Biopeptide-EL | Elastin |

ACE, angiotensin I-converting enzyme; ECM, extracellular matrix; HA, hyaluronic acid; HGF, hepatocyte growth factor; IgG, immunoglobulin G; MMP, matrix metalloproteinases; MSH, melanocyte-stimulating hormone; SNARE, soluble *N*-ethylmaleimide sensitive factor attachment receptor; TGF-β, transforming growth factor-β; TIMP, tissue inhibitor of MMP.
Companies: Atrium Biotechnologies (Quebec City, QC, Canada); Grant industries (Elmwood, NJ, USA); Lipotec (Barcelona, Spain), Pentapharm (Basel, Switzerland); Procyte (Photomedix, Montgomeryville, PA, USA); Sederma (Le Perray en Yvelines, France).

**Figure 4.22** Summary of bioactive peptides currently marketed for inclusion as active ingredients in skin care products. *Source*: Reprinted with permission from Reference 148.

inhibitors or modulators of enzyme inhibitors. These applications well illustrate the importance of the development of alternative methods for enzyme inhibition, as to date the focus has been primarily placed on small molecule-based drug discovery.

## 4.13 STRATEGIES FOR THE DISCOVERY OF NOVEL PEPTIDE INHIBITORS

This chapter has described peptide-based inhibitors for the majority of enzyme classifications. Successful approaches for obtaining these inhibitors are varied and highly creative. The most straightforward approach is the exploration of substrate analogs (as observed for MMPs, HIV-1 protease, kinases, glycosyltransferases, telomerases, PPIases, and HDM), including substrate-based transition-state or suicide inhibitors (for MMPs, glycosyltransferases, and HDM). One can also screen with substrate fragments/products as potential inhibitors [86]; captopril is a product-based inhibitor of ACE. Peptidomimetics based on substrates and products have been developed as HIV-1 protease and MMP inhibitors. Peptide-based inhibitors targeted outside of the active site, such as those that disrupt interactions of enzyme binding partners or docking sites, mimic autoinhibitory domains, or perturb allosteric/secondary binding sites, have proven effective for MMPs, anthrax LF, kinases, and tyrosinase. Novel peptide inhibitors have been identified for tyrosinase, PPIase, and HDAC through screening of peptide libraries. Libraries can be produced by chemical or phage display methodologies, or obtained from natural sources. Library screening may proceed using the entire enzyme, catalytic subunits/domains, or regulatory subunits (docking sites, anchoring sites, scaffold-interacting subunits). Overall, a plethora of options exist for the identification and design of peptide-based enzyme inhibitors, and one is likely to see continued growth in this area of probe and pharmaceutical development.

## ACKNOWLEDGMENTS

Research in the Fields' laboratory on proteases and protease inhibitors has been supported by the National Institutes of Health (CA098799, AR063795, MH078948, and NHLBI Contract 268201000036C-0-0-1), the Multiple Sclerosis National Research Institute, Texas Higher Education STAR Award Program, the Robert A. Welch Foundation, and the Torrey Pines Institute for Molecular Studies.

## REFERENCES

1. Rishton GM. Failure and success in modern drug discovery: guiding principles in the establishment of high probability of success drug discovery organizations. Med Chem 2005;1:519–527.
2. Abbenante G, Fairlie DP. Protease inhibitors in the clinic. Med Chem 2005;1:71–104.
3. Turk B. Targeting proteases: successes, failures and future prospects. Nat Rev Drug Discovery 2006;5:785–799.
4. Cushman DW, Ondetti MA. Design of angiotensin converting enzyme inhibitors. Nat Med 1999;5:1110–1113.
5. Zaman MA, Oparil S, Calhoun DA. Drugs targeting the renin-angiotensin-aldosterone system. Nat Rev Drug Discov 2002;1(8):621–636.
6. Ferreira SH. A bradykinin-potentiating factor (Bpf) present in the venom of bothrops jararca. Br J Pharmacol Chemother 1965;24:163–169.
7. Aiken JW, Vane JR. Inhibition of converting enzyme of the renin-angiotensin system in kidneys and hindlegs of dogs. Circ Res 1972;30(3):263–273.
8. Case DB.. et al. Estimating renin participation in hypertension: superiority of converting enzyme inhibitor over saralasin. Am J Med 1976;61:790–796.
9. Byers LD., Wolfenden R. A potent reversible inhibitor of carboxypeptidase A. J Biol Chem 1972;247:606–608.
10. Acharya KR.. et al. Ace revisited: a new target for structure-based drug design. Nat Rev Drug Discov 2003;2(11):891–902.
11. Natesh R. et al. Crystal structure of the human angiotensin-converting enzyme-lisinopril complex. Nature 2003;421(6922):551–554.
12. Jandeleit-Dahm KA. Dual ACE/NEP inhibitors - more than playing the ACE card. J Hum Hypertens 2006;20(7):478–481.

13. Kerr MA, Kenny AJ. The purification and specificity of a neutral endopeptidase from rabbit kidney brush border. Biochem J 1974;137(3):477–488.
14. Kerr MA, Kenny AJ. The molecular weight and properties of a neutral metallo-endopeptidase from rabbit kidney brush border. Biochem J 1974;137(3): 489–495.
15. Roques BP. et al. The enkephalinase inhibitor thiorphan shows antinociceptive activity in mice. Nature 1980;288(5788):286–288.
16. Roques BP. Zinc metallopeptidases: active site structure and design of selective and mixed inhibitors: new approaches in the search for analgesics and anti-hypertensives. Biochem Soc Trans 1993;21((Pt 3) 3):678–685.
17. Shirotani K. et al. Neprilysin degrades both amyloid beta peptides 1-40 and 1-42 most rapidly and efficiently among thiorphan- and phosphoramidon-sensitive endopeptidases. J Biol Chem 2001;276(24):21895–21901.
18. Iwata N. et al. Metabolic regulation of brain Abeta by neprilysin. Science 2001;292(5521):1550–1552.
19. Burnett JC. Jr. Vasopeptidase inhibition: a new concept in blood pressure management. J Hypertens Suppl 1999;17(1):S37–S43.
20. Roques BP. et al. Novel approaches in the development of new analgesics. Neurophysiol Clin 1990;20(5):369–387.
21. Oefner C. et al. Structural studies of a bifunctional inhibitor of neprilysin and DPP-IV. Acta Crystallogr D Biol Crystallogr 2007;63(Pt 9):975–981.
22. Robl JA. et al. Dual metalloprotease inhibitors: mercaptoacetyl-based fused heterocyclic dipeptide mimetics as inhibitors of angiotensin-converting enzyme and neutral endopeptidase. J Med Chem 1997;40(11):1570–1577.
23. Jeang KT. World AIDS Day 2007: AIDS at 26, are we there yet? Retrovirology 2007;4:86.
24. Wainberg MA, Jeang KT. 25 years of HIV-1 research - progress and perspectives. BMC Med 2008;6:31.
25. Roberts NA. et al. Rational design of peptide-based HIV proteinase inhibitors. Science 1990;248(4953):358–361.
26. Montaner JS. et al. Association of highly active antiretroviral therapy coverage, population viral load, and yearly new HIV diagnoses in British Columbia, Canada: a population-based study. Lancet 2010;376:532–539.
27. De Clercq E. Strategies in the design of antiviral drugs. Nat Rev Drug Discov 2002;1:13–25.
28. De Clercq E. Anti-HIV drugs: 25 compounds approved within 25 years after the discovery of HIV. Int J Antimicrob Agents 2009;33:307–320.
29. Condra JH. et al. In vivo emergence of HIV-1 variants resistant to multiple protease inhibitors. Nature 1995;374:569–571.
30. Schock HB, Garsky VM, Kuo LC. Mutational anatomy of an HIV-1 protease variant conferring cross-resistance to protease inhibitors in clinical trials. Compensatory modulations of binding and activity. J Biol Chem 1996;271:31957–31963.
31. Ghosh AK. et al. Design of HIV protease inhibitors targeting protein backbone: an effective strategy for combating drug resistance. Acc Chem Res 2008;41:78–86.
32. Altman MD. et al. Computational design and experimental study of tighter binding peptides to an inactivated mutant of HIV-1 protease. Proteins 2008;70:678–694.

33. Plaza A. et al. Mirabamides A-D, depsipeptides from the sponge Siliquariaspongia mirabilis that inhibit HIV-1 fusion. J Nat Prod 2007;70:1753–1760.
34. Woessner JF. Matrix metalloproteinases and their inhibitors in connective tissue remodeling. FASEB J 1991;5:2145–2154.
35. Page-McCaw A, Ewald AJ, Werb Z. Matrix metalloproteinases and the regulation of tissue remodelling. Nat Rev Mol Cell Biol 2007;8:221–233.
36. Chang C, Werb Z. The many faces of metalloproteases: cell growth, invasion, angiogenesis and metastasis. Trends Cell Biol 2001;11:S37–S43.
37. Egeblad M, Werb Z. New functions for the matrix metalloproteinases in cancer progression. Nat Rev Cancer 2002;2:161–174.
38. Ishiguro N. et al. Relationships of matrix metalloproteinases and their inhibitors to cartilage proteoglycan and collagen turnover and inflammation as revealed by analyses of synovial fluids from patients with rheumatoid arthritis. Arthritis Rheum 2001;44(11):2503–2511.
39. Skiles JW, Gonnella NC, Jeng AY. The design, structure, and therapeutic application of matrix metalloproteinase inhibitors. Curr Med Chem 2001;8(4):425–474.
40. Verma RP, Hansch C. Matrix metalloproteinases (MMPs): chemical-biological functions and (Q)SARs. Bioorg Med Chem 2007;15(6):2223–2268.
41. Overall CM, Lopez-Otin C. Strategies for MMP inhibition in cancer: innovations for the post-trial era. Nat Rev Cancer 2002;2:657–672.
42. Saghatelian A. et al. Activity-based probes for the proteomic profiling of metalloproteases. Proc Natl Acad Sci U S A 2004;101:10000–10005.
43. Rao BG. Recent developments in the design of specific matrix metalloproteinase inhibitors aided by structural and computational studies. Curr Pharm Design 2005;11:295–322.
44. Lauer-Fields JL. et al. Application of topologically constrained mini-proteins as ligands, substrates, and inhibitors. Methods Mol Biol 2007;386:125–166.
45. Sawa M. et al. New type of metalloproteinase inhibitor: design and synthesis of new phosphonamide-based hydroxamic acids. J Med Chem 2002;45(4):919–929.
46. Dennis MS. et al. Peptide exosite inhibitors of factor VIIa as anticoagulants. Nature 2000;404:465–470.
47. Roberge M. et al. A novel exosite on coagulation factor VIIa and its molecular interactions with a new class of peptide inhibitors. Biochemistry 2001;40:9522–9531.
48. Izaguirre G, Rezale AR, Olson ST. Engineering functional antithrombin exosites in $\alpha$1-proteinase inhibitor that specifically promote the inhibition of factor Xa and factor IXa. J Biol Chem 2008;284:1550–1558.
49. Scheer JM, Romanowski MJ, Wells JA. A common allosteric site and mechanism in caspases. Proc Natl Acad Sci U S A 2006;103:7595–7600.
50. Hauske P. et al. Allosteric regulation of proteases. ChemBioChem 2008;9:2920–2928.
51. Mikkelsen JH. et al. Inhibition of the proteolytic activity of pregnancy-associated plasma protein-A by targeting substrate exosite binding. J Biol Chem 2008;283:16772–16780.
52. Selent J. et al. Selective inhibition of the collagenase activity of cathepsin K. J Biol Chem 2007;282:16492–16501.
53. Engel CK. et al. Structural Basis for the Highly Selective Inhibition of MMP-13. Chem Biol 2005;12:181–189.

54. Johnson AR. et al. Discovery and characterization of a novel inhibitor of matrix metalloprotease-13 that reduces cartilage damage in vivo without joint fibroplasia side effects. J Biol Chem 2007;282(38):27781–27791.

55. Gooljarsingh LT. et al. Characterization of an exosite binding inhibitor of matrix metalloprotease 13. Protein Sci 2008;17:66–71.

56. Baragi VM. et al. A new class of potent matrix metalloproteinase 13 inhibitors for potential treatment of osteoarthritis: evidence of histologic and clinical efficacy without musculoskeletal toxicity in rat models. Arthritis Rheum 2009;60:2008–2018.

57. Lauer-Fields JL, Fields GB, Development of selective, exosite binding matrix metalloproteinase (MMP) inhibitors, in peptides: breaking away. In: Lebl M, editor. Proceedings of the Twenty-First American Peptide Symposium; American Peptide Society, Albuquerque, NM; 2009 p 239–241.

58. Jüngel A. et al. Effect of the oral application of a highly selective MMP-13 inhibitor in three different animal models of rheumatoid arthritis. Ann Rheum Dis 2010;69:898–902.

59. Roth J. et al. *Discovery of a novel, exosite-binding matrix metalloproteinase-13 scaffold.* Bioorg. Med Chem Lett 2011;21:7180–7184.

60. Babine RE, Bender SL. Molecular recognition of protein-ligand complexes: applications to drug design. Chem Rev 1997;97:1359–1472.

61. Gall AL. et al. Crystal structure of the stromelysin-3 (MMP-11) catalytic domain complexed with a phosphinic inhibitor mimicking the transition-state. J Mol Biol 2001;307:577–586.

62. Lauer-Fields JL. et al. Identification of specific hemopexin-like domain residues that facilitate matrix metalloproteinase collagenolytic activity. J Biol Chem 2009;284:24017–24024.

63. Lauer-Fields JL, Fields GB. Triple-helical peptide analysis of collagenolytic protease activity. Biol Chem 2002;383:1095–1105.

64. Fan C-Y. et al. Production of multivalent protein binders using a self-trimerizing collagen-like peptide scaffold. FASEB J 2008;22:3795–3804.

65. Lauer-Fields JL. et al. Screening of potential ADAMTS-4 inhibitors utilizing a collagen-model FRET substrate. Anal Biochem 2008;373:43–51.

66. Lauer-Fields JL. et al. Triple-helical transition-state analogs: a new class of selective matrix metalloproteinase inhibitors. J Am Chem Soc 2007;129:10408–10417.

67. Bhowmick M. et al. Efficient synthesis of Fmoc-protected phosphinic pseuedodipeptides: building blocks for the synthesis of matrix metalloproteinase inhibitors (MMPIs). Biopolymers Pept Sci 2011;96:1–3.

68. Niyibizi C. et al. A 92 kDa gelatinase (MMP-9) cleavage site in native type V collagen. Biochem Biophys Res Commun 1994;202:328–333.

69. Overall CM, Kleifeld O. Towards third generation matrix metalloproteinase inhibitors for cancer therapy. Br J Cancer 2006;94:941–946.

70. Minond D. et al. Matrix metalloproteinase triple-helical peptidase activities are differentially regulated by substrate stability. Biochemistry 2004;43:11474–11481.

71. Minond D. et al. The roles of substrate thermal stability and P2 and P1′ subsite identity on matrix metalloproteinase triple-helical peptidase activity and collagen specificity. J Biol Chem 2006;281:38302–38313.

72. Lauer-Fields JL. et al. Substrate conformation modulates aggrecanase (ADAMTS-4) affinity and sequence specificity: suggestion of a common topological specificity of functionally diverse proteases. J Biol Chem 2007;282:142–150.
73. Holmgren SK. et al. Code for collagen's stability deciphered. Nature 1998;392:666–667.
74. Holmgren SK. et al. A hyperstable collagen mimic. Chem Biol 1999;6:63–70.
75. Malkar NB. et al. Modulation of triple-helical stability and subsequent melanoma cellular responses by single-site substitution of fluoroproline derivatives. Biochemistry 2002;41:6054–6064.
76. Lauer-Fields JL. et al. Selective modulation of matrix metalloproteinase 9 (MMP-9) functions via exosite inhibition. J Biol Chem 2008;283:20087–20095.
77. Xu X. et al. Inhibition of MMP-2 gelatinolysis by targeting exodomain-substrate interactions. Biochem J 2007;406:147–155.
78. Ganz T. Defensins: antimicrobial peptides of vertebrates. C R Biol 2004;327(6): 539–549.
79. Lehrer RI. Primate defensins. Nat Rev Microbiol 2004;2(9):727–738.
80. Tang YQ. et al. A cyclic antimicrobial peptide produced in primate leukocytes by the ligation of two truncated alpha-defensins. Science 1999;286(5439):498–502.
81. Trabi M, Craik DJ. Circular proteins–no end in sight. Trends Biochem Sci 2002;27(3):132–138.
82. Craik DJ. Chemistry. Seamless proteins tie up their loose ends. Science 2006;311(5767): 1563–1564.
83. Kim C. et al. Human alpha-defensins neutralize anthrax lethal toxin and protect against its fatal consequences. Proc Natl Acad Sci U S A 2005;102(13):4830–4835.
84. Wang W. et al. Retrocyclins kill bacilli and germinating spores of Bacillus anthracis and inactivate anthrax lethal toxin. J Biol Chem 2006;281(43):32755–32764.
85. Kaidanovich-Beilin O, Eldar-Finkelman H. Peptides targeting protein kinases: strategies and implications. Physiology (Bethesda) 2006;21:411–418.
86. Bogoyevitch MA, Barr RK, Ketterman AJ. Peptide inhibitors of protein kinases-discovery, characterisation and use. Biochim Biophys Acta 2005;1754(1-2): 79–99.
87. Cheng HC. et al. A potent synthetic peptide inhibitor of the cAMP-dependent protein kinase. J Biol Chem 1986;261(3):989–992.
88. Glass DB. et al. Differential and common recognition of the catalytic sites of the cGMP-dependent and cAMP-dependent protein kinases by inhibitory peptides derived from the heat-stable inhibitor protein. J Biol Chem 1986;261(26):12166–12171.
89. Knighton DR. et al. Structure of a peptide inhibitor bound to the catalytic subunit of cyclic adenosine monophosphate-dependent protein kinase. Science 1991;253(5018):414–420.
90. Knighton DR. et al. Crystal structure of the catalytic subunit of cyclic adenosine monophosphate-dependent protein kinase. Science 1991;253(5018):407–414.
91. Plotkin B. et al. Insulin mimetic action of synthetic phosphorylated peptide inhibitors of glycogen synthase kinase-3. J Pharmacol Exp Ther 2003;305(3):974–980.
92. Kemp BE, Pearson RB, House CM. Pseudosubstrate-based peptide inhibitors. Methods Enzymol 1991;201:287–304.
93. Kemp BE, Pearson RB. Design and use of peptide substrates for protein kinases. Methods Enzymol 1991;200:121–134.

94. Ho DT. et al. A docking site in MKK4 mediates high affinity binding to JNK MAPKs and competes with similar docking sites in JNK substrates. J Biol Chem 2003;278(35):32662–32672.

95. Bardwell AJ. et al. A conserved docking site in MEKs mediates high-affinity binding to MAP kinases and cooperates with a scaffold protein to enhance signal transmission. J Biol Chem 2001;276(13):10374–10386.

96. Roux PP, Richards SA, Blenis J. Phosphorylation of p90 ribosomal S6 kinase (RSK) regulates extracellular signal-regulated kinase docking and RSK activity. Mol Cell Biol 2003;23(14):4796–4804.

97. Guan QH. et al. Neuroprotection against ischemic brain injury by a small peptide inhibitor of c-Jun N-terminal kinase (JNK) via nuclear and non-nuclear pathways. Neuroscience 2006;139(2):609–627.

98. Bonny C. et al. Cell-permeable peptide inhibitors of JNK: novel blockers of beta-cell death. Diabetes 2001;50(1):77–82.

99. Kaneto H. et al. Possible novel therapy for diabetes with cell-permeable JNK-inhibitory peptide. Nat Med 2004;10(10):1128–1132.

100. Niv MY. et al. Sequence-based design of kinase inhibitors applicable for therapeutics and target identification. J Biol Chem 2004;279(2):1242–1255.

101. Licht T. et al. Induction of pro-angiogenic signaling by a synthetic peptide derived from the second intracellular loop of S1P3 (EDG3). Blood 2003;102(6):2099–2107.

102. Rosenmund C. et al. Anchoring of protein kinase A is required for modulation of AMPA/kainate receptors on hippocampal neurons. Nature 1994;368(6474):853–856.

103. Colledge M, Scott JD. AKAPs: from structure to function. Trends Cell Biol 1999;9(6):216–221.

104. Vijayaraghavan S. et al. Protein kinase A-anchoring inhibitor peptides arrest mammalian sperm motility. J Biol Chem 1997;272(8):4747–4752.

105. Jamaluddin MF. et al. Polypeptide binding specificities of Saccharomyces cerevisiae oligosaccharyltransferase accessory proteins Ost3p and Ost6p. Protein Sci 2011;20(5):849–855.

106. Peluso S. et al. Neoglycopeptides as inhibitors of oligosaccharyl transferase: insight into negotiating product inhibition. Chem Biol 2002;9(12):1323–1328.

107. Culyba EK. et al. Protein native-state stabilization by placing aromatic side chains in N-glycosylated reverse turns. Science 2011;331(6017):571–575.

108. Bause E. et al. Epoxyethylglycyl peptides as inhibitors of oligosaccharyltransferase: double-labelling of the active site. Biochem J 1997;322 (Pt 1):95–102.

109. Bause E. Structural requirements of N-glycosylation of proteins. Studies with proline peptides as conformational probes. Biochem J 1983;209(2):331–336.

110. Bause E. Active-site-directed inhibition of asparagine N-glycosyltransferases with epoxy-peptide derivatives. Biochem J 1983;209(2):323–330.

111. Blackburn EH. Telomere states and cell fates. Nature 2000;408(6808):53–56.

112. Chen H, Li Y, Tollefsbol TO. Strategies targeting telomerase inhibition. Mol Biotechnol 2009;41(2):194–199.

113. Phatak P, Burger AM. Telomerase and its potential for therapeutic intervention. Br J Pharmacol 2007;152(7):1003–1011.

114. Ahmed A, Tollefsbol T. Telomeres and telomerase: basic science implications for aging. J Am Geriatr Soc 2001;49(8):1105–1109.
115. Holt SE, Shay JW, Wright WE. Refining the telomere-telomerase hypothesis of aging and cancer. Nat Biotechnol 1996;14(7):836–839.
116. Shay JW, Bacchetti S. A survey of telomerase activity in human cancer. Eur J Cancer 1997;33(5):787–791.
117. Kim NW. et al. Specific association of human telomerase activity with immortal cells and cancer. Science 1994;266(5193):2011–2015.
118. Morin GB. Is telomerase a universal cancer target? J Natl Cancer Inst 1995;87(12):859–861.
119. Fordyce CA. et al. Telomere content correlates with stage and prognosis in breast cancer. Breast Cancer Res Treat 2006;99(2):193–202.
120. Mitchell M. et al. Structural basis for telomerase catalytic subunit TERT binding to RNA template and telomeric DNA. Nat Struct Mol Biol 2010;17(4):513–518.
121. Villa R. et al. Inhibition of telomerase activity by a cell-penetrating peptide nucleic acid construct in human melanoma cells. FEBS Lett 2000;473(2):241–248.
122. Hamilton SE. et al. Cellular delivery of peptide nucleic acids and inhibition of human telomerase. Chem Biol 1999;6(6):343–351.
123. Daquinag AC. et al. Primary structure of a potent endogenous dopa-containing inhibitor of phenol oxidase from Musca domestica. Proc Natl Acad Sci U S A 1995;92(7):2964–2968.
124. Carsberg CJ, Warenius HM, Friedmann PS. Ultraviolet radiation-induced melanogenesis in human melanocytes. Effects of modulating protein kinase C. J Cell Sci 1994;107(Pt 9):2591–2597.
125. Schurink M. et al. Improvement of lipoxygenase inhibition by octapeptides. Peptides 2007;28(12):2268–2275.
126. Schurink M. et al. Novel peptides with tyrosinase inhibitory activity. Peptides 2007;28(3):485–495.
127. Suzuki I. et al. Binding of melanotropic hormones to the melanocortin receptor MC1R on human melanocytes stimulates proliferation and melanogenesis. Endocrinology 1996;137(5):1627–1633.
128. Burchill SA, Virden R, Thody AJ. Regulation of tyrosinase synthesis and its processing in the hair follicular melanocytes of the mouse during eumelanogenesis and phaeomelanogenesis. J Invest Dermatol 1989;93(2):236–240.
129. Park HY. et al. Topical application of a protein kinase C inhibitor reduces skin and hair pigmentation. J Invest Dermatol 2004;122(1):159–166.
130. Noh JM. et al. Kojic acid-amino acid conjugates as tyrosinase inhibitors. Bioorg Med Chem Lett 2009;19(19):5586–5589.
131. Kim H. et al. Solid-phase synthesis of kojic acid-tripeptides and their tyrosinase inhibitory activity, storage stability, and toxicity. Bioorg Med Chem Lett 2004;14(11):2843–2846.
132. Etzkorn FA, Noel JP, Zhang Y, Wang XJ. Pin1: Inhibitors and mechanism. In: Blondelle SE, editor. *Understanding Biology Using Peptides*. San Diego: Springer, Inc.; 2006. p 759–762.
133. Duncan KE. et al. Discovery and characterization of a nonphosphorylated cyclic peptide inhibitor of the peptidylprolyl isomerase, Pin1. J Med Chem 2011;54(11):3854–3865.

134. Etzkorn FA. Pin1 flips Alzheimer's switch. ACS Chem Biol 2006;1(4):214–216.
135. Pastorino L. et al. The prolyl isomerase Pin1 regulates amyloid precursor protein processing and amyloid-beta production. Nature 2006;440(7083):528–534.
136. Liu T. et al. Membrane permeable cyclic peptidyl inhibitors against human Peptidylprolyl Isomerase Pin1. J Med Chem 2010;53(6):2494–2501.
137. Piekarz RL. et al. Inhibitor of histone deacetylation, depsipeptide (FR901228), in the treatment of peripheral and cutaneous T-cell lymphoma: a case report. Blood 2001;98(9):2865–2868.
138. Yurek-George A. et al. The first biologically active synthetic analogues of FK228, the depsipeptide histone deacetylase inhibitor. J Med Chem 2007;50(23):5720–5726.
139. Greshock TJ. et al. Improved total synthesis of the potent HDAC inhibitor FK228 (FR-901228). Org Lett 2008;10(4):613–616.
140. Culhane JC. et al. A mechanism-based inactivator for histone demethylase LSD1. J Am Chem Soc 2006;128(14):4536–4537.
141. Culhane JC. et al. Comparative analysis of small molecules and histone substrate analogues as LSD1 lysine demethylase inhibitors. J Am Chem Soc 2010;132(9):3164–3176.
142. Ganz T. Microbiology: gut defence. Nature 2003;422(6931):478–479.
143. Shay JW, Wright WE. Telomerase therapeutics for cancer: challenges and new directions. Nat Rev Drug Discov 2006;5(7):577–584.
144. Weiler J. et al. Hybridisation based DNA screening on peptide nucleic acid (PNA) oligomer arrays. Nucleic Acids Res 1997;25(14):2792–2799.
145. Kobayashi T. et al. Tyrosinase related protein 1 (TRP1) functions as a DHICA oxidase in melanin biosynthesis. EMBO J 1994;13(24):5818–5825.
146. Wang XJ. et al. Conformationally locked isostere of phosphoSer-cis-Pro inhibits Pin1 23-fold better than phosphoSer-trans-Pro isostere. J Am Chem Soc 2004;126(47):15533–15542.
147. Kim HJ, Bae SC. Histone deacetylase inhibitors: molecular mechanisms of action and clinical trials as anti-cancer drugs. Am J Transl Res 2011;3(2):166–179.
148. Zhang L, Falla TJ. Cosmeceuticals and peptides. Clin Dermatol 2009;27(5):485–494.
149. Overton HA. et al. Effect of two novel inhibitors of the human immunodeficiency virus protease on the maturation of the HIV gag and gag-pol polyproteins. Virology 1990;179:508–511.
150. Danner SA. et al. A short-term study of the safety, pharmacokinetics, and efficacy of ritonavir, an inhibitor of HIV-1 protease. European-Australian Collaborative Ritonavir Study Group. N Engl J Med 1995;333:1528–1533.
151. Kaldor SW. et al. Viracept (nelfinavir mesylate, AG1343): a potent, orally bioavailable inhibitor of HIV-1 protease. J Med Chem 1997;40:3979–3985.
152. Falcoz C. et al. Pharmacokinetics of GW433908, a prodrug of amprenavir, in healthy male volunteers. J Clin Pharmacol 2002;42:887–898.
153. Baba M. et al. Potent and selective inhibition of human immunodeficiency virus type 1 transcription by piperazinyloxoquinoline derivatives. Antimicrob Agents Chemother 1997;41:1250–1255.
154. Bold G. et al. New aza-dipeptide analogues as potent and orally absorbed HIV-1 protease inhibitors: candidates for clinical development. J Med Chem 1998;41:3387–3401.

155. Koh Y. et al. Novel bis-tetrahydrofuranylurethane-containing nonpeptidic protease inhibitor (PI) UIC-94017 (TMC114) with potent activity against multi-PI-resistant human immunodeficiency virus in vitro. Antimicrob Agents Chemother 2003;47:3123–3129.
156. Wild C, Greenwell T, Matthews T. A synthetic peptide from HIV-1 gp41 is a potent inhibitor of virus-mediated cell-cell fusion. AIDS Res Hum Retrov 1993;9:1051–1053.
157. Chen G. et al. Glycogen synthase kinase 3beta (GSK3beta) mediates 6-hydroxydopamine-induced neuronal death. FASEB J 2004;18(10):1162–1164.
158. Kaidanovich-Beilin O, Eldar-Finkelman H. Long-term treatment with novel glycogen synthase kinase-3 inhibitor improves glucose homeostasis in ob/ob mice: molecular characterization in liver and muscle. J Pharmacol Exp Ther 2006;316(1):17–24.
159. Theodore L. et al. Intraneuronal delivery of protein kinase C pseudosubstrate leads to growth cone collapse. J Neurosci 1995;15(11):7158–7167.
160. Verhoeven AJ. et al. Inhibition of neutrophil NADPH oxidase assembly by a myristoylated pseudosubstrate of protein kinase C. J Biol Chem 1993;268(25):18593–18598.
161. Laudanna C. et al. Evidence of zeta protein kinase C involvement in polymorphonuclear neutrophil integrin-dependent adhesion and chemotaxis. J Biol Chem 1998;273(46):30306–30315.
162. Brickey DA. et al. Mutational analysis of the autoinhibitory domain of calmodulin kinase II. J Biol Chem 1994;269(46):29047–29054.
163. Soderling TR, Chang B, Brickey D. Cellular signaling through multifunctional Ca2+/calmodulin-dependent protein kinase II. J Biol Chem 2001;276(6):3719–3722.
164. Fantz DA. et al. Docking sites on substrate proteins direct extracellular signal-regulated kinase to phosphorylate specific residues. J Biol Chem 2001;276(29):27256–27265.
165. Zheleva DI. et al. Highly potent p21(WAF1)-derived peptide inhibitors of CDK-mediated pRb phosphorylation: delineation and structural insight into their interactions with cyclin A. J Pept Res 2002;60(5):257–270.
166. Thomas GM. et al. A GSK3-binding peptide from FRAT1 selectively inhibits the GSK3-catalysed phosphorylation of axin and beta-catenin. FEBS Lett 1999;458(2):247–251.
167. Franca-Koh J. et al. The regulation of glycogen synthase kinase-3 nuclear export by Frat/GBP. J Biol Chem 2002;277(46):43844–43848.
168. Zhang F. et al. Inhibitory phosphorylation of glycogen synthase kinase-3 (GSK-3) in response to lithium. Evidence for autoregulation of GSK-3. J Biol Chem 2003;278(35):33067–33077.
169. Johnson JA. et al. A protein kinase C translocation inhibitor as an isozyme-selective antagonist of cardiac function. J Biol Chem 1996;271(40):24962–24966.
170. Liu GS. et al. Protein kinase C-epsilon is responsible for the protection of preconditioning in rabbit cardiomyocytes. J Mol Cell Cardiol 1999;31(10):1937–1948.
171. Ron D, Luo J, Mochly-Rosen D. C2 region-derived peptides inhibit translocation and function of beta protein kinase C in vivo. J Biol Chem 1995;270(41):24180–24187.
172. Coghlan VM. et al. A-kinase anchoring proteins: a key to selective activation of cAMP-responsive events? Mol Cell Biochem 1993;127-128:309–319.
173. Newhall KJ. et al. Dynamic anchoring of PKA is essential during oocyte maturation. Curr Biol 2006;16(3):321–327.

# 5

# DISCOVERY OF PEPTIDE DRUGS AS ENZYME INHIBITORS AND ACTIVATORS

Jeffrey-Tri Nguyen

*Department of Medicinal Chemistry, Kyoto Pharmaceutical University, Kyoto, Japan*

Yoshiaki Kiso

*Department of Molecular Pharmacy, Kobe Gakuin University, Kobe, Japan*

## 5.1 INTRODUCTION

There are many reasons why peptides are not ideal drug candidates when compared to small drug molecules. Most people's general impression of peptides is that they are large molecules that are easily digested by proteases. These two important drawbacks mean that peptides are often considered as being less stable with low oral bioavailability. Consequently, administration by injection is often required, and the peptides are readily inactivated by peptidases and cleared from the body. Moreover, they have a relatively higher risk of immunogenic effects when compared with small drug molecules. Owing to their relatively larger size and often hydrophilic nature, solubilization, and the delivery of peptide drugs across membranes become challenges. Cost-wise, the synthesis of peptide drugs is often considered more expensive if one associates cost with molecular size.

Viewed from the opposite side, where a detriment is reinterpreted as a benefit, peptide drugs hold several key advantages over small drug molecules. Peptide drugs are

*Peptide Chemistry and Drug Design*, First Edition. Edited by Ben M. Dunn.

large molecules and thus would only be recognized by a targeted receptor or active site. Therefore, peptide drugs are expected to be very potent because when compared to small drug molecules, peptide drugs have high activity and high specificity due to very few unspecific bindings to nontargeted sites. On the same line of reasoning, peptide drugs would have fewer drug–drug interactions, although, as previously stated, they have an increased risk of immunogenic effects. Considering that peptide drugs are large molecules composed mainly of natural amino acids with high target specificity and are easily degraded by peptidases, peptide drugs would in theory exhibit lower toxicity than small drug molecules. Likewise, considering that peptide drugs have difficulty crossing membranes, they are less likely to accumulate in tissues and thus have a lower risk of adverse drug reactions over time. The sheer largeness of peptide drugs also means that they are more biologically and chemically diverse.

In actual practice, however, peptide drugs are often used to derive small nonpeptide drug molecules. Doing so offers the benefits from both classes and the fine line that differentiates between a peptide drug and small drug molecule becomes faded. Indeed, after a lengthy process of rational drug design where residues are changed from natural amino acids to nonnatural amino acids then to nonamino acids, it becomes rather challenging at times to classify if a drug is peptide or nonpeptide. Although we would like to classify a nonpeptide drug as a compound that does not possess any amino acid, out of respect for the developers of the drugs, in this chapter, we will keep the nonpeptide or peptide assignments that the drug developers have chosen, and will thus avoid any debate over semantics. We will focus on successful stories of peptide-derived drugs that are processed by enzymes. To restrict our scope, we will mainly narrow our discussion to drugs that have received marketing approval by the US Food and Drug Administration (FDA), because it is one of the larger agencies that are responsible for the safety regulation of drugs. We will try to be as up-to-date as possible in the information that we provide at the time that this chapter is being written.

It should be noted that, in this chapter, most comparisons done between different drugs are restricted to our own personal viewpoint; because of legal reasons and personal pride, the drug developers would claim originality to their own discoveries. Hence, we would like the readers to read with an open mind and come up with their own interpretations of the information that we provide.

### 5.1.1 Peptide Residue Nomenclature

Peptides are composed of amino acid residues. During the process of changing a peptide drug to a peptide-like drug and eventually to a nonpeptide drug, the naming of each residue becomes confusing because two or more residues may be merged into one functional structure. We will be using the Schechter and Berger [1] nomenclature that assumes that the substrate binds to the active site of an enzyme in an extended backbone conformation. Peptide substrates are proteins or peptides that are cleaved by the enzyme. Within the active site, subsites, also referred to as *pockets*, are denoted as $S_n$ and $S'_n$, where $n$ represents the number of subsite away from the catalytic $S_1$ subsite, with the prime symbol denoting the opposite direction. Accordingly, $P_n$ and $P'_n$ residues in the substrates are accommodated by their respective $S_n$

and $S'_n$ subsites. Often, N-terminal residues are referred as $P_n$, whereas C-terminal residues are referred as $P'_n$. The naming of peptide drugs follows the same rules as that of peptide substrates. For example, $P_2$–$P_1$–$P'_1$–$P'_2$ is a tetrapeptide drug with a scissile bond between the $P_1$ and $P'_1$ residues. For peptide inhibitors, the inhibitory unit, which is the unit that prevents enzyme cleavage, is assigned to the $P_1$ residue. One should keep in mind that because the numbering is based on the subsites of the active site rather than the sequential order of the residues of the peptide drug, and that the chemical structures of the enzyme and peptide drug are three-dimensional by nature, that in some cases, the numbering of the residues of the peptide drug may not follow a sequential order. In simpler words, there are cases where the peptide drug does not bind to the active site in an extended backbone conformation. An example of an irregular order numbering is argatroban, a direct thrombin inhibitor, which has a $P_3$–$P_1$–$P_2$ sequence (Section 5.4.3). For homodimeric enzymes such as the human immunodeficiency virus (HIV) type 1 protease (Section 5.8), some authors may refer to a residue of a nonpeptide drug as $P_1$, whereas others may refer to the same residue as $P'_1$, which is completely acceptable due to the symmetrical nature of the enzyme.

### 5.1.2 Common Methods of Drug Design

As with most things in life, it is easier to either maintain the *status quo* or destroy, than to create. Hence, it is often easier to commercialize natural enzymes or activators of enzymes found in nature, and to develop inhibitors of enzymes, than to create more potent enzyme activators. A philosophical reasoning for this observation could be that nature has selected the *best* enzymes and their activators, whereas man can only copy or destroy nature's refinements. Despite the previous statement, researchers have designed a few enzyme activators, such as α-methyldopa and droxidopa (Section 5.3.3). Here, we are loosely equating the term *enzyme activator* to *substrate*, because as far as we are aware, there is no allosteric activator in the pharmaceutical market.

Peptide drugs have traditionally been discovered by natural product screening. Most activators of enzymes, or the enzymes themselves, are developed via either extraction of pharmacologically active natural substances from a crude inexpensive natural source or by replicating the natural substances by synthetic means. On the contrary, most potent inhibitors of enzymes are derived from natural lead compounds, or from natural substrates that have been corrupted to become enzyme inhibitors.

From our own experience, the first step in substrate-based drug design of modulators is to establish an assaying system for enzyme activity. A modulator is either an activator or inhibitor, which in our case, applies to a substrate or its peptide inhibitor. As the initial step, a reproducible enzyme activity assay system must be developed from a substrate and enzyme that both must be stable and pure. It is noteworthy that the enzyme often can process several different substrates and the choice of substrate, especially in substrate-based design of enzyme inhibitors, will determine the structural outcome of the derived modulators. For example, in designing inhibitors against the HIV (Section 5.8) and human T-cell lymphotrophic/leukemia virus type 1 (HTLV-I) (Section 5.9.6), we kept in mind that the general Xaa-Pro cleavage sequence is more specific to retroviral than mammalian proteases [2], and specifically chose substrates that contained Xaa-Pro at the cleavage site. Consequently, most of our inhibitors have Pro or one of its isostere at the $P_1'$ position. As discussed in Section 5.5.3, dipeptidyl peptidase-4 (DPP-IV) is an exception in that it is a mammalian protease that recognizes the Xaa-Pro sequence. In an enzyme inhibition assay, the substrate is used as a comparative control. In order to improve the processing efficiency of the substrate by the enzyme, the substrate and enzyme may be structurally altered by synthetic means to improve purity and stability, so as to reduce variations between experiment results. Often, the final substrate used in the assay is a shortened yet active version of a natural substrate, and the enzyme is modified from its natural form to prevent self-digestion. For example, in our research on HTLV-I protease inhibitors (Section 5.9.6), we used a more proteolytic L40I mutant HTLV-I protease, where Leu40 was modified to an Ile, to cleave an 11-residue substrate [3]. Any drastic change from the natural substrate or enzyme could be viewed by the scientific community as a huge leap from the substrate and the natural form of the enzyme, and thereby negatively reflecting on the research as a false image of nature.

A common method of substrate-based design of inhibitors entails the introduction of an inhibitory unit near the scissile bond, between the $P_1$ and $P_1'$ residues of the substrate. The inhibitory unit is a modified version of the $P_1$ residue of the substrate such that the enzyme can recognize and bind to the inhibitory unit at the catalytic site, but the enzyme cannot readily cleave the inhibitor. A common mechanistic feature of protease inhibitors is the presence of a transition state isostere, as a part of the inhibitory unit, to simulate the transition state of amide bond hydrolysis, as depicted in Figure 5.1. Using three-dimensional structure data from nuclear magnetic resonance (NMR) and X-ray diffraction crystallography, our research group

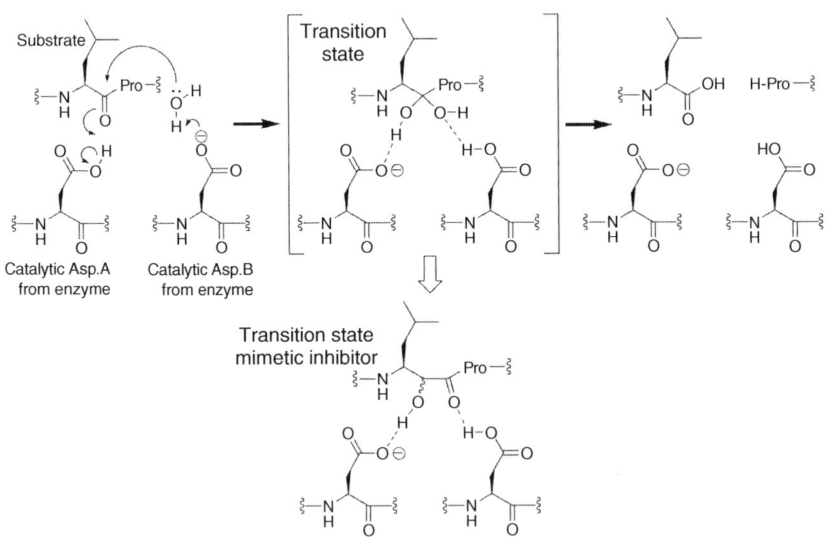


**Figure 5.1** A hydroxymethylcarbonyl isostere, found in an aspartic protease inhibitor, acts as a mimic of the intermediary tetrahedral transition state formed during peptide hydrolysis. Dashed lines represent hydrogen bond interactions.

was first to prove the hydrogen bond network formed by the inhibitory unit [4]. Our recent studies combined neutron diffraction crystallography to conclusively provide direct experimental evidence of the catalytic mechanism of the protease and its inhibition by the inhibitory unit [5]. The actual positions of deuterated protons can be observed. In the initial design of protease inhibitors, other than the central inhibitory unit, the remaining residues of the inhibitor are kept similar to that of the substrate. In simpler words, the inhibitor is a mimic of the substrate and cannot be processed by the enzyme. Hence, the inhibitor competes against natural substrates for the enzyme. If the enzyme can cleave the inhibitor, albeit at a slower rate, or if the inhibitor can be washed out over time, the inhibition is considered reversible. If the inhibitor forms strong interactions with the enzyme to the extent that the inhibitor cannot be removed until the enzyme is degraded, then the inhibition is irreversible. If an unforeseen adverse drug effect is observed with an inhibitor, the adverse effect is expected to be more prolonged in an irreversible inhibitor than a reversible inhibitor. Hence, due to safety concerns associated with mammalian enzymes, the design of reversible inhibitors is often preferred over that of irreversible inhibitors. However, when it comes to nonmammalian enzymes, such as those of viruses and parasites, irreversible inhibitors may be favored over reversible inhibitors, in order to eliminate completely and quickly the viral or parasitic threat, once it has been ascertained that there is absolutely no chance of recognition by other mammalian host enzymes.

Following the introduction of the inhibitory unit in the design, several attempts are performed to minimize the peptide nature of the molecule to avoid most peptide-associated problems that we have discussed in the introduction (Section 5.1). Of course, for the case of substrate-based design of activators, an inhibitory unit is obviously not introduced. During the ensuing rational drug optimization process, quantitative structure–activity relationship studies are performed to statistically confirm and suggest any potency trend observed in modulatory activity. The peptide drug is truncated to reduce size-related pharmacodynamic and pharmacokinetic problems. In consideration that the enzyme can most likely be able to process several different substrates, natural amino acid substitution studies are done on each amino acid residue of the peptide drug to improve inhibitory activity against the enzyme. Nonnatural amino acids are also substituted to avoid recognition and premature degradation by other enzymes. Our work on HTLV-I protease inhibitors serves as an elegant example of this rational drug optimization process (Section 5.9.6) [6]. Generally speaking, amino acids serve as simple units that can somewhat be readily assembled, to probe the active site of the enzyme and obtain valuable information on the nature of the subsites [7]. Further structural changes to the drug are performed to improve several aspects, which may include balancing hydrophilicity and hydrophobicity so as to improve blood–brain barrier permeation, oral bioavailability, and duration of action, or reducing adverse drug reactions and cost of synthesis. During the process of drug optimization, these modifications progressively decrease the peptide nature of the molecule. After the peptide bonds of the peptide drug are altered, the final drug is then reclassified by its inventors as being a nonpeptide. Of interest, we note that most FDA-approved peptide-derived drugs consist of three to five residues, or to be more exact, the drugs are anchored to three to five subsites.

Substrate-based drug design has been substantially improved with the availability of three-dimensional structural information from such sources as X-ray diffraction crystallography and NMR studies, that is, information from solid and liquid state three-dimensional data, respectively. Three-dimensional structural information provides a computer image of a complex of an enzyme and its inhibitor. It is noteworthy that the shape of the enzyme in complex with an inhibitor is completely different from that of an unbound enzyme. Hence, examining a three-dimensional depiction of an unbound enzyme is an exercise in futility. Moreover, it is obviously practically difficult to obtain a substrate-enzyme complex because peptide hydrolysis of the substrate would occur before any data could be gathered. Inspecting the coordinates of an inhibitor bound to an enzyme provides information about the nature of the subsites including pocket shapes and sizes, presences of sub-pockets, hydrophilic and hydrophobic surfaces, and potential sites for hydrogen bond, van der Waals, or hydrophobic interactions. Moreover, because we believe that inhibitor-enzyme binding follows an induced-fit model, when several complexes of different inhibitors in the same enzyme are available, the flexibility of the subsites to accommodate for different residues can be deduced. From studies aimed at improving the cleavage efficiency of a substrate, researchers can also obtain valuable information about the shape, size, hydrophobicity, and accommodating nature of the subsites, although with less details than three-dimensional structural data. Substrate optimization studies on HIV (Section 5.8) and HTLV-I proteases (Section 5.9.6) serve as excellent examples of probing the subsites for information [8]. It is noteworthy that because the final desired drug is a small molecule, complexes of small inhibitors in the enzyme are preferred over larger ones. Complexes of small inhibitors focus on the specific subsites that are in close proximity to the catalytic subsite, whereas complexes of large inhibitors may induce distortions in the enzyme and lead to misinterpretations on the nature of the active site. Taken together what we have discussed, several three-dimensional structural coordinates of the derived small and potent inhibitors in complex with the enzyme are used to clarify the bound form of the active site of the enzyme. Knowing the flexibility, shape, and electronic properties of the active site means that novel modulators, that is, inhibitors or substrates, can be designed without peptide drawbacks.

High throughput screening is often used to develop nonpeptide modulators. At this stage of research, three-dimensional information of inhibitors bound to the enzyme along with information pertaining to the flexibility of the active site have provided sufficient data to search for potential nonpeptide lead compounds. From a generic chemical library, compounds that can fit and favorably electronically interact with the active site are searched through computer-assisted docking simulations, namely, virtual high throughput screening. These potential lead compounds are then synthesized and processed by high throughput assay screening to verify for activating or inhibitory activity toward or against the enzyme. Essentially, high throughput assay screening is an automated assaying method of a large library of potential lead compounds in microtiter plates. Once lead compounds are identified, the compounds are structurally refined under rational drug optimization to derive potent compounds with desired pharmacodynamic and pharmacokinetic properties. Cellular and animal experiments are performed to confirm the expected pharmacodynamics and pharmacokinetics, as well as to examine for any unexpected adverse drug effects.

### 5.1.3 Phases of Drug Development

Development of new medicines goes through five different phases. We have just described the preclinical stage (Section 5.1.2), in which the drug is discovered, developed, and tested on nonhuman subjects, after which the drug can advance to clinical testing. Clinical trials are divided into four phases in which the drug is administered to volunteer trial participants. Because the tests are ethically conducted on living humans, there are extensive rules and standards governing the trials and their evaluations. Throughout the clinical phases, safety, effectiveness, adverse risks, and adverse reactions associated with the investigational drug in human are continuously monitored. In other words, the pharmacodynamic properties of the drug are diligently kept under close watch.

In phase I clinical trials, low doses of the investigational new drug are given to healthy individuals and gradually increased to investigate for the safety and tolerability of the drug. In certain cases, the drug is given to patients with the targeted disease. The investigators examine for pharmacokinetic properties in healthy individuals to assess drug bioavailability and isolate potential drug distribution problems, so as to determine safe and tolerable dosage levels.

Phase II clinical trials are usually restricted to patients with the targeted disease. Up to several hundred patients are involved. The main focus of the trials is to determine the most appropriate method of drug delivery and its associated therapeutic dosage. Hence, this phase looks at the pharmaceutics of the drug in patients afflicted with the targeted disease.

Phase III clinical trials test the results of the two previous phases in larger populations from several hundred to several thousand patients. Investigators and patients are randomized and double-blinded to provide the primary basis for the benefit-versus-risk assessment for the new drug, while comparing the drug with conventional treatments.

Following successful phase III clinical trials, the drug is registered to the national agency that is responsible for the safety regulation of drugs, such as the FDA. Once the manufacturing process and clinical trials are reviewed by the agency, the drug may be approved for marketing. Phase IV clinical trials, commonly known as *post-marketing studies*, are conducted for an indefinite length of time to evaluate for long-term risks, benefits, and optimal use of the drug. It should be noted that even after a drug has received marketing approval, the FDA is known to revoke its drug approval after reviewing negative post-marketing reports.

In this chapter, we will primarily focus on peptide drugs that have been approved by the FDA and are in post-marketing studies, that is, phase IV clinical trials. However, investigational peptide drugs that have not received FDA approval will be used as illustrative examples, to explain rational drug designs and compare drug pharmacophores. Although different countries will have different phases of clinical trials than the US system that we have described, the main stages of drug developments remain as preclinical, clinical, and post-marketing.

## 5.2 ENZYME TYPES THAT PROCESS PEPTIDES

Among all the different types of enzymes that process peptides, we will examine protein kinases, protein phosphatases, and peptidases.

Phosphates are important in signal transduction because they regulate the proteins to which they are attached. Protein kinases modify peptides or proteins by attaching a phosphate group to one of the three amino acids that have a free hydroxyl group, namely, serine, threonine, and tyrosine. Certain protein kinases, such as histidine kinase, may phosphorylate other amino acids. Owing to their important effect on cell growth, movement, and death, the activity of protein kinases is highly regulated by several mechanisms. A deregulation of protein kinase activity often causes cell proliferation diseases such as cancer. Several tyrosine kinase inhibitors are marketed as anticancer agents. These inhibitors are either of monoclonal antibody or small molecule class, and none of them seems to have been derived from peptides.

While protein kinases add a phosphate group to serine, threonine, tyrosine, or histidine, protein phosphatases remove the phosphate group. Protein phosphatases catalyze the removal of the phosphate thus reversing the regulatory effect of phosphates. As far as we are aware, there is currently no protein phosphatase modulating drug on the pharmaceutical market.

Belonging to the hydrolase category, peptidases, as their names suggest, catalyze the hydrolysis of a peptide bond. Most developed peptide drugs modulate the action of peptidases. Semantically, considering that proteins are longer peptides, the terms *protease* and *peptidase* are often used interchangeably to denote an enzyme that breaks down a protein or peptide. Proteases are classified as serine, threonine, cysteine, aspartic acid, glutamic acid, and metalloprotease, to denote the participation of the active site amino acid(s) within the enzyme or metal ion, during the hydrolysis of a substrate.

### 5.2.1 Enzymes as Chemicals in Consumer and Medical Products

In consideration of the importance of proteases in the human body, it is not inconceivable that the enzymes themselves are commercially and therapeutically useful. We will examine a few of these enzymes that are used in common assay tests, commercial products and as therapeutic agents. Use of serine protease as therapeutic agents and their importance in blood coagulation will be discussed separately in Section 5.4.

Bromelain refers to two cysteine proteases, fruit and stem bromelain, found in the respective parts of the pineapple, *Ananas comosus*. Bromelain is a popular digestive aid and meat tenderizing agent. Classified as a herbal product with anti-inflammatory activity by the FDA, bromelain is approved by the German Commission E panel, the German herbal medicine regulatory body, to treat sinusitis caused by injury or surgery.

Chymosin, also known as rennin (with two *n*'s), is an aspartic protease found in rennet, a natural complex of enzymes produced in infant mammalian stomach to curdle ingested mother's milk thus allowing longer residence in the bowels, and thereby improving absorption of the milk-product. Renin (with one *n*) is an aspartic protease that is involved with the regulation of blood pressure and body fluids (Section 5.6). Chymosin catabolyses K-casein between Phe105 and Met106 to produce an insoluble 1–105 fragment, which forms a curd in the presence of calcium, and a soluble fragment 106–169 fragment, which becomes a part of the whey. The catabolytic effects of chymosin are exploited in the making of cheese, curd, and junket.

Collagenase, derived from *Clostridium histolyticum* bacteria, is a metalloprotease that is approved by the FDA to be used as a sterile enzymatic debriding ointment to digest collagen in necrotic tissue. In other words, collagenase is used to break down the collagen that binds dead tissues together.

Papain is a cysteine protease found in papaya, *Carica papaya*, and mountain papaya, *Vasconcellea cundinamarcensis*. In the laboratory, papain is used to dissociate cells in the first step of cell culture preparation. For thousands of years in South America, papain is used to break down tough meat fibers and is also currently marketed as a meat tenderizer, digestive aid and used in breweries. Topically papain is used as a home remedy treatment to digest protein toxins in the venom of jellyfish, bee, wasp stings, and mosquito bites. Papain is also found as an ingredient in various enzymatic debriding preparations to remove dead or contaminated skin tissues for medical and cosmetic purposes, as shampoo, as enzyme cleaners for soft contact lenses, as tooth whitener in much diluted form and as dental caries removal in more concentrated form. In 2008, the FDA stated that no topical drug product containing papain has been approved by them and urged consumers to cease using papain-containing topical products following reports of allergic reactions, hypotension, and tachycardia.

Pepsin is a major stomach enzyme that digests food proteins into peptides. Discovered in 1836 by Theodor Schwann, the aspartic protease has found success as a digestive aid [9]. The discovery of a potent inhibitor of pepsin, pepstatin, would popularize substrate-based inhibitory peptide drug design (Section 5.2.2).

Pregnancy-associated plasma protein A, also known as pappalysin 1, is a metalloprotease used as a biochemical marker for Down's syndrome in prenatal screening. Low plasma level of pregnancy-associated plasma protein A has been positively correlated with aneuploid fetuses that may develop to babies with Down's syndrome.

Prostate specific antigen is also known as kallikrein III, seminin, semenogelase, $\gamma$-seminoprotein and P-30 antigen. It is a serine protease that is often elevated in blood in the presence of prostate-related disorders. Consequently, it has been approved by the FDA as an effective marker for early detection of prostate cancer.

Subtilisin is a bacteria-derived serine protease that has found its use as a general purpose peptidase in laundry and dishwashing detergents, skin-care ointments, and contact lens cleaners to breakdown unwanted proteins. Subtilisin is also used in food processing.

Trypsin is a serine protease found in the digestive system of many animals. It is used for numerous biotechnological processes, including assays, to cleave many peptide chains at the carboxyl side of Lys and Arg that are not followed by Pro. In the consumer food industry, trypsin is used to predigest baby food by breaking down large protein molecules, so that the developing babies' stomach can easily absorb the smaller nutrients. Serum trypsinogen, the precursor form of trypsin, is found at high levels in acute pancreatitis and cystic fibrosis. It should be noted that serum amylase and/or lipase screening is more standard for the diagnosis of acute pancreatitis, and that serum trypsinogen may return false positives for cystic fibrosis.

### 5.2.2 Nonspecific Enzyme Inhibitors

$\alpha_1$-Antitrypsin, the short name for $\alpha_1$-*proteinase inhibitor*, is a natural plasma glycoprotein that inhibits a wide variety of proteases, such as trypsin (Section 5.2.1) and elastase [10]. Elastase is divided in two subgroups, pancreatic and neutrophil elastases. Emphysema is a result of an overt activity of neutrophil elastase that breaks down elastin, resulting in a decrease in lung elasticity. As an inhibitor of neutrophil elastase, $\alpha_1$-antitrypsin is approved by the FDA as an injectable in the management of emphysema. Several other inhibitors of neutrophil elastase are under investigation (Section 5.9.4).

Pepstatin is a potent aspartic protease inhibitor that was originally isolated from *Actinomyces* bacteria and named after its ability to potently inhibit pepsin (Section 5.2.1) [11]. Pepstatin is a hexapeptide with a statine moiety, (3*S*,4*S*)-4-amino-3-hydroxy-6-methylheptanoic acid, also known as a hydroxyethylcarbonyl unit. Pepstatin inhibits nearly all acid proteases with high potency, due to the statine inhibitory unit (Section 5.1.2). The mechanism of action of the statine unit is similar to that of the hydroxymethylcarbonyl isostere that is illustrated in Figure 5.1. Derivatives of the statine unit sparked interests in the development of several aspartic protease inhibitors such as HIV (Section 5.8) and HTLV-I (Section 5.9.6) for proteases, plasmepsins (Section 5.9.6), β-secretase (Section 5.9.3), and direct renin inhibitors (Section 5.6.2) used to combat acquired immunodeficiency syndrome (AIDS), HTLV-associated diseases, malaria, Alzheimer's disease, and hypertension, respectively. Moreover, pepstatin is a common constituent of the protease inhibitor combination therapy used to treat AIDS (Section 5.8).

## 5.3 AMINO ACID DRUGS

If one would loosely use the term *peptide*, the shortest *peptide* is an amino acid. The single amino acid residue is then processed by enzymes to derive precursor and active neurotransmitters, such as dopamine, norepinephrine, epinephrine, tryptophan, 5-hydroxytryptophan, serotonin, and $\gamma$-aminobutyric acid (GABA). Many drugs have been designed from such amino acids. Although a fair amount of amino acid drugs target specific receptors, we will restrict our discussion to amino acid drugs that modulate the activity of enzymes. Blood coagulating amino acids, aminocaproic acid, and tranexamic acid are examined in their own Section 5.4.1.

### 5.3.1 Thyroid Hormones

According to chemical nomenclature, the thyroid hormones thyroxine ($T_4$) and triiodothyronine ($T_3$) are α-amino acids (Figure 5.2). $T_4$ is the major prohormone secreted by the follicular cells of the thyroid gland that is activated as the $T_3$

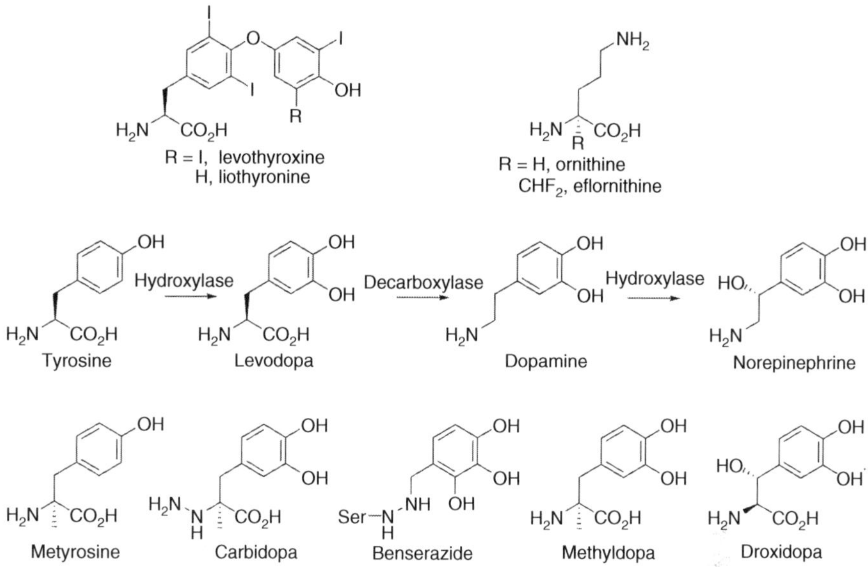


**Figure 5.2** Amino acid-based drug design. Thyroid hormones, an ornithine decarboxylase inhibitor, catecholamines and catecholamine modulators.

hormone. Although to a lesser extent than $T_4$, $T_3$ is also be secreted. As the most powerful thyroid hormone, $T_3$ affects almost every process in the body including body temperature, growth, and heart rate. Indeed, $T_3$ stimulates the production of the enzyme RNA polymerase I and II, to increase the rate of protein synthesis. Conversely, $T_3$ can also increase the rate of protein degradation. Levothyroxine, a synthetic L-$T_4$, is widely marketed for the management of hypothyroidism and enlarged thyroid gland [12]. Moreover, liothyronine sodium, a synthetic salt of L-$T_3$, being more potent than L-$T_4$, is indicated for severe hypothyroidism as well as the treatment of myxedema coma, a state of decompensated hypothyroidism. Hence, levothyroxine and liothyronine are used in thyroid hormone supplementation therapy as general activators of enzymes and regulators of biophysical processes.

### 5.3.2 An Ornithine Decarboxylase Inhibitor

Ornithine is a product of the action of the enzyme arginase on arginine (Figure 5.2). Eflornithine is an α-difluoromethyl derivative of ornithine that acts as an irreversible inhibitor of ornithine decarboxylase [13]. As its name suggests, ornithine decarboxylase removes the carboxylate function from α-amino acid ornithine to yield putrescine. Cofactor pyridoxal phosphate, the activated form of vitamin $B_6$, is also involved. The reaction is the first and rate limiting step for the production of polyamines that are required for cell division. In human, the drug eflornithine has a very short half-life and is degraded much faster than in the parasite *Trypanosoma brucei gambiense*. The parasite causes African trypanosomiasis, which is more commonly referred as *sleeping sickness*. Hence, due to the lower bioavailability of eflornithine in humans than in the parasite, eflornithine pharmacokinetically favors harming the parasite. However, because sleeping sickness mainly affects Africa, a developing continent where patients cannot afford the drug, the original manufacturer of eflornithine does not deem the inhibitor to be profitable and have reduced its production. The WHO subsequently pressured the manufacturer to resume production and to eventually transfer manufacturing to other willing companies. Interestingly, eflornithine as a topical agent, manufactured by another company, is approved by the FDA as an effective hair-growth retardant in women, due to its inhibition of polyamine production.

### 5.3.3 Catecholamines

From tyrosine, mammals can synthesize several important catecholamine neurotransmitters, namely, dopamine, epinephrine, and norepinephrine. L-Tyrosine is hydroxylated by the enzyme tyrosine hydroxylase to L-dihydroxyphenylalanine (L-DOPA or levodopa) that is then processed to dopamine by DOPA decarboxylase and cofactor pyridoxal phosphate (Figure 5.2). Dopamine can be further processed by dopamine β-hydroxylase and cofactor ascorbate to form norepinephrine. Epinephrine can be produced from norepinephrine following a methylation of the distal amine by phenylethanolamine *N*-transferase in the cytosol of adrenergic neurons and chromaffin cells of the adrenal medulla.

In the first step, the process from tyrosine to levodopa is catalyzed by tyrosine hydroxylase. Metyrosine is an amino acid drug that inhibits the enzyme tyrosine hydroxylase to deplete levels of catecholamines [14]. Although metyrosine is rarely used in clinical settings, the drug was approved by the FDA in 1979 for the treatment of pheochromocytoma, a neuroendocrine tumor that induces excessive secretion of catecholamines. From its chemical structure, it is apparent that metyrosine is an α-methyl derivative of tyrosine that competes against tyrosine for tyrosine hydroxylase.

Tyrosine hydroxylase adds an extra hydroxyl function to the aromatic ring of tyrosine to form levodopa, which is then decarboxylated by DOPA decarboxylase to yield dopamine, which is subsequently *β*-hydroxylated to norepinephrine. Levodopa, technically an α-amino acid, can cross the blood–brain barrier whereas dopamine cannot [15]. In 1957, Nobel Prize winner Arvid Carlsson discovered that the administration of levodopa to animals with Parkinsonian symptoms would reduce the symptoms. Eventually, levodopa, as a drug, was approved by the FDA for the management of Parkinson's disease.

In consideration that levodopa can be converted to dopamine both inside and outside of the brain, and the inability of dopamine to cross the blood–brain barrier, DOPA decarboxylase inhibitors are used as an adjunct to levodopa treatment to slow down peripheral conversion of levodopa to dopamine. A retardation of peripheral levodopa conversion raises central nervous system levels of dopamine to manage Parkinsonian symptoms, and decreases peripheral nervous system levels of dopamine, which would result in fewer and less severe adverse drug effects.

Carbidopa is a DOPA decarboxylase inhibitor that is approved by the FDA as a levodopa mimic that competes for the enzyme [16]. Carbidopa is an α-methyl and hydrazinyl derivative of levodopa. Carbidopa and levodopa are often combined in a single tablet for the management of Parkinsonism. Benserazide, a dopamine-Ser derivative, is a DOPA decarboxylase inhibitor that is approved as a combination with levodopa in the United Kingdom and Canada, but not in the United States where carbidopa is used for the same purpose [17]. When compared, both peripheral DOPA decarboxylase inhibitors, carbidopa and benserazide, contain a hydrazinyl function. Vitamin $B_6$, usually provided in the form of pyridoxine, is often included in the combination therapy, so that it can be activated to cofactor pyridoxal phosphate and assist in the conversion of levodopa to dopamine in the brain by DOPA decarboxylase.

Interestingly, α-methyldopa, is an α-methyl derivative of levodopa without the hydrazinyl function found in carbidopa [18]. α-Methyldopa is readily metabolized in the intestines and liver to α-methylnorepinephrine that acts as a central $\alpha_2$-receptor agonist. Similar to norepinephrine and epinephrine, α-methylnorepinephrine exerts $\alpha_2$-receptor negative feedback that results in antihypertensive effects. Consequently, the FDA has approved α-methyldopa for the treatment of hypertension. Its use is now deprecated following introduction of alternative safer classes of antihypertensive agents. It should be noted that, although α-methyldopa was originally designed to be a DOPA decarboxylase inhibitor, the actual pharmacological effects suggest that α-methyldopa may worsen Parkinsonism. In a broad sense, α-methyldopa is a substrate, that is, enzyme activator, of several enzymes and can be considered as a precursor of α-methylnorepinephrine, an agonist analog of norepinephrine. Of interest, from a drug design perspective, whereas metyrosine is an α-methyl derivative of tyrosine, carbidopa and α-methyldopa are α-methyl derivatives of levodopa.

Droxidopa is an analog of both levodopa and norepinephrine that was approved since 1989 in Japan for the treatment of neurogenic hypotension associated with Parkinson's disease [19]. As of 2008, the drug is under clinical trials in Australia, Europe, Canada, and the United States. Contrary to α-methyldopa, which is metabolized to a norepinephrine analog, droxidopa is a prodrug of norepinephrine. In a sense, droxidopa, as a substrate of DOPA decarboxylase, can be considered as an activator of the enzyme, because the removal of the carboxylate function in droxidopa by the enzyme would by-pass the formation of dopamine and directly form norepinephrine. Droxidopa can either cross the blood–brain barrier or remain in the periphery where it is converted to norepinephrine. Moreover, droxidopa can be used in combination with a DOPA decarboxylase inhibitor, such as carbidopa, to increase brain norepinephrine levels while maintaining peripheral levels.

## 5.4 SERINE PROTEASES AND BLOOD CLOTTING

Coagulation is a complex process by which blood forms clots. The primary pathway to initiate blood coagulation is defined by a series of reactions, in which an inactive serine protease precursor, that is, a zymogen, and its cofactor become activated components that then catalyze the next reaction in the cascade, eventually resulting in cross-linked fibrin. In this section, we will touch on serine proteases and serine protease inhibitors that modulate blood coagulation.

### 5.4.1 Blood Coagulating Agents

Peptide drug aprotinin is a naturally occurring broad-spectrum serine protease inhibitor isolated from bovine lung tissue [20]. The monomeric globular 58 residue polypeptide is known to inhibit several serine proteases, namely, trypsin, chymotrypsin, plasmin, and kallikrein. Although aprotinin action on plasmin slows fibrinolysis, its effect on kallikrein leads to the inhibition of the formation of factor XIIa, and consequently stops coagulation and fibrinolysis. Hence, as an injectable drug, aprotinin was used since 1964 to reduce bleeding during complex surgery. However, in 2008, the FDA pressured the manufacturer to discontinue marketing the drug due to an increase risk of complications or death.

As an alternative to aprotinin, aminocaproic acid, and tranexamic acid were developed from lysine (Figure 5.3) [21]. These drugs are $\varepsilon$-amino acids. Both drugs inhibit zymogens and enzymes that particularly bind to lysine. In particular, both drugs bind reversibly to zymogen plasminogen, so that it cannot be activated to plasmin. Without plasmin, fibrinolysis does not occur. Orally administered aminocaproic acid, is FDA-approved to treat excessive postoperative bleeding, in such situations such as dental surgery. Injectable tranexamic acid has roughly eight times the antifibrinolytic activity of its older analog, aminocaproic acid. Tranexamic also inhibits serine proteases known as plasminogen activators, which activates plasminogen to plasmin, as their names suggest (Section 5.4.2). Approved by the FDA, tranexamic acid is the drug of choice as a nonhormonal treatment of menstrual bleeding, namely, dysfunctional bleeding and heavy bleeding associated with uterine fibroids. The drug is also commonly used in cardiac, dental, obstetric, and orthopedic surgery. Of interest, although both aminocaproic acid and tranexamic acid were available in both oral and injectable forms, the manufacturers decided to only support one dosage form.

Several serine proteases, namely, blood clotting factors II, VII, IX, and X, are combined as a prothrombin complex concentrate. Fresh frozen plasma and the prothrombin complex concentrate are used to medically correct for prothrombin deficiencies. When the factors are activated, blood coagulates. Hence, the prothrombin complex concentrate is often used to resolve intractable bleeding caused by the anticoagulant drug warfarin.

Interestingly, the knowledge of the sequence of activated factor II, also known as factor IIa or thrombin, is valuable in constructing recombinant fusion proteins. The

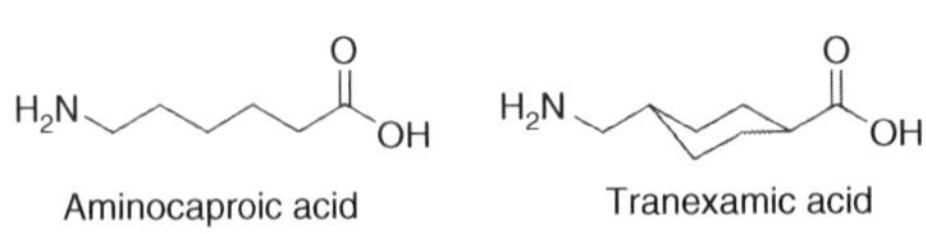


**Figure 5.3** Fibrinolytic amino acid drugs.

thrombin cleavage site sequence of Leu-Val-Pro-Arg//Gly-Ser is commonly included as a linker region between recombinant fusion protein constructs for ease of purification. Once purified, thrombin is used to specifically cleave between Pro-Arg and Gly of the cleavage site sequence, thereby removing the purification tag from the protein of interest.

### 5.4.2 Enzymes as Blood Anticoagulants

Plasminogen activators are serine proteases that catalyze the conversion of plasminogen to plasmin, a major anticoagulant serine protease [22]. In general, plasminogen activators are injected enzymes that are indicated for clotting-related conditions including pulmonary embolism, myocardial infarction and stroke. The activators are classified according to the source where the enzyme drugs are obtained, such as recombinant tissue plasminogen activators, alteplase, monteplase, reteplase, and tenecteplase, from endothelial cells; urokinase from urine; streptokinase and anistreplase from *Streptococcus* bacteria. As previously discussed (Section 5.4.1), tranexamic acid is an amino acid drug that inhibits plasminogen activators to prevent excessive bleeding.

Protein C is a vitamin K-dependent serine protease that is activated by thrombin into activated protein C. Once activated, the enzyme is a major physiological anticoagulant and exhibits both anti-inflammatory and antiapoptotic activities. Activated drotrecogin alfa is an injectable recombinant activated protein C that is approved by the FDA to be used in intensive care medicine as a treatment for severe sepsis [23].

### 5.4.3 Direct Thrombin Inhibitors as Blood Anticoagulants

Current anticoagulation therapies often involve antiplatelet agents such as aspirin, clopidogrel, dipyridamole, and ticlopidine and anticoagulants such as heparin and warfarin. In particular, although coumarin derivatives such as warfarin are very useful and powerful anticoagulant drugs, they are plagued with problems associated with drug–drug and drug–disease interactions, and a very narrow therapeutic window that requires very careful therapeutic drug monitoring. To resolve these unspecific therapeutic interactions, direct thrombin inhibitors were developed as anticoagulants to bind at the active site of thrombin and inhibit its blood coagulating activity. There are two classes of direct thrombin inhibitors: bivalent and univalent [24]. Although the bivalent inhibitors bind at both the active site and exosite 1 of thrombin, the univalent inhibitors bind more specifically to the active site. Although few direct thrombin inhibitors are available to patients, further development and market accessibility to direct thrombin inhibitors would provide very excellent and most likely safer alternatives to coumarin derivatives.

In 1884, John Haycraft demonstrated that medicinal leeches, *Hirudo medicinalis*, secreted a substance, hirudin, with potent anticoagulant properties [25]. Until the discovery of heparin, these leeches were the only mean of preventing blood from clotting. Hirudin is a bivalent direct thrombin inhibitor. Structurally, hirudin is a

65-amino acid residue protein with a compact N-terminal domain with three disulfide bonds and a disordered C-terminal domain [26]. Unfortunately, natural hirudin exists in various isoforms and is difficult to extract in sufficient therapeutic amount from natural sources, that is, leeches. However, homogeneous hirudin can be produced using recombinant techniques [27]. Lepirudin and desirudin are marketed as recombinant hirudins to be used when heparin is contraindicated, such as in unstable angina or for patients with a risk of developing heparin-induced thrombocytopenia (HIT). Bivalirudin is a synthetic 20-amino acid peptide derivative of hirudin, containing a tripeptide active direct thrombin inhibitor (D-Phe)-Pro-Arg attached to Pro and a tetrapeptide Gly linker, followed by a dodecapeptide analog of the C-terminus of hirudin, that is (D-Phe)-Pro-Arg//Pro-Gly-Gly-Gly-Gly-(hirudin C-terminus) [28]. Of interest, the Pro-Arg sequence is shared between bivalirudin and the aforementioned linker chain used in recombinant fusion protein construct, so that thrombin can recognize and process the respective peptide (Section 5.4.1). Although hirudin is an irreversible inhibitor of thrombin, bivalirudin is a reversible inhibitor that is slowly processed by thrombin between the Pro-Arg and Pro sequences of the drug. As a result of bivalirudin reversibility, there is a low risk of severe bleeding associated with bivalirudin when compared with standard combination heparin therapy. Unfortunately, due to their large sizes, bivalent direct thrombin inhibitors, lepirudin, desirudin, and bivalirudin, need to be administered by injection, and thereby limiting their use for long-term treatment.

Univalent direct thrombin inhibitors are considerably much smaller than bivalent inhibitors. The (D-Phe)-Pro-Arg sequence represents the $P_3-P_2-P_1$ residues of bivalent thrombin inhibitor bivalirudin. The (D-Phe)-Pro-Arg motif was heavily modified in the design of univalent inhibitor melagatran [29]. The drug was marketed outside of the FDA's jurisdiction as an oral prodrug, ximelagatran, to increase oral bioavailability (Figure 5.4). In the body, the ethyl ester moiety in ximelagatran is hydrolyzed, whereas the hydroxyl group is removed to uncover the main binding portion of the inhibitor. Unfortunately, due to liver toxicity in a subpopulation of patients found during clinical trials, the FDA rejected the drug's application and the manufacturer discontinued developing the drug.

Argatroban is an extensively modified derivative of the (D-Phe)-Pro-Arg motif [30]. It is noteworthy that argatroban does not fit in the active site of thrombin in an extended backbone conformation, resulting in a nonsequential numbering of the residues as $P_3-P_1-P_2$. Argatroban was approved by the FDA for patients with HIT undergoing prophylaxis or treatment for thrombosis or percutaneous coronary intervention. However, the univalent direct thrombin inhibitor must be administered intravenously because of the highly basic $P_1$ Arg side-chain that interferes with gastrointestinal absorption [31]. To partially alleviate the alkalinity of the $P_1$ residue and improve drug tolerability, the carboxylic acid at the 2-position of the $P_2$ piperidine amide acts as an internal counter-ion for the $P_1$ guanidine function.

Dabigatran etexilate is an orally administered prodrug that is metabolized to univalent direct thrombin inhibitor dabigatran through ethyl de-esterification and the removal of a long hexyloxycarbonyl function from a benzamidine moiety [32]. Using

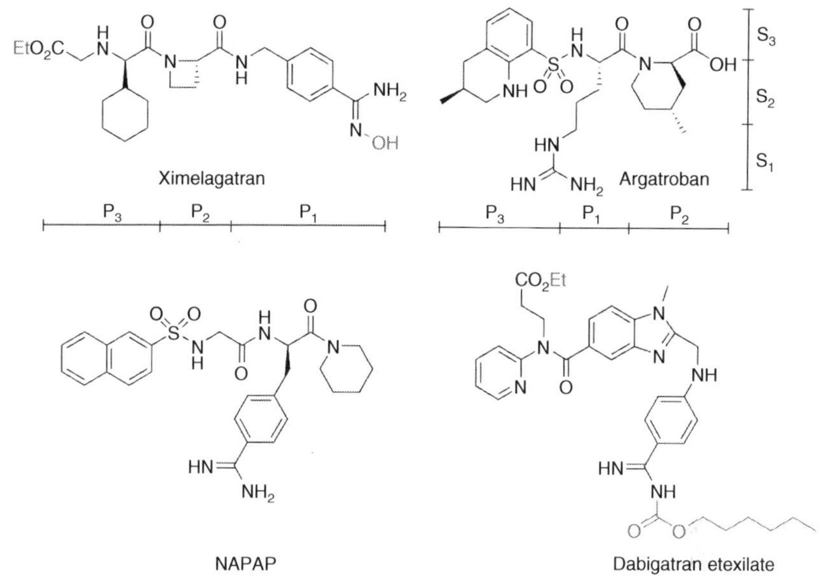


**Figure 5.4** Univalent direct thrombin inhibitors. The leaving groups of prodrugs ximelagatran and dabigatran etexilate are drawn in gray.

X-ray crystallography and computer-assisted docking experiments, dabigatran etexilate was derived from NAPAP ($N^{\alpha}$-(2-naphthylsulfonylglycyl)-4-amidinophenylalala nine piperidide), a powerful nonpeptide inhibitor of various serine proteases, specifically thrombin and trypsin [33]. Despite its strong activity, NAPAP is not intestinally absorbed and rapidly eliminated from the circulation by hepatic uptake. Much like argatroban, NAPAP and dabigatran etexilate have a nonsequential $P_3$–$P_1$–$P_2$ numbering convention. Heavily modified from NAPAP, dabigatran etexilate has a benzimidazole central core that holds the $P_1$ benzamidine function found in NAPAP and an aromatic $P_3$ residue. In contrast to NAPAP and argatroban, dabigatran etexilate is an excellent oral candidate and could serve as a replacement for the often problematic anticoagulant drug, warfarin. Interestingly, dabigatran etexilate does not take advantage of potential interactions with the $S_2$ subsite because it lacks a $P_2$ residue. Considering that dabigatran etexilate has received market authorization by the European Medicines Agency, National Health Services in Britain and Health Canada in 2008, it is expected that the FDA too would soon approve the drug.

Common structural features are present in the (D-Phe)-Pro-Arg motif found in bivalent thrombin inhibitor bivalirudin and the univalent thrombin inhibitors ximelagatran, argatroban, NAPAP, and dabigatran etexilate. As mentioned, the motif and ximelagatran have a conventional sequential $P_3$–$P_2$–$P_1$ numbering whereas the residues of argatroban, NAPAP and dabigatran etexilate are laid out as $P_3$–$P_1$–$P_2$, because subsite numbering takes precedence over residue numbering (Section 5.1.1). As a part of the critical $P_1$ residue, all motif and inhibitors share a similar amidine or guanidine functional group that interacts with thrombin

Asp189B. Benzamidine function in ximelagatran, NAPAP, and dabigatran etexilate is an isostere of the guanidine side-chain of arginine in the motif and argatroban. Interestingly, the molecule benzamidine itself is a reversible competitive inhibitor of trypsin, trypsin-like enzymes and serine proteases. The $P_2$ residue, which is absent in dabigatran etexilate, exists as a nonaromatic cyclic 4-, 5-, or 6-member ring system in the motif or other univalent thrombin inhibitors, respectively. According to X-ray diffraction crystallography data, the lipophilic region of the $P_2$ residue provides the majority of the binding energy through favorable van der Waals contact with the largely lipophilic $S_2$ subsite. Any $P_2$ polar groups were merely tolerated by the $S_2$ subsite. The $S_3$ subsite of thrombin is unconventional in that it is only accessible by a compound with a $P_3$ D-configuration, such as D-phenylalanine of the motif and D-α-cyclohexylglycine of ximelagatran. In the natural fibrinogen substrate, the pocket is accessible when $P_8$-Leu and $P_9$-Phe loops around to make the interactions. This loop suggests that the $S_3$ subsite can accommodate for large cyclic functions, such as those found in the motif and univalent thrombin inhibitors. In the design of dabigatran etexilate, following a report that highly lipophilic thrombin inhibitors would exhibit less activity in the blood plasma due to protein binding [34], a butyric acid function was attached to the $P_3$ amide nitrogen to increase hydrophilicity. X-ray diffraction crystallography data of dabigatran in complex with thrombin reveal that this attachment did not greatly interfere with drug-enzyme binding because the $P_3$ amide nitrogen projected into bulk solvent without forming further interaction with the enzyme. Interestingly, ximelagatran similarly possessed an acetic function that is attached to the $P_3$ nitrogen. Moreover, both ximelagatran and dabigatran etexilate are orally bioavailable prodrugs of melagatran and dabigatran, respectively. Both drugs have similar protection points that are needed to improve their oral bioavailability, namely, a protected $P_1$ amidine or guanidine function, and a protective $P_3$ ethyl ester function. Bivalirudin, argatroban, and NAPAP are not orally bioavailable.

## 5.5 DIABETES MELLITUS

Diabetes mellitus is a syndrome of a metabolic disorder that results in abnormally high blood–sugar levels. Blood glucose levels are controlled by a complex interaction of multiple chemicals and hormones, including the peptide hormone insulin. Defects in either insulin secretion or insulin action lead to diabetes mellitus. When food is present in the lumen of the small intestine, the gastrointestinal hormone incretins are released to enhance insulin secretion, even before blood glucose levels become elevated. We will touch on peptide drug strategies in the management of diabetes mellitus that entail the administration of incretin derivatives and inhibition of DPP-IV, an enzyme that breaks down incretins.

### 5.5.1 Peptide Hormones and Blood Glucose Regulation

Insulin is a 51-amino acid peptide hormone with extensive effects on metabolism and several other body systems, such as vascular compliance. Insulin is known to modify the activity of numerous enzymes. It is released from the β-cells of the islets of

Langerhans in the pancreas to cause most of the body's cells to take up glucose from the blood, storing it as glycogen in the liver and muscle, and stops the use of fat as an energy source. Patients with type 1 diabetes mellitus depend on subcutaneously injected insulin for their survival because their bodies cannot produce insulin. Most synthetic human insulins for therapeutic use are manufactured as recombinant proteins, and exist as insulin analogs with different absorption and duration of action profiles. To meet the patient's varying insulin requirements, the derivatives are classified as rapid-, short-, intermediate-, and long-acting insulin analogs.

Glucagon is a 29-amino acid peptide hormone that is involved in carbohydrate metabolism. The binding of glucagon to glucagon receptors leads to a cascade of enzyme activations. When blood glucose level is low, glucagon, which is produced from the pancreas, is released to cause the liver to convert stored glycogen into glucose that is subsequently released into the bloodstream. The action of glucagon is opposite to that of insulin. In severe hypoglycemia when the victim cannot take glucose orally, glucagon is used as a peptide drug that is given intramuscularly, intravenously, or subcutaneously by injection to quickly raise blood glucose levels.

Secretin is the first hormone to be identified [35]. Discovered by William Bayliss and Ernest Starling in 1902, secretin is a 27-amino acid peptide hormone that has some sequence similarity with glucagon. Its primary effect is to regulate the pH of the duodenal contents through the control of gastric acid secretion and buffering with bicarbonate, by several means, including modulating digestive enzyme activity such as the inhibition of gastrin and stimulation of pepsin. Secretin also regulates blood glucose levels by triggering increase insulin release from the pancreas, or conversely by stimulating the release of glucagon [36]. Secretin is used in the medical field in pancreatic function tests to detect abnormalities in the pancreas such as gastrinoma, pancreatitis, or pancreatic cancer.

### 5.5.2 Glucagon-like Peptide-1 and Analogs

Glucagon-like peptide-1 (GLP-1) functions as a gastrointestinal hormone incretin that enhances insulin secretion prior to blood glucose levels elevation. GLP-1 also decreases glucagon secretion from pancreatic $\alpha$-cells, and increases $\beta$-cell mass and insulin gene expression. Moreover, GLP-1 inhibits acid secretion and gastric emptying of the stomach, and increases satiety resulting in decrease food intake. Although these characteristics suggest that GLP-1 is a good agent to combat diabetes mellitus, GLP-1 must be administered by continuous subcutaneous infusion. The reason behind such a tedious administration method is because GLP-1 is a large peptide hormone with a half-life of less than 2 min due to a rapid degradation by DPP-IV.

Exenatide is the only analog of GLP-1 that is currently approved by the FDA for the treatment of diabetes mellitus type 2. Exenatide was discovered from the venom of a lizard, the Gila monster, *Heloderma suspectum* [37]. As an analog of GLP-1 with 53% homology and longer half-life, exenatide is a 39-residue peptide and therefore must be administered by subcutaneous injection twice daily. Although several other long acting GLP-1 analogs such as albiglutide and liraglutide are currently in clinical trials, they too must be administered by subcutaneous injection.

### 5.5.3 Dipeptidyl Peptidase-4 Inhibitors

Another approach to promote the actions of GLP-1 is to inhibit DPP-IV, the enzyme that inactivates GLP-1. DPP-IV also degrades another incretin, glucose-dependent insulinotropic peptide (GIP). Hence, inhibition of DPP-IV would prevent the degradation of two incretins that induce insulin secretion. The major advantage of DPP-IV inhibitors is that they can be designed to be orally bioavailable, due to their smaller sizes. Since its discovery in 1967, DPP-IV was known to cleave two amino acids from the N-terminus of a substrate after an alanine residue or, more importantly, a proline residue using a catalytic serine residue on the enzyme [38]. The active site of DPP-IV is fairly indiscriminant with at least 62 known substrates and thus permitted for the discovery of diverse inhibitors [39]. This lack of discrimination also means that the inhibitors could also have affinity for other DPPs, such as DPP-II, DPP-VIII, and DPP-IX, and thereby could result in drug-induced toxicity [40].

Substrate mimic inhibitors were designed from a general Xaa-Pro substrate sequence where cleavage occurs after Pro. Cyanopyrrolidine inhibitors, as the name suggests, contain a nitrile function that forms a reversible covalent bond with the catalytic serine residue and a pyrrolidine moiety of DPP-IV to mimic the proline residue of the substrate. The interaction of the nitrile function with the catalytic serine is critical for both potent inhibitory activity against DPP-IV, and selectivity over other DPPs [40, 41]. However, the cyanopyrrolidine moiety can undergo intramolecular cyclization that deactivates the inhibitor (Figure 5.5). The *trans*-rotamer of the drug, which is the binding rotamer at the active site, can convert to the *cis*-rotamer that then undergoes intramolecular cyclization to form a structurally different compound.

DPP-IV inhibitor vildagliptin introduced a steric bulk as the $P_2$ residue's capping moiety (the R function in Figure 5.5) to slow intramolecular cyclization (Figure 5.6) [42]. The bulky hydroxyl adamantyl ring of the inhibitor relatively slowed intramolecular cyclization by 30 times, resulting in an orally bioavailable drug with long-lasting pharmacodynamics and higher inhibitory potency against DPP-IV. In 2008, vildagliptin was approved by the European Medicines Agency for use within the EU as a single pharmaceutical ingredient, or in combination with metformin, an antidiabetic drug. However, within the same year, the manufacturer withdrew its application to the FDA because the agency had requested that further clinical studies be made.

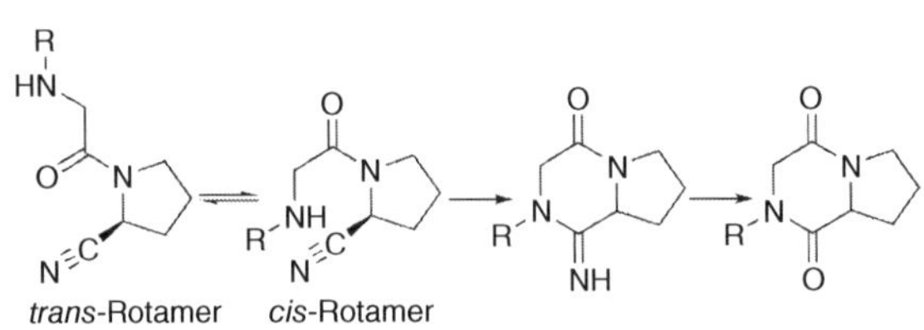


**Figure 5.5** Intramolecular cyclization of cyanopyrrolidine dipeptidyl peptidase-4 inhibitors.

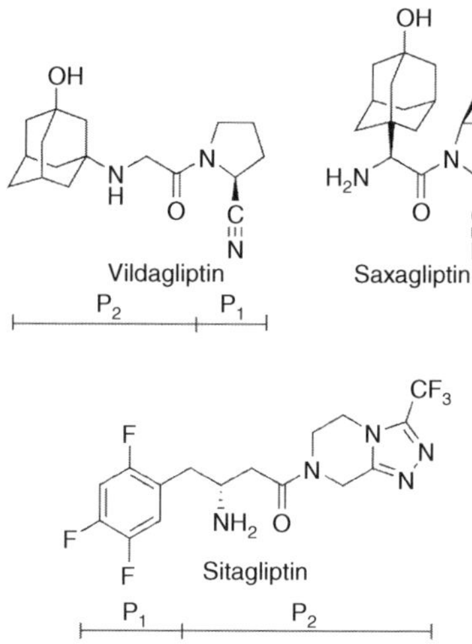


**Figure 5.6** Dipeptidyl peptidase-4 inhibitors.

Developed by a competing pharmaceutical company, saxagliptin is another DPP-IV inhibitor with a bulky hydroxyl adamantyl ring at the $P_2$ position [43]. However, the bulk was placed as a side-chain instead of a capping group as found in vildagliptin. To further stabilize saxagliptin against intramolecular cyclization, a *cis*-4,5-methylene function was appended to the $P_1$ pyrrolidine ring so as to reduce molecular flexibility. Saxagliptin is in phase III clinical trials.

From X-ray diffraction crystallography studies of substrate-based DPP-IV inhibitors in complex with the enzyme, three-dimensional information pertaining to the flexibility, shape, and electronic properties of the active site of the enzyme were elucidated (Section 5.1.2). From such information, high throughput screening was performed to discover lead compounds that were followed-up by rational drug optimization to improve on inhibitory activity and selectivity against DPP-IV. As a result, several nonsubstrate-based inhibitors such as sitagliptin, alogliptin, and linagliptin were obtained [44]. Considering that the lead compounds were semirandomly screened and that the enzyme itself is fairly indiscriminant toward substrates, the mode of interactions of these nonsubstrate-based DPP-IV inhibitors are, as expected, different from one drug to another. For example, when the mode of binding of nonsubstrate-based inhibitor sitagliptin is compared to that of substrate-based inhibitors vildagliptin and saxagliptin, one realizes that the N-terminus of the β-amino acid of sitagliptin occupies the $S_1$ subsite instead of the $S_2$ (Figure 5.6). Sitagliptin is currently the only FDA-approved DPP-IV inhibitor for type 2 diabetes mellitus, while alogliptin and linagliptin are under phase III clinical trials. While sitagliptin is a β-amino acid, it would take quite a stretch of the imagination to describe alogliptin and linagliptin as peptide drugs since they either have a dihydropyrimidine or dihydropurinedione, respectively, as the central core.

## 5.6 RENIN–ANGIOTENSIN–ALDOSTERONE SYSTEM

The renin–angiotensin–aldosterone system is a complex regulator of blood pressure, renal hemodynamics and volume homeostasis in normal physiology. Simply put, the system aims at increasing blood pressure in response to hypotension, decrease sodium concentration in the distal tubule of the kidney, renal sympathetic nerve stimulation and decrease blood volume. When the system is activated, the kidneys release renin (with one *n*) that cleaves the liver-derived peptide angiotensinogen into angiotensin I. Angiotensin-converting enzyme (ACE), found in the pulmonary circulation and the endothelium of blood vessels, converts angiotensin I to angiotensin II by removing two residues from the C-terminus of angiotensin I. In consideration that angiotensin II is a potent vasoconstrictor and that ACE is also involved in the inactivation of bradykinin, a potent vasodilator, ACE, essentially increases blood pressure. Angiotensin II also stimulates aldosterone production that causes the tubules of the kidneys to increase re-uptake of sodium and water, increasing the plasma volume and blood pressure. Hence, ACE regulates fluid volume homeostasis. We will discuss two classes of enzyme-targeting peptide-derived drugs, namely, ACE inhibitors and renin inhibitors.

### 5.6.1 ACE Inhibitors

ACE is a metalloprotease. ACE inhibitors antagonize the actions of ACE. Consequently, they are primarily indicated for the management of hypertension and congestive heart failure. ACE inhibitors are also indicated for left ventricular dysfunction, the prevention of cardiovascular disorders, and the prevention of nephropathy in diabetes mellitus. Development of ACE inhibitors is of historical significance because the research demonstrated early on that peptides could be developed as oral bioavailable drugs.

In 1965, the research team of Sérgio H. Ferreira [45] discovered bradykinin potentiating factor, a family of peptides that were isolated from the venom of the Brazilian pit viper, *Bothrops jararaca*. The team elucidated a small pentapeptide, Glu-Lys-Trp-Ala-Pro [46], that, although had little to no hypotensive effect, potentiated the hypotensive effects evoked by bradykinin [47]. From the bradykinin-potentiating factor family of peptides, Ondetti and colleagues isolated a more potent nonapeptide named *teprotide*, Glu-Trp-Pro-Arg-Pro-Glu-Ile-Pro-Pro [48], that, however, had limited clinical value because it lacked oral activity [49]. Although further studies on the sequences did not lead to a marketed drug, the discoveries highlighted the importance of ACE in hypertension.

In 1973, Larry D. Byers and Richard Wolfenden [50] discovered that carboxypeptidase A, a bovine pancreatic enzyme similar to ACE, was potently inhibited by L-benzylsuccinic acid (Figure 5.7). The carboxylate moiety of the compound binds to the catalytic zinc ion that is present at the active site. The active site encompasses to the C-terminal region of a peptide substrate. Both carboxypeptidase A and ACE are zinc metalloexopeptidases. In contrast to carboxypeptidase A, ACE removes two residues instead of one residue from a peptide substrate. Having considered that

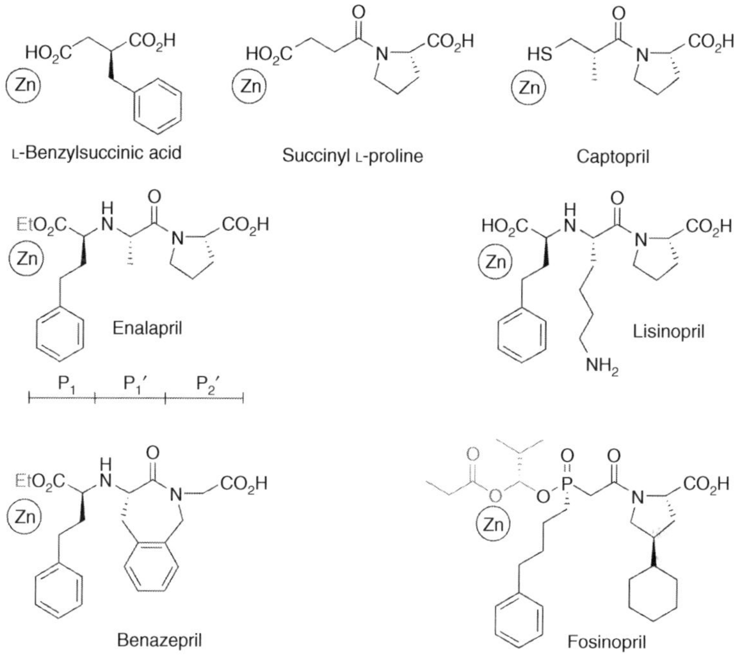


**Figure 5.7** L-Benzylsuccinic acid and angiotensin converting enzyme inhibitors. The leaving groups of prodrugs enalapril, benazepril, and fosinopril are drawn in gray. The zinc ion found at the catalytic site is illustrated to mark the proximity of the chelating group of the inhibitor.

ACE removes a dipeptide and teprotide contained proline at the C-terminus, David Cushman and Miguel Ondetti [51] derived a weak ACE inhibitor, succinyl L-proline. Methyl branching and chain elongation studies on the succinyl moiety, followed by a replacement of the carboxylate moiety by sulfhydryl as a zinc-chelating group with greater affinity, eventually led to the first marketed ACE inhibitor, captopril.

The most common adverse effects of captopril are skin rash and loss of taste, which are believed to be caused by the sulfhydryl moiety [52]. Designers of captopril opted to switch back to a less potent carboxylate moiety as the zinc-chelating group found in succinyl L-proline, while improving binding to the active site in other sections of the drug to compensate for the expected reduction in inhibitory potency against ACE. The results from the studies were peptide drugs enalaprilat and lisinopril. The researchers adopted a three-residue model. Proline was kept as the $P_2'$ end anchor. In enalaprilat, a $P_1'$ alanine residue was adopted to mimic the methyl functional found in captopril. Alternatively, in lisinopril, an adjacent basic amino acid $P_1'$ lysine residue improved potency as a result of a hydrogen bond network formed with several intermediary water molecules, as suggested by X-ray diffraction crystallography data [53]. As for the $P_1$ residue containing the chelating group, a hydrophobic residue could be accommodated by the $S_1$ pocket. Interestingly, we would like to point out that the chemical structures for enalaprilat and lisinopril are similar to the sequences for

Phe-Ala-Pro and Phe-Lys-Pro, where the $P_1$ L-homophenylalanine residue is similar to the $P_1$ L-phenylalanine found in angiotensin I near the scissile bond. Moreover, the removed dipeptide sequence of angiotensin I is composed of a basic and a nonpolar amino acid, His-Leu, that corresponds to the polarity pattern of the $P'_1$–$P'_2$ residues of lisinopril, namely, Lys-Pro.

Enalaprilat is only suitable for intravenous administration because, being dicarboxylated, it exhibits unfavorable ionization characteristics to allow sufficient stability for oral administration. Consequently, enalapril was developed as a prodrug of enalaprilat, in which the chelating carboxylate moiety was converted to the ethyl ester. The prodrug is metabolized in the body by various esterases to afford the parent compound, enalaprilat. As previously mentioned, ximelagatran and dabigatran etexilate, two univalent direct thrombin inhibitor prodrugs, also used ethyl esterification to improve oral bioavailability (Section 5.4.3). Since the introduction of enalapril, at least six other ethyl ester ACE inhibitor prodrugs with a carboxylate chelating moiety, namely, benazepril, moexipril, perindopril, quinapril, ramipril, and trandolapril, were approved by the FDA for the US market. Of interest, benazepril has a benzodiazepine core moiety that encompasses the $P'_1$–$P'_2$ Ala-Pro sequence in enalapril. Hence, one could infer that the $S'_1$–$S'_2$ subsites form a wide cavity. Fosinopril is a prodrug that is metabolized *in vivo* to fosinoprilat, in order to overcome similar oral bioavailability problems associated with enalaprilat. Contrary to enalaprilat, fosinoprilat has a phosphonate functional group as the zinc-chelating moiety. Fosinoprilat has a chemical structure similar to a Phe-Gly-Pro sequence.

The hot bed of ACE inhibitor research now lies in selective domain inhibition. ACE has two active sites [52]. Although the C-domain is mainly responsible for the regulation of blood pressure by converting angiotensin I, the N-domain is principal for the processing of Ac-Ser-Asp-Lys-Pro, a natural hemoregulatory peptide hormone. Bradykinin is hydrolyzed at both domains. Current ACE inhibitors are nonselective toward the two active sites, which means that bradykinin degradation is also inhibited, resulting in vasodilator-related adverse effects such as angioedema. Researchers are currently attempting to develop C-domain selective ACE inhibitors that would permit some degradation of bradykinin at the N-domain, thereby reducing the chance of angioedemic attacks.

### 5.6.2 Renin Inhibitors

After chronic use of drugs that interfere with angiotensin, for example, ACE inhibitors, and aldosterone, the body compensates by increasing renin production. In order to overcome the negative feedback mechanism, researchers have focused on renin inhibitors as antihypertensive agents.

Renin is a highly specific aspartic protease that selectively cleaves angiotensinogen to generate angiotensin I, using two aspartic acids at the active site of the enzyme. An important sequence of angiotensinogen is Pro-Phe-His-Leu//Val within which the scissile bond is between $P_1$ Leu and $P'$ Val (Figure 5.8). From the sequence and after several studies, in 1990, our research group derived a potent and orally bioactive renin inhibitor, KRI-1314 [54]. In the compound, the $P_4$ Pro residue was

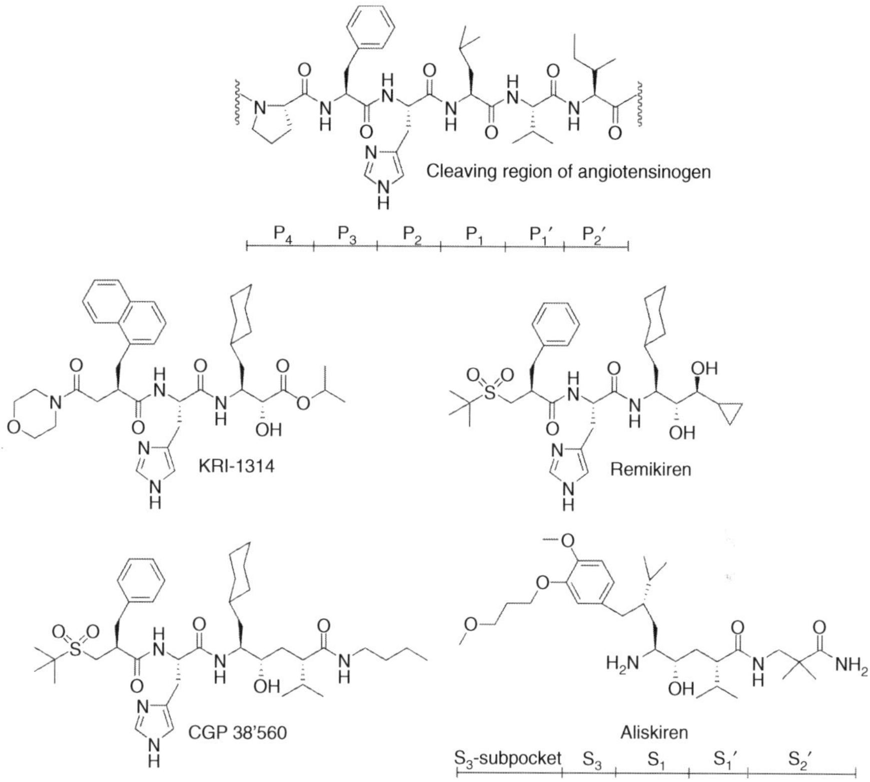


**Figure 5.8** A segment of angiotensinogen and renin inhibitors.

enlarged and optimized to a morpholinyl moiety. A succinic acid residue, having a $P_3$–$P_4$ retro-inverso amide bond, was introduced to reduce the risk of premature degradation by other proteases [55]. The $P_3$ Phe side-chain was replaced by a 1-methylnaphthyl moiety to avoid recognition and degradation by chymotrypsin, a digestive enzyme that particularly cleaves at Phe-His amide bonds. Moreover, a larger bicyclic aromatic $P_3$ moiety is better accommodated by the $S_3$ pocket than the $P_3$ Phe residue of the angiotensinogen. $P_2$ His residue was kept as in angiotensinogen to maintain hydrogen bond interactions with Ser233 of the renin. As for the critical $P_1$ inhibitory residue, a cyclohexylnorstatine moiety, (2*R*,3*S*)-3-amino-4-cyclohexyl-2-hydroxybutyric acid could more efficiently inhibit renin than several other hydroxymethylcarbonyl isosteric inhibitory units that we have substituted. Hydroxymethylcarbonyl structure may also be referred as a norstatine unit. As briefly mentioned in Section 5.1.2, the design of the hydroxymethylcarbonyl isosteres as inhibitory units is based on a tetrahedral transition state formed during the general acid–base hydrolysis mechanism of a substrate (Figure 5.1). Indeed, as proven by X-ray diffraction crystallography, the hydroxyl group and, to a lesser extent, the carbonyl oxygen, participate in hydrogen bond interactions with the two catalytic aspartic acids at the active site, and thereby anchoring the inhibitor inside the active site [4a]. In KRI-1314, the $P_1'$ moiety was set as an isopropyl ester to resemble angiotensinogen's $P_1'$ Val residue [54]. Oral administration of KRI-1314 to salt-depleted monkeys resulted in a fall of 10–20 mm Hg of mean blood pressure and reduction of the plasma-renin activity for a 5-h period. Thus, renin inhibitor KRI-1314 is an oral bioavailable, effective and long-lasting antihypertensive agent.

Remikiren is a renin inhibitor with a dihydroxyethylene isostere as the inhibitory transition state mimic [56]. Comparisons between KRI-1314 and remikiren reveal that, in the case of remikiren, the large $S_4$ subsite is accommodated by a $P_4$ *tert*-butyl sulfonyl moiety, the $P_3$ residue is structurally similar to the $P_3$ Phe residue of angiotensinogen, while the $P_2$ residue is kept as the histidine in angiotensinogen. The side-chain of the $P_1$ transition state mimic residue shares a common cyclohexylmethyl function between the two compounds. The $P_1'$ residues in both peptide drugs were designed to resemble angiotensinogen's $P_1'$ Val residue. The proton acceptor carbonyl group of KRI-1314's $P_1'$ residue was replaced by a weaker proton acceptor hydroxyl group in remikiren. Although remikiren is a potent orally active inhibitor, its overall oral bioavailability was low due to hepatic clearance.

Another research group developed peptide inhibitor CGP 38′560 that shares several features with KRI-1314 and remikiren [57]. The $P_4$–$P_1$ residues in CGP 38′560 are similar to remikiren. The $P_1$ inhibitory unit of CGP 38′560 is a hydroxyethylene isostere. As with the other peptide renin inhibitors, the $P_1'$ residue resembles angiotensinogen's $P_1'$ Val residue. As a longer molecule than KRI-1314 and remikiren, CGP 38′560 has a butyl retroamide as the $P_2'$ residue. Although CGP 38′560 is a potent and specific inhibitor of human renin, it exhibits weak blood-pressure lowering effect in salt-depleted monkeys after oral dosing, due to mainly very limited overall bioavailability [58].

From CGP 38′560, aliskiren, a nonpeptide renin inhibitor, was developed [59]. After analyzing the shape and chemical properties of the active site, the developers of CGP 38′560 came to realize that the $S_1$ and $S_3$ pockets essentially form a contiguous and large hydrophobic cavity [60]. They went on to design through computer-assisted molecular modeling methods several hydroxyethylene transition state mimetic inhibitors with a directly linked $P_1$–$P_3$ moiety that is large. Doing so, the $P_1$–$P_4$ spanning backbone of the aforementioned peptide inhibitors was eliminated. Although not directly mentioned by the researchers in their reports, we believe these computer-assisted modeling methods also involved virtual high throughput screening that we have briefly touched on (Section 5.1.2). In a similar manner as KRI-1314, remikiren and CGP 38′560, a phenyl ring is present within the $P_1$–$P_3$ moiety of aliskiren to mimic the $P_3$ Phe residue of angiotensinogen. Interestingly, the $P_1$ cyclohexylmethyl function found in the aforementioned peptide inhibitors is replaced by an isopropyl function in aliskiren that occupies the same $S_1$ subsite. We believe that this computer-assisted design was possible because the $S_1$–$S_3$ cavity is a large hydrophobic area in which improving fit (e.g., van der Waals contacts and hydrophobic interactions) is a stronger determinant of renin inhibitory activity in the $S_1$–$S_3$ region than hydrogen bond interactions. Moreover, the researchers had the resources to perform several X-ray diffraction crystallography studies that greatly accelerated their discoveries. Indeed, X-ray diffraction crystallography data permitted the team to exploit a 9 Å deep, narrow, and well-defined hydrophobic $S_3$ subpocket by implementing a long hydrophobic ether side-chain to the $P_1$–$P_3$ moiety of aliskiren. Neither substrates nor peptide inhibitors bind to this subpocket. In nonpeptide aliskiren, the $S_3$ subpocket interactions, along with improved interactions at the $S_2'$ pocket, could sufficiently compensate for the lack of $S_2$ and $S_4$ subsites interactions.

Aliskiren is currently the only marketed renin inhibitor. It was approved in 2007 by the FDA for the treatment of primary hypertension [61]. Orally active aliskiren, designed through molecular modeling techniques, is an octanamide transition state renin inhibitor with good water solubility and low lipophilicity. It exhibits potent and specific *in vitro* inhibition of human renin ($IC_{50}$ in the nanomolar range) with a plasma half-life of around 24 h. Despite an amide bond, aliskiren is resistant to protease biodegradation. As with ACE inhibitors, angioedema may occur with aliskiren, although the risk is theoretically lower. We expect that future renin inhibitors will have several structural features that are similar to aliskiren so as to take advantage of its nonpeptide benefits.

## 5.7 PENICILLIN AND CEPHALOSPORIN ANTIBIOTICS

The discovery of antibiotic benzylpenicillin, commonly known as penicillin G, from *Penicillium notatum* is attributed to by Nobel laureate Alexander Fleming in 1928, while its medicinal development is attributed to Nobel laureate Howard Walter Florey [62]. The discovery of cephalosporin compounds from *Cephalosporium acremonium* is attributed to Giuseppe Brotzu in 1948, and cephalosporin C was subsequently isolated at the University of Oxford [63]. In consideration that the biosynthesis of penicillin G and cephalosporin C begin from a tripeptide comprising of L-α-aminoadipic acid, L-cysteine, and D-valine, in a sense, most antibiotics in the penicillin and cephalosporin classes are peptide drugs (Figure 5.9). Penicillin antibiotics are known to inhibit a bacterial enzyme, DD-transpeptidase, also known as penicillin-binding protein and serine-type D-Ala D-Ala carboxypeptidase, and interfere with the cross-linking of peptidoglycan chains of the enzyme to form rigid bacterial cell walls. Cephalosporin antibiotics share the same mode of action as penicillin.

Bacteria that are resistant to penicillins secrete an enzyme, *β*-lactamase, which breaks the drug's *β*-lactam ring, and thereby inactivating the drug. Using a boosting peptide inhibitor to prevent *β*-lactamase from metabolizing *β*-lactam antibiotics is common practice. Clavulanic acid is a biosynthetic product of amino acid arginine and sugar glyceraldehyde-3-phosphate isolated from *Streptomyces clavuligerus* [64]. Potassium clavulanate is available as combination penicillin products of amoxicillin-clavulanate and ticarcillin-clavulanate, to inhibit bacterial *β*-lactamase and subsequently overcome bacterial resistance. Likewise, sulbactam and tazobactam

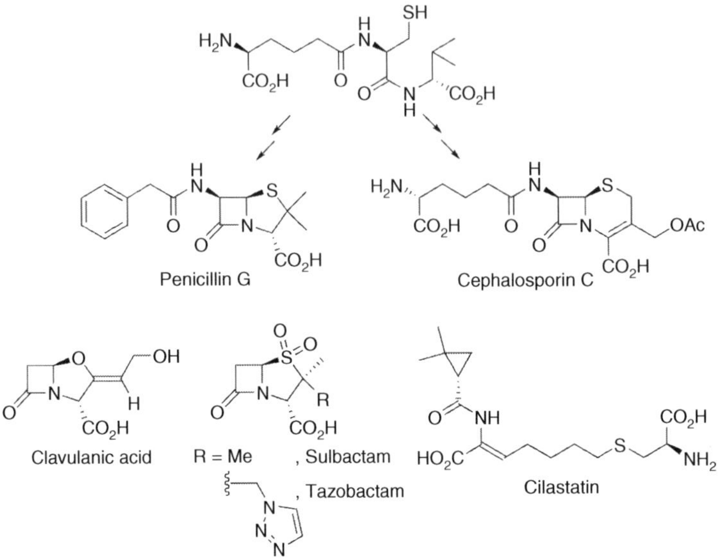


**Figure 5.9** Biosynthesis of penicillin and cephalosporin and $\beta$-lactam antibiotic boosters.

are found in ampicillin-sulbactam and piperacillin-tazobactam penicillin antibiotics combinations [65]. Although clavulanate, sulbactam and tazobactam share the $\beta$-lactam ring that is characteristic of $\beta$-lactam antibiotics, for example, penicillins and cephalosporins, these compounds are not exploited for their antibacterial properties but more for their activity as competitive inhibitors of $\beta$-lactamase. In a similar manner, cilastatin protects $\beta$-lactam antibiotic imipenem from being degraded by human renal enzyme dehydropeptidase, and prolongs the antibacterial effect of imipenem [66]. Lacking a $\beta$-lactam ring, cilastatin is a cysteine analog that does not itself have any antibacterial activity.

## 5.8 HIV PROTEASE

HIV causes AIDS. In their weakened state, HIV-infected immunodeficient patients are susceptible to opportunistic infections and cancers from which they eventually pass away. Considering the severity of AIDS and high mutation rate of the virus, an aggressive combination of several antiretroviral drugs, referred as *highly active antiretroviral therapy* (HAART), is strongly recommended for symptomatic patients. The drug cocktail contains several different classes of antiretroviral drugs that act at different stages of the HIV life cycle. Pepstatin, a general aspartic protease inhibitor (Section 5.2.2), and aspartic protease inhibitors that are more specific to HIV are included in the cocktail to target viral assembly. A role that HIV protease plays in the life cycle of the virus is to cleave precursor polyproteins into proteins that are subsequently used to assemble and form new virions. Inhibiting HIV protease would essentially stop the propagation of the virus.

Physically, HIV-1 protease is a small homodimeric enzyme composed of two identical peptide chains that form into a pincer shape, where the tip of the pincer is referred as the flap region, and the central tunnel-like cavity as the active site region. A protein is introduced within the active site, where its cleavage is coordinated by a water molecule and two aspartic acid residues found at the base of the active site of the protease. For several cases of aspartic protease inhibitors, another water molecule anchors the inhibitor to the flap. As previously mentioned in Sections 5.2.2 and 5.6.2, pepstatin and renin inhibitors are aspartic protease inhibitors with a central inhibitory unit that form hydrogen bonds with the two catalytic aspartic acid residues of the enzyme as described in Section 5.1.2. In the design of HIV-1 protease inhibitors, the participation of the water molecules and symmetrical nature of the enzyme are also exploited in the design of HIV-1 protease inhibitors.

### 5.8.1 HIV-Specific Protease Inhibitors

The protease of HIV and other retroviruses recognize the Xaa-Pro sequence as the cleavage site, whereas most mammalian aspartic proteases do not [2]. Taking advantage of this selectivity toward retroviruses, our research group based several of our inhibitor studies on the MA/CA (matrix/capsid, also known as p17/p24) and TF/PR (transframe/protease) cleavage regions of HIV-1 polyproteins that have Phe-Pro as the scissile residues (Figure 5.10). One of our more potent HIV-1 protease inhibitor, KNI-227, possessed a 5-isoquinolinyloxyacetyl moiety as the $P_3$ moiety that was optimized from the $P_3$ Phe of the TF/PR sequence [67]. As the $P_2$ residue, a methylcysteine was preferred by the $S_2$ subsite over the asparagine of the substrate. The $P_1$ inhibitory unit of KNI-227 contained an allophenylnorstatine, (2*S*,3*S*)-3-amino-2-hydroxy-4-phenylbutyric acid, as a hydroxymethylcarbonyl isostere. The TF/PR sequence's $P_1'$ Pro moiety was improved to a bulkier L-5,5-dimethylthiazolidine-4-carboxylic acid moiety and was simply capped as a *tert*-butyl amide. In our design, we attempted to increase steric bulk in order to balance hydrophilicity and hydrophobicity, in order to increase desolvation entropy [68]. In other words, a more hydrophobic drug would be more entropically favored to release water molecules as the drug and active site undergo complete or partial desolvation upon binding. Moreover, the increase steric bulk would also reduce flexibility to the drug molecule. Indeed, our engineered conformation constraints were designed to make the free conformation of the inhibitor similar to its bound conformation, so as to minimize loss of conformation entropy. Although a certain level of hydrophilicity is required for water solubilization in body fluids, polar functions can be strategically placed in noninteracting sections of the inhibitor, as exemplified by the design of direct thrombin inhibitor dabigatran (Section 5.4.3). KNI-227 is superpotent and highly selective toward HIV-1 protease with excellent antiviral activity in cells [69].

Saquinavir is the first HIV-1 protease inhibitor approved by the FDA in 1996. Although our KNI-227 has a $P_3$ 5-isoquinoline structure, saquinavir has a similar quinoline structure at the $P_3$ position. In saquinavir, asparagine was kept as the $P_2$ residue found in the MA/CA and TF/PR substrates, and introduced an inhibitory unit,

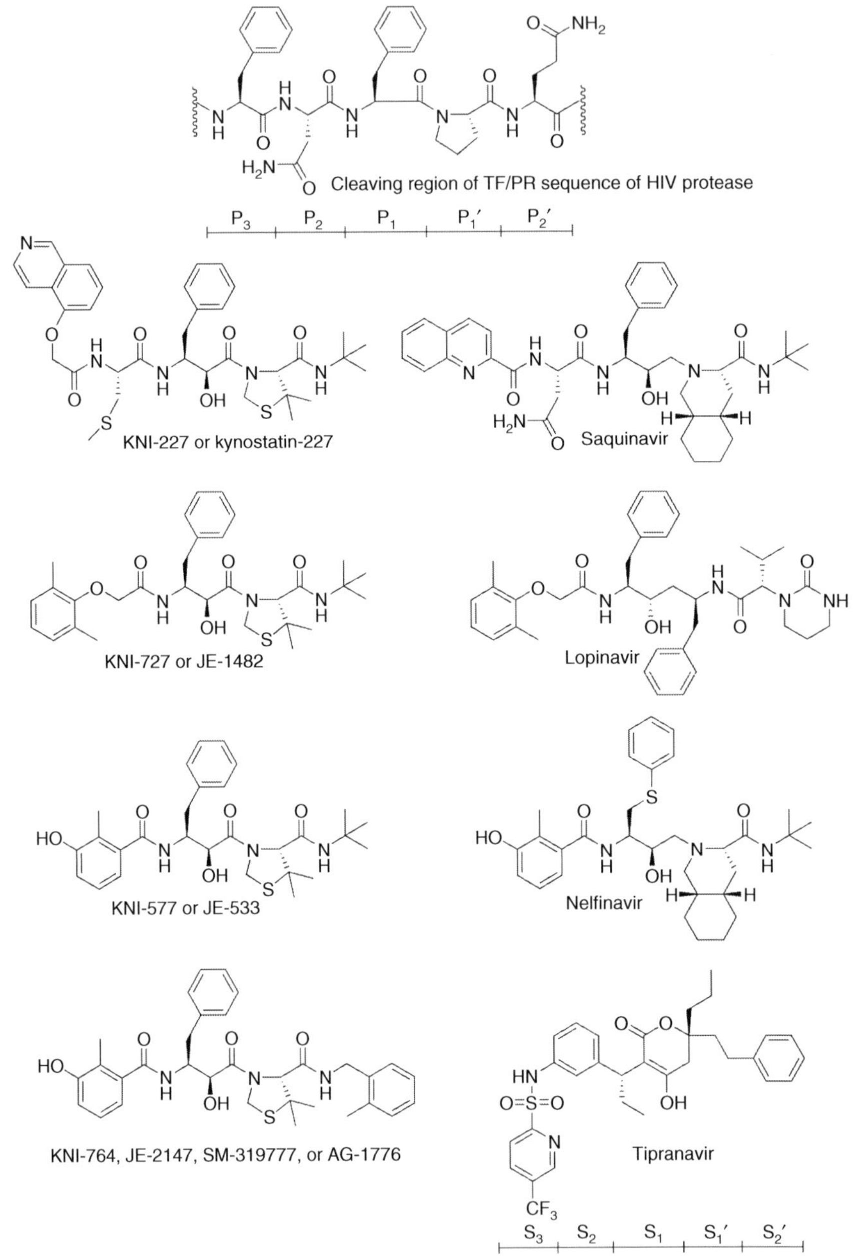


**Figure 5.10** Cleaving region of HIV-1 TF/PR sequence and HIV protease inhibitors.

a hydroxyethylamine isostere, that was similar to the allophenylnorstatine found in KNI-227. Moreover, similar to KNI-227, saquinavir's $P_1'$ residue was also bulkier than the $P_1'$ Pro found in the polyprotein sequences, and was also capped as a *tert*-butyl amide. However, due to its low oral bioavailability as a result of poor absorption and extensive hepatic metabolism, the efficacy of saquinavir must be boosted with the second FDA-approved protease inhibitor, ritonavir. It should be noted that ritonavir is hardly used for its antiviral activity, and more for its ability to inhibit a liver enzyme, cytochrome P450 3A4 (CYP3A4), that normally metabolizes protease inhibitors [70]. Hence, ritonavir-boosting elevates plasma drug levels of saquinavir. We have previously described this strategy of using a boosting peptide inhibitor to prevent an enzyme from prematurely metabolizing the main drug in combination antibiotics therapy (Section 5.7).

FDA-approved lopinavir was developed to improve on the HIV resistance and serum protein-binding properties of ritonavir [71]. Resistance to monotherapy with HIV-1 protease inhibitor is attributed to amino acid substitutions in the protease. Much like saquinavir, lopinavir is marketed as a mixture with ritonavir as a booster to enhance oral bioavailability. Lopinavir is an improved derivative that retained the $P_1$–$P_2'$ residues of ritonavir. The $P_1$–$P_1'$ residues in ritonavir and lopinavir were designed while keeping in mind the symmetrical nature of the homodimeric HIV-1 protease. Structurally, the $P_1$ and $P_1'$ residues in lopinavir are nearly symmetrical to each other with two Phe-like moieties and symmetrical peptide chain directions. This inhibitory unit is commonly referred as a hydroxyethylene dipeptide isostere. Compared with KNI-227 and saquinavir, the large $P_2$ and $P_3$ moieties were replaced by a smaller six-member ring structure that symmetrically flanked both sides of the $P_1$–$P_1'$ region of lopinavir.

Our own study suggested that the $P_2$ to $P_2'$ residues form the pharmacophore of potent inhibitors [72]. Consequently, we adapted the 2,6-dimethylphenoxyacetyl moiety found in lopinavir as the $P_2$ residue in our KNI-727 inhibitor design [73]. Although KNI-727 exhibited very high HIV-1 protease inhibitory activity, its cellular antiviral activity was less than desirable.

On the other hand, when a 3-hydroxy-2-methylbenzoyl group was used as the $P_2$ residue in our inhibitor KNI-577, the drug exhibited both potent HIV-1 protease inhibition and cellular antiviral activity [73]. The $P_2$ residue was borrowed from nelfinavir, an HIV protease inhibitor that boasted of inhibiting both HIV-1 and HIV-2 proteases [74]. Nelfinavir was approved by the FDA as a mesylate ester prodrug. Structurally, the $P_2$ residues of lopinavir and nelfinavir are substituted phenyl moieties. The $P_1$–$P_1'$ sequence of nelfinavir is similar to that of saquinavir with the exception of a longer sulfanyl $P_1$ side-chain.

In consideration of the symmetrical nature of the enzyme and the design of lopinavir, inhibitor KNI-577 was improved at the $P_2'$ position with a methylphenyl moiety to uncover inhibitor KNI-764 [75]. In other words, the $P_2$ and $P_2'$ residues of the inhibitor contain phenyl rings that reside in the respective symmetrical $S_2$ and $S_2'$ subsites. KNI-764 is a highly potent inhibitor of HIV-1 protease that exerts potent cellular antiviral activity with moderate oral bioavailability and low cytotoxicity profiles. Moreover, the inhibitor is effective against both HIV-1 and HIV-2 proteases along with HIV-1 resistant strains.

High throughput screening of a large library of compounds afforded a lead compound that was optimized to derive a nonpeptide HIV protease inhibitor, tipranavir [76]. As described with DPP-IV inhibitor sitagliptin (Section 5.5.3) and renin inhibitor aliskiren (Section 5.6.2), lead compounds discovered by high throughput screening and assay, and followed by rational drug optimization often result in drugs with fascinating structural features. Tipranavir has an interesting structure with several symmetrical pharmacophoric features that are recognize by the symmetrical HIV protease dimer. As previously described (Section 5.1.2), a water molecule is involved in the hydrolysis of a substrate through the catalytic amino acid residues of the enzyme. Most substrate-based HIV protease inhibitors interact with the leucine flap residues through another water molecule. In tipranavir, the main inhibitory unit, a 5,6-dihydro-4-hydroxy-2-pyrone ring system, interacts with both the catalytic aspartic acid residues and the leucine flap residues of the enzyme directly without the mediation of a water molecule. Moreover, the inhibitory unit could potentially undergo keto–enol tautomerism. Other symmetrical features include the ethyl and *n*-propyl structures that are accommodated by the respective $S_1$ and $S_1'$ subsites, and phenyl rings that fit in the $S_2$ and $S_2'$ subsites, respectively. The presence of the $P_3$ trifluoromethylpyridinylsulfonamide moiety is necessary to enhance inhibitory potency against HIV protease. Approved by the FDA in 2005, tipranavir is administered with ritonavir to treat patients who are resistant to other treatments. However, side effects of tipranavir can be more severe than other HIV protease inhibitors.

Other FDA-approved peptide HIV protease inhibitors include amprenavir, atazanavir, darunavir, and indinavir sulfate. Fosamprenavir calcium is marketed as a slow-release $P_1$ phosphate ester prodrug that undergoes cleavage by phosphatase in the body to provide the parent drug, amprenavir.

## 5.9 PEPTIDE DRUGS UNDER DEVELOPMENT

Many protease inhibitors are currently under development. Being in development, the inhibitors are either in the preclinical or clinical testing phases. We will provide a brief overview of enzymes that are being investigated as drug targets. Readers are encouraged to seek further details in the literature.

### 5.9.1 Cathepsins

Cathepsins form a family of enzymes that are activated at the low pH found in lysosomes. The enzymes play a vital role in mammalian cellular turnover, such as bone resorption, and thus are implicated in diseases, in which biological structures are destroyed and formed, such as cancer, stroke, Alzheimer's disease, arthritis, and chronic obstructive pulmonary disease.

Cathepsin B is a cysteine protease that when overexpressed is associated with tumor metastasis, inflammation, bone resorption, and myocardial infarction.

Cathepsin B inhibitor CA-074 was shown to delay death in the monkey neurons after an ischemic insult, whereas its methyl ester prodrug prevented bone resorption in rat osteoclast, as reported in preclinical studies [77].

Cathepsin D is an intracellular aspartic protease that is overexpressed in breast cancer cells and associated with an increased risk of metastasis due to enhanced cell growth. Moreover, the enzyme may be involved in the formation of $\beta$-amyloid peptides in Alzheimer's disease. However, it is uncertain whether inhibition of cathepsin D would be beneficial, because the roles of cathepsin D in the disease states are not yet well defined. Consequently, no attempt at discovering cathepsin D-specific inhibitors has been reported.

Cathepsin K is a cysteine protease that is highly expressed in osteoclasts, and catabolizes elastin, collagen, and gelatin to break down bone and cartilage. Odanacatib is in phase III clinical trials as a cathepsin K inhibitor to prevent bone resorption [78]. Odanacatib is a nonpeptide drug, originally derived from a peptide origin, being developed for the treatment of postmenopausal osteoporosis.

Cathepsin L is a cysteine protease that has similar bone resorption roles as cathepsin K, and like cathepsin B, is implicated in tumor metastasis. Several cathepsin L inhibitors are under preclinical development for osteoporosis and cancer [79].

### 5.9.2 Cysteine Proteases

Calpains comprise a family of at least six distinct members, whose precise functions are unclear. Nevertheless, calpains are believed to be involved in the pathology of stroke, Alzheimer's disease, muscular dystrophy, cataracts, and arthritis. Calpain inhibitors, such as calpeptin, were synthesized to determine the roles of calpains [80]. Owing to limited information that is available on the enzyme, the development of calpain inhibitors is still in its infancy.

Caspase-1, formerly known as interleukin-1 (IL-1) converting enzyme, processes pro-IL-1 to IL-1$\beta$, a key inflammatory mediator. Inhibition of caspase-1 would slow the inflammatory response. In a mouse and rat models with an inflammatory disease, an inhibitor of caspase-1, VE-13045, was reported to delay the onset as well as reduce the severity of the respective disease [81]. Caspase-3, also known as apopain, is a key executioner in apoptosis and has been implicated in neurodegenerative diseases such as Alzheimer's disease. Peptide inhibitors that are selective for either caspase-1 or caspase-3 have been reported [82].

### 5.9.3 Secretases in Alzheimer's Disease

The *amyloid hypothesis* suggests that the etiology of Alzheimer's disease begins with a transmembrane glycoprotein, $\beta$-amyloid precursor protein (APP). In the normal nonpathogenic pathway, APP is processed by two enzymes, $\alpha$-secretase and $\gamma$-secretase. In the pathogenic pathway, APP is cleaved by $\beta$-secretase and $\gamma$-secretase, while $\alpha$-secretase is not involved. The products of the pathogenic path are amyloid $\beta$ peptides (A$\beta$) ranging from 38 to 43 residues that readily form oligomers due to their hydrophobic nature in an aqueous cerebral environment. A$\beta$42 is the more aggregenic species.

In the development of secretases in Alzheimer's disease, there has been very little progress in α-secretase activators [83]. More research has been focused on the inhibition of $\beta$-secretase due to the principal pathogenic role of the enzyme in Alzheimer's disease. One of our peptidomimetic inhibitor, KMI-429, was found to reduce A$\beta$ production in both APP-transgenic and wild-type mice [84]. Another more advanced peptidomimetic inhibitor from our research, KMI-574, was shown to change the conformation of $\beta$-secretase, thereby shifting the enzyme from lipid raft membranes to nonraft membranes, resulting in a disruption of protein transport [85]. Along with other research groups, we have expanded our research to nonpeptide $\beta$-secretase inhibitors [86]. Considering that several pharmaceutical companies have expressed much interest in nonpeptide $\beta$-secretase inhibitors, we expect that $\beta$-secretase inhibitors will soon be in clinical trials. Pharmaceutical companies have also invested much research in $\gamma$-secretase inhibition, resulting in a $\gamma$-secretase inhibitor, LY-450139, arriving at phase III clinical trials [87].

### 5.9.4 Trypsin-Like Serine Proteases

Activated factor X, namely, factor Xa, cleaves prothrombin to form thrombin that would in its turn activate several reaction cascades to build blood clots (Section 5.4.2). A number of natural product inhibitors have been reported such as antistatin, ecotin, and tick anticoagulant peptide. However, there is very little report on factor Xa inhibitors being developed for medicinal purpose.

Human neutrophil elastase belongs to the same family as chymotrypsin. The elastase breaks down connective tissues such as collagen, elastin, laminin, fibronectin, and progeoglycan in lung structures and thereby increases airspaces. The damaging effects lead to inflammation of the lungs and causes pulmonary diseases such as adult respiratory distress syndrome (ARDS), chronic bronchitis, pulmonary emphysema, along with cystic fibrosis and rheumatoid arthritis. Several inhibitors against neutrophil elastase are in clinical evaluation [88].

Tryptase is found in mast cells that are involved with inflammatory and allergic responses. Inhibition of tryptase could alleviate the symptoms of asthma, conjunctivitis and rhinitis. Peptide inhibitor APC-366 is undergoing clinical trials for the treatment of asthma [89].

### 5.9.5 Zinc Metalloproteases

As discussed in Section 5.6.1, ACE is a zinc metalloprotease. Numerous ACE inhibitors are approved by the FDA to manage hypertension. Inhibitors of several others zinc metalloproteases are being investigated as therapeutic agents on their own terms.

Matrix metalloproteases (MMPs) are responsible for the reconstruction of the extracellular matrix by remodeling structural proteins such as membrane collagens, aggrecan, fibronectin, and laminin. An over-expression of the proteases leads to inflammatory diseases, cancer, and muscular dystrophy. At least 10 MMP inhibitors, such as marimastat, are being assessed in clinical trials as agents against arthritis, cancer, and multiple sclerosis [90].

Neprilysin, also known as neutral endoprotease, degrades atrial natriuretic peptide, a protein hormone secreted by atrial myocytes in response to high blood pressure. Inhibition of neprilysin would elevate levels of atrial natriuretic peptides and reduce blood pressure. Candoxatril is the orally active ester prodrug of candoxatrilat, an inhibitor of neprilysin [91]. Candoxatril has a potential therapeutic role in the management of hypertension, especially in congestive heart failure patients, and is in clinical trials. During the development of neprilysin selective inhibitors, compounds that are active against both neprilysin and ACE (Section 5.6.1) were discovered. Among these dual-acting inhibitors, fasidotril, mixanpril, and sampatrilat are in clinical trials [92]. Omapatrilat, a dual-inhibitor, was not approved by the FDA due to angioedema safety concerns [93].

A disintegrin and metalloprotease domain 17 (ADAM17), also known as tumor necrosis factor-α (TNF-α) converting enzyme (TACE), is involved in the processing of TNF-α at the surface of cells. Because TNF-α is a potent and pivotal mediator in the inflammatory process, inhibition of TACE would reduce the severity of inflammatory responses in several disease states such as arthritis and multiple sclerosis. Research on TACE inhibition is in preclinical phase [94].

### 5.9.6 Non-mammalian Proteases

The Candida yeast strains *Candida albicans*, *Candida tropicalis*, and *Candida parapsilosis* exist in small colonies in a healthy intestinal tract. In immunocompromised patients, *C. albicans* is the cause of opportunistic oral and genital infections, whereas *C. tropicalis* is the predominant cause of fungal infections in neutropenic cancer patients, and *C. parapsilosis* is associated with sepsis along with wound and tissue infections. These yeasts release secreted aspartic proteases of broad specificity that is linked with the virulence of the strains. HIV-1 protease inhibitor indinavir was found to be a weak inhibitor of secreted aspartyl proteinases and could reduce the viability and growth of *C. albicans* [95]. Very little inhibitor designs have been reported [96].

Hepatitis C is a blood-borne infectious liver disease that is caused by the hepatitis C flavivirus. Hepatitis C NS3/4A serine protease, encoded by the nonstructural 3 and 4A regions of the viral genome, is essential for viral replication. Boceprevir and telaprevir are two prospective inhibitors of the protease that are undergoing clinical trials [97].

HTLV-I causes adult T-cell lymphoma/leukemia that may worsen to HTLV-associated myelopathy (HAM)/tropical spastic paraparesis (TSP) along with other inflammatory diseases and opportunistic infections. Despite the fact that HTLV-I belongs to the same *Retroviridae* family as HIV, HTLV-I protease inhibition by HIV protease inhibitors has been unsuccessful in managing the infection in the clinics. Using the rational drug design method that we have described in this chapter (Section 5.1.2), we are developing several small and potent peptide HTLV-I protease inhibitors [6].

Among malarial parasites, the *Plasmodium falciparum* species contributes to the highest incidence of death. Plasmepsins form a group of at least nine aspartic proteases and a histo-aspartic protease, plasmepsin III, found in malarial parasites. While cleavage by aspartic proteases involves the aspartic acid residues of the respective enzymes, cleavage by the histo-aspartic protease is coordinated by a histidine and aspartic acid residue at the catalytic site. Plasmepsins I to IV are essential for digestion of the protozoan parasite's major food source, human hemoglobin. Our research group has designed and synthesized potent peptide inhibitors that are active against all plasmepsins I–IV, and will continue to refine and vary our inhibitor designs [98].

Rhinovirus 3C protease is responsible for viral replication in the common cold virus. Inhibition of the protease would stop the propagation of the virus. A peptidomimetic inhibitor, AG7008, against the protease is in clinical trials [99].

In 2003, severe acute respiratory syndrome (SARS) became a global crisis that was eventually resolved by isolating the infected from the general populace. A coronavirus (CoV) was found to be the causative agent. SARS-CoV 3CL$^{pro}$ is the chymotrypsin-like cysteine protease that plays a pivotal role in the replication of the virus. Several SARS-CoV 3CL$^{pro}$ inhibitors are currently under development by our research group [100].

## 5.10 DISCUSSION

In this chapter, we hope to have provided some insights on peptide drugs that affect proteases. A peptide drug can be as small as a single amino acid residue or as large as an enzyme. Enzymes, activating peptide substrates, and peptide inhibitors can all be considered as peptide drugs when they are used for commercial or therapeutic purposes. Most enzymes are exploited for their ability to break down proteins, and are thus used as digestive and debridement agents. Not many activator peptide drugs have reached the pharmaceutical market, and of those, most are used in supplementation therapy. Indeed, although there are far fewer examples of synthetic peptide enzyme activators than enzymes or peptide inhibitors that are used as drugs, these activating substrates are nonetheless important for researchers to study the nature of the enzymes. Such studies would provide further elaborations on the specificity of the enzyme, as well as its roles in the healthy and disease processes. Once the nature of the enzyme is clarified, inhibitory peptide drugs can be more easily designed.

From a peptide substrate, especially a small one, a substrate mimic that competes for the enzyme can be designed, by replacing the scissile residue of the substrate with an inhibitory unit that cannot be cleaved by the enzyme. The derivation of substrate-based peptide inhibitors is aided by the likelihood that substrates interact with the active site in an extended backbone conformation. The substrate mimicking inhibitors are further refined by truncation and natural amino acid substitution studies, followed by nonnatural amino acid studies and an eventual replacement of the peptide bonds by nonpeptide bonds. Indeed throughout the rational drug design process, researchers use peptide inhibitors to probe the active site of the enzyme to understand the flexibility, topology, and charge distribution of the individual binding subsites. Of interest, most inhibitors that have reached the pharmaceutical market interact with three to five subsites. In other words, most peptide-derived inhibitors are composed of three to five residues. Once multiple three-dimensional images of the active site are obtained to illustrate the dynamics of the inhibitor-bound enzyme, these four-dimensional data are used to virtually screen for potential nonpeptide lead compounds that can fit and form favorable hydrogen bond interactions with the active site. The potential leads are synthesized and assayed for inhibitory activity against the enzyme. The active leads are then structurally refined for desired pharmacodynamic and pharmacokinetic properties. Ideally, the resulting drug should be specific for the targeted enzyme, should have high overall bioavailability, and could be conveniently administered, preferably by oral route.

One must not forget that the pharmaceutical market is a business that relies on potential monetary profit to fuel drug development. A drug would not be marketed if there is insufficient profit. A pharmaceutical company is not likely to venture against a competing company that has established a clear dominancy with an unsurpassable drug. A pharmaceutical company would not be interested in developing drugs for consumers who cannot afford the drug, for example, diseases affecting developing countries. A pharmaceutical company may even stop a drug development when the drug approval agency, being pressured by current socioeconomic situations and the media, places too many demands on the company. Although we have mainly restricted our scope of discussion to peptide drugs that have been approved by the FDA, it is evident that we have also left out a fair number of peptide drugs that have been approved by various drug approval agencies throughout the world. In consideration that we, the authors, are also restricted by our personal resources to explore the many assortments of peptide drugs, we strongly encourage the readers to further expand their learning through literature and their own research, and share their discoveries with the international scientific community.

## ACKNOWLEDGMENTS

The writing of this chapter was supported in part by the Frontier Research Program, the twenty-first century COE (Centers of Excellence) Program from the Ministry of Education, Culture, Sports, Science and Technology (MEXT) – Japan; and the unyielding efforts from many researchers and patients to find reliefs and cures to various ailments and intractable conditions.

## REFERENCES

1. Schechter I, Berger A. On the size of the active site of proteases. I. Papain. Biochem Biophys Res Commun 1967;27:157–162.
2. Hellen CUT, Krausslich HVK, Wimmer E. Proteolytic processing of polyproteins in the replication of RNA viruses. Biochemistry 1989;28:9881–9890.
3. Nguyen J-T, Zhang M, Kumada H-O, Itami A, Nishiyama K, Kimura T, Cheng M, Hayashi Y. Kiso. Y. Truncation and non-natural amino acid substitution studies on HTLV-I protease hexapeptidic inhibitors. Bioorg Med Chem Lett 2008;18:366–370.
4. (a) Baldwin ET, Bhat TN, Gulnik S, Liu B, Topol IA, Kiso Y, Mimoto T, Mitsuya H, Erickson JW. Structure of HIV-1 protease with KNI-272, a tight-binding transition-state

analog containing allophenylnorstatine. Structure 1995;3:581–590; (b) Wang Y-X, Freedberg DI, Yamazaki T, Wingfield PT, Stahl SJ, Kaufman JD, Kiso Y, Torchia DA. Solution NMR evidence that the HIV-1 protease catalytic aspartyl groups have different ionization states in the complex formed with the asymmetric drug KNI-272. Biochemistry 1996;35:9945–9950; (c) Wang Y-X, Freedberg DI, Wingfield PT, Stahl SJ, Kaufman JD, Kiso Y, Bhat TN, Erickson JW, Torchia DA. Bound water molecules at the interface between the HIV-1 protease and a potent inhibitor, KNI-272, determined by NMR. J Am Chem Soc 1996;118:12287–12290.

5. Adachi M, Ohhara T, Kurihara K, Tamada T, Eijiro H, Nobuo O, Shigeki A, Shoyama Y, Kimura K, Matsumura H, Sugiyama S, Adachi H, Takano K, Mori Y, Hidaka K, Kimura T, Hayashi Y, Kiso Y, Kuroki R. Structure of HIV-1 protease in complex with potent inhibitor KNI-272 determined by high-resolution x-ray neutron crystallography. Proc Natl Acad Sci U S A 2009;106:4641–4646.
6. Zhang M, Nguyen J-T, Kumada H-O, Kimura T, Cheng M, Hayashi Y, Kiso Y. Locking the two ends of tetrapeptidic HTLV-I protease inhibitors inside the enzyme. Bioorg Med Chem 2008;16:6880–6890.
7. Chin GJ. Searching adaptive space. Science 2003;301:143.
8. Tözsér J, Zahuczky G, Bagossi P, Louis JM, Copeland TD, Oroszlan S, Harrison RW, Weber IT. Comparison of the substrate specificity of the human T-cell leukemia virus and human immunodeficiency virus proteanases. Eur J Biochem 2000;267:6287–6295.
9. Florkin M. Discovery of pepsin by Theodor Schwann. Rev Med Liege 1957;12:139–144.
10. Gettins PG. Serpin structure, mechanism, and function. Chem Rev 2002;102: 4751–47804.
11. Umezawa H, Aoyagi T, Morishima H, Matsuzaki M, Hamada M. Pepstatin, a new pepsin inhibitor produced by Actinomycetes. J Antibiot 1970;23:259–262.
12. (a) Vaidya B, Pearce SH. Management of hypothyroidism in adults. BMJ Clin Res 2008;337:a801; (b) Svensson J, Ericsson UB, Nilsson P, Olsson C, Jonsson B, Lindberg B, Ivarsson S-A. Levothyroxine treatment reduces thyroid size in children and adolescents with chronic autoimmune thyroiditis. J Clin Endocr Metab 2006;91:1729–1734.
13. Van Nieuwenhove S, Schechter PJ, Declercq J, Bone G, Burke J, Sjoerdsma A. Treatment of gambiense sleeping sickness in the Sudan with oral DFMO (DL-α-difluoromethylornithine), an inhibitor of ornithine decarboxylase; first field trial. Trans R Soc Trop Med Hyg 1985;79:692–698.
14. Green KN, Larsson SK, Beevers DG, Bevan PG, Hayes B. Alpha-methyltyrosine in the management of phaeochromocytoma. Thorax 1982;37:632–633.
15. (a) Carlsson A, Lindqvist M, Magnusson T. 3,4-Dihydroxyphenylalanine and 5-hydroxytryptophan as reserpine antagonists. Nature 1957;180:1200; (b) Abbott A. Neuroscience: the molecular wake-up call. Nature 2007;447:368–370.
16. Gilbert JA, Frederick LM, Ames MM. The aromatic-L-amino acid decarboxylase inhibitor carbidopa is selectively cytotoxic to human pulmonary carcinoid and small cell lung carcinoma cells. Clin Cancer Res 2000;6:4365–4372.
17. Shen H, Kannari K, Yamato H, Arai A, Matsunaga M. Effects of benserazide on L-DOPA-derived extracellular dopamine levels and aromatic L-amino acid decarboxylase activity in the striatum of 6-hydroxydopamine-lesioned rats. Tohoku J Exp Med 2003;199:149–159.

18. Fernandez PG, Zachariah PK, Bryant DG, Missan SS. Antihypertensive efficacy of α-methyldopa, chlorothiazide and Supres-150 (α-methyldopa-chlorothiazide). Can Med Assoc J 1980;123:284–287.

19. Goldstein DS. L-Dihydroxyphenylserine (L-DOPS): a norepinephrine prodrug. Cardiovasc Drug Rev 2006;24:189–203.

20. Tice DA, Worth MH Jr, Clauss RH, Reed GH. The inhibition of trasylol of fibrinolytic activity associated with cardiovascular operations. Surg Gynecol Obstet 1964;119:71–74.

21. (a) Royston D. Blood-sparing drugs: Aprotinin, tranexamic acid, and epsilon-aminocaproic acid. Int Anesthesiol Clin 1995;33:155–179; (b) Mannucci P. Hemostatic Drugs. N Engl J Med 1998;339:245–253.

22. Sheehan JJ, Tsirka SE. Fibrin-modifying serine proteases thrombin, tPA, and plasmin in ischemic stroke: a review. Glia 2005;50:340–350.

23. PROWESS Study Group. Efficacy and safety of recombinant human activated protein C for severe sepsis. N Engl J Med 2001;344:699–709.

24. Di Nisio M, Middeldorp S, Buller H. Direct thrombin inhibitors. N Engl J Med 2005;353:1028–1040.

25. Markwardt F. The development of hirudin as an antithrombotic drug. Thromb Res 1994;74:1–23.

26. (a) Folkers PJM, Clore GM, Driscoll PC, Dodt J, Koehler S, Gronenborn AM. Solution structure of recombinant hirudin and the Lys-47-Glu mutant: a nuclear magnetic resonance and hybrid distance geometry-dynamical simulated annealing study. Biochemistry 1989;28:2601–2617; (b) Haruyama H, Wuthrich K. Conformation of recombinant desulfatohirudin in aqueous solution determined by nuclear magnetic resonance. Biochemistry 1989;28:4301–4312.

27. Rydell TJ, Tulinsky A, Bode W, Huber R. Refined structure of the hirudin-thrombin complex. J Mol Biol 1991;221:583–601.

28. REPLACE-2 Study Group. Bivalirudin and provisional glycoprotein IIb/IIIa blockade compared with heparin and planned glycoprotein IIb/IIIa blockade during percutaneous coronary intervention: REPLACE-2 randomized trial. JAMA 2003;289:853–863.

29. Ho SJ, Brighton TA. Ximelagatran: direct thrombin inhibitor. Vasc Health Risk Manag 2006;2:49–58.

30. (a) Okamoto S, Hijikata A. Potent inhibition of thrombin by the newly synthesized arginine derivative no. 805. The importance of stereo-structure of its hydrophobic carboxamide portion. Biochem Biophys Res Commun 1981;101:440–446; (b) Okamoto S, Kinjo K, Hijikata A. Thrombin inhibitors. 1. Ester derivatives of $N^{\alpha}$-(arylsulfonyl)-L-arginine. J Med Chem 1980;23:827–830.

31. Rewinkel JBM, Adang AEP. Strategies and progress towards the ideal orally active thrombin inhibitor. Curr Pharm Design 1999;5:1043–1075.

32. Hauel NH, Nar H, Priepke H, Ries U, Stassen J-M, Wienen W. Structure-based design of novel potent nonpeptide thrombin inhibitors. J Med Chem 2002;45:1757–1766.

33. Sturzebecher J, Markwardt F, Voight B, Wagner G, Walsmann P. Cyclic amides of *N*-α-arylsulfonylaminoacylated 4-amidinophenylalanine-tight binding inhibitors of thrombin. Thromb Res 1983;29:635–642.

34. Tucker TJ, Lumma WC, Lewis SD, Gardell SJ, Lukas BJ, Baskin EP, Woltmann R, Lynch JJ, Lyle EA, Appleby SD, Chen I-W, Dancheck KB, Vacca JP. Potent noncovalent thrombin inhibitors that utilize the unique amino acid D-dicyclohexylalanine in the $P_3$ position. Implications on oral bioavailability and antithrombotic efficacy. J Med Chem 1997;40:1565–1569.
35. Bayliss W, Starling EH. The mechanism of pancreatic secretion. J Physiol Lond 1902;28:325–353.
36. Kraegen EW, Chisholm DJ, Young JD, Lazarus L. The gastrointestinal stimulus to insulin release. II. A dual action of secretin. J Clin Invest 1970;49:524–529.
37. Knudsen LB. Glucagon-like peptide-1: The basis of a new class of treatment for type 2 diabetes. J Med Chem 2004;47:4128–4134.
38. (a) Green B, Flatt P, Bailey C. Dipeptidyl peptidase IV (DPP IV) inhibitors: a newly emerging drug class for the treatment of type 2 diabetes. Diab Vasc Dis Res 2006;3:159–165; (b) Yaron A, Naider F. Proline-dependent structural and biological properties of peptides and proteins. Crit Rev Biochem Mol Biol 1993;28:31–81; (c) Lu I-L, Tsai K-C, Chiang Y-K, Jiaang W-T, Wu S-H, Mahindroo N, Chien C-H, Lee S-J, Chen X, Chao Y-S, Wu S-Y. A three-dimensional pharmacophore model for dipeptidyl peptidase IV inhibitors. Eur J Med Chem 2008;43:1603–1611.
39. (a) Chen X. Biochemical properties of recombinant prolyl dipeptidases DPP-IV and DPP8. Adv Exp Med Biol 2006;575:27–32; (b) Kuhn B, Hennig M, Mattei P. Molecular recognition of ligands in dipeptidyl peptidase IV. Curr Top Med Chem 2007;7:609–619.
40. Veken PVD, Haemers A, Augustyns K. Prolyl peptidase related to dipeptidyl peptidase IV: potential of specific inhibitors in drug discovery. Curr Top Med Chem 2007;7:621–635.
41. Peters J-U. 11 years of cyanopyrrolidines as DPP-IV inhibitors. Curr Top Med Chem 2007;7:579–595.
42. Villhauer EB, Brinkman JA, Naderi GB, Burkey BF, Dunning BE, Kapa P, Mangold BL, Russell ME, Hughes TE. 1-{[(3-Hydroxy-1-adamantyl)amino]acetyl}-2-cyano-(*S*)-pyrrolidine: A potent, selective, and orally bioavailable dipeptidyl peptidase IV inhibitor with antihyperglycemic properties. J Med Chem 2003;46:2774–2789.
43. Augeri DJ, Robl JA, Betebenner DA, Magnin DR, Khanna A, Robertson JG, Wang A, Simpkins LM, Taunk P, Huang Q, Han S-P, Abboa-Offei B, Cap M, Xin L, Tao L, Tozzo E, Welzel GE, Egan DM, Marcinkevicienne J, Chang SY, Biller SA, Kirby MS, Parker RA, Hamann LG. Discovery and preclinical profile of saxagliptin (BMS-477118): a highly potent, long-acting, orally active dipeptidyl peptidase IV inhibitor for the treatment of type 2 diabetes. J Med Chem 2005;48:5025–5037.
44. (a) Kim D, Wang L, Beconi M, Eiermann GJ, Fisher MH, He H, Hickey GJ, Kowalchick JE, Leiting B, Lyons K, Marsilio F, McCann ME, Patel RA, Petrov A, Scapin G, Patel SB, Roy RS, Wu JK, Wyvratt MJ, Zhang BB, Zhu L, Thornberry NA, Weber AE. (2*R*)-4-oxo-4-[3-(trifluoromethyl)-5,6-dihydro[1,2,4]triazolo[4,3-a]pyrazin-7(8*H*)-yl]-1-(2,4,5-trifluorophenyl)butan-2-amine: a potent, orally active dipeptidyl peptidase IV inhibitor for the treatment of type 2 diabetes. J Med Chem 2005;48:141–151; (b) Feng J, Zhang Z, Wallace MB, Stafford JA, Kaldor SW, Kassel DB, Navre M, Shi L, Skene RJ, Asakawa T, Takeuchi K, Xu R, Webb DR, Gwaltney SL II. Discovery of alogliptin: A potent, selective, bioavailable, and efficacious inhibitor of dipeptidyl peptidase IV. J Med Chem 2007;50:2297–2300; (c) Eckhardt M, Langkopf E, Mark M, Tadayyon M,

Thomas L, Nar H, Pfrengle W, Guth B, Lotz R, Sieger P, Fuchs H, Himmelsbach F. 8-(3-(*R*)-Aminopiperidin-1-yl)-7-but-2-ynyl-3-methyl-1-(4-methyl-quinazolin-2-ylmethyl)-3,7-dihydropurine-2,6-dione (BI 1356), a highly potent, selective, long-acting, and orally bioavailable DPP-4 inhibitor for the treatment of type 2 diabetes. J Med Chem 2007;50:6450–6453.

45. Ferreira SH. A bradykinin-potentiating factor, (BPF) present in the venom of *Bothrops jararaca*. Br J Pharmacol Chemother 1965;24:163–169.
46. Ferreira SH, Bartelt DC, Greene LJ. Isolation of bradykinin-potentiating peptides from *Bothrops jararaca* venom. Biochemistry 1970;9:2583–2593.
47. Stewart JM, Ferreira SH, Greene LJ. Bradykinin potentiating peptide PCA-Lys-Trp-Ala-Pro. An inhibitor of the pulmonary inactivation of bradykinin and conversion of angiotensin I to II. Biochem Pharmacol 1971;20:1557–1567.
48. (a) Ondetti MA, Williams NJ, Sabo EF, Pluscec J, Weaver ER, Kocy O. Angiotensin-converting enzyme inhibitors from the venom of *Bothrops jararaca*. Isolation, elucidation of structure, and synthesis. Biochemistry 1971;10:4033–4038; (b) Engel SL, Schaeffer TR, Gold BI, Rubin B. Inhibition of pressor effects of angiotensin I and augmentation of depressor effects of bradykinin by synthetic peptides. Proc Soc Exp Biol Med 1972;140:240–244.
49. Crantz FR, Swartz SL, Hollenberg NK, Moore TJ, Dluhy RG, Williams GH. Differences in response to the peptidyldipeptide hydrolase inhibitors SQ 20,881 and SQ 14,225 in normal-renin essential hypertension. J Am Heart Assoc 1980;2:604–609.
50. Byers LD, Wolfenden R. Binding of the by-product analog benzylsuccinic acid by carboxypeptidase A. Biochemistry 1973;12:2070–2078.
51. Ondetti MA, Rubin B, Cushman DW. Design of specific inhibitors of angiotensin-converting enzyme: new class of orally active antihypertensive agents. Science 1977;196:441–444.
52. (a) Acharya KR, Sturrock ED, Riodan JK, Ehlers MR. ACE revisited: a new target for structure-based drug design. Nat Rev 2003;2:891–902; (b) Patchett AA, Harris E, Tristram EW, Wyvratt MJ, Wu MT, Taub D, Peterson ER, Ikeler TJ, ten Broeke J, Payne LG, Ondeyka DL, Thorsett ED, Greenlee WJ, Lohr NS, Hoffsommer RD, Joshua H, Ruyle WV, Rothrock JW, Aster SD, Maycock AL, Robinson FM, Hirschmann R, Sweet CS, Ulm EH, Gross DM, Vassil TC, Stone CA. A new class of angiotensin-converting enzyme inhibitors. Nature 1980;288:280–283.
53. Natesh R, Schwager SLU, Sturrock ED, Acharya KR. Crystal structure of the human angiotensin-converting enzyme-lisinopril complex. Nature 2003;421:551–554.
54. Iizuka K, Kamijo T, Kubota T, Akahane K, Umeyama H, Kiso Y. New human renin inhibitors containing an unnatural amino acid, norstatine. J Med Chem 1988;31:704–706.
55. Iizuka K, Kamijo T, Harada H, Akahane K, Kubota T, Umeyama H, Ishida T, Kiso Y. Orally potent human renin inhibitors derived from angiotensinogen transition state: design, synthesis, and mode of interaction. J Med Chem 1990;33:2707–2714.
56. Jensen C, Herold P, Brunner HR. Aliskiren: the first renin inhibitor for clinical treatment. Nat Rev Drug Discov 2008;7:399–410.
57. Buehlmayer P, Caselli A, Fuhrer W, Goeschke R, Rasetti V, Rueger H, Stanton JL, Criscione L, Wood JM. Synthesis and biological activity of some transition-state inhibitors of human renin. J Med Chem 1988;31:1839–1846.

58. Wood JM, Criscione L, de Gasparo M, Buhlmayer P, Rueger H, Stanton JL, Jupp RA, Kay J. CGP 38560: orally active, low molecular weight renin inhibitor with high potency and specificity. J Cardiovasc Pharmacol 1989;14:221–226.
59. Rahuel J, Rasetti V, Maibaum J, Rueger H, Goschke R, Cohen N-C, Stutz S, Cumin F, Fuhrer W, Wood JM, Grütter MG. Structure-based drug design: the discovery of novel nonpeptide orally active inhibitors of human renin. Chem Biol 2000;7:493–504.
60. Rasetti V, Cohen NC, Rueger H, Goschke R, Maibaum J, Cumin F, Fuhrer W, Wood JM. Bioactive hydroxyethylene dipeptide isosteres with hydrophobic ($P_3$–$P_1$)-moieties. A novel strategy towards small nonpeptide renin inhibitors. Bioorg Med Chem Lett 1996;6:1589–1594.
61. (a) Gradman A, Schmieder R, Lins R, Nussberger J, Chiang Y, Bedigian M. Aliskiren, a novel orally effective renin inhibitor, provides dose-dependent antihypertensive efficacy and placebo-like tolerability in hypertensive patients. Circulation 2005;111:1012–1018; (b) Straessen JA, Li Y, Richart T. Oral renin inhibitors. Lancet 2006;368:1449–1456; (c) Ingelfinger JR. Aliskiren and dual therapy in type 2 diabetes mellitus. N Engl J Med 2008;358:2503–2505.
62. (a) Fleming A. On the antibacterial action of cultures of a penicillium, with special reference to their use in the isolation of *B. influenzæ*. Br J Exp Pathol 1929;10:226–236; (b) Drews J. Drug discovery: a historical perspective. Science 2000;287:1960–1964.
63. Bo G. Giuseppe Brotzu and the discovery of cephalosporins. Clin Microbiol Infect 2000;6(Suppl. 3):6–9.
64. Tahlan K, Park HU, Wong A, Beatty PH, Jensen SE. Two sets of paralogous genes encode the enzymes involved in the early stages of clavulanic acid and clavam metabolite biosynthesis in *Streptomyces clavuligerus*. Antimicrob Agents Chemother 2004;48:930–939.
65. (a) Totir MA, Helfand MS, Carey MP, Sheri A, Buynak JD, Bonomo RA, Carey PR. Sulbactam forms only minimal amounts of irreversible acrylate-enzyme with SHV-1 beta-lactamase. Biochemistry 2007;46:8980–8987; (b) The Piperacillin/Tazobactam Skin and Skin Structure Study Group. Treatment of hospitalized patients with complicated skin and skin structure infections: double-blind, randomized, multicenter study of piperacillin-tazobactam versus ticarcillin-clavulanate. Antimicrob Agents Chemother 1993;37:1580–1586.
66. Keynan S, Hooper NM, Felici A, Amicosante G, Turner AJ. The renal membrane dipeptidase (dehydropeptidase I) inhibitor, cilastatin, inhibits the bacterial metallo-beta-lactamase enzyme CphA. Antimicrob Agents Chemother 1995; 39:1629–1631.
67. Mimoto T, Imai J, Kisanuki S, Enomoto H, Hattori N, Akaji K, Kiso Y. Kynostatin (KNI)-227 and −272, highly potent anti-HIV agents: Conformationally constrained tripeptide inhibitors of HIV protease containing allophenylnorstatine. Chem Pharm Bull (Tokyo) 1992;40:2251–2253.
68. Freire E. Do enthalpy and entropy distinguish first in class from best in class? Drug Discov Today 2008;13:869–874.
69. Mimoto T, Kato R, Takaku H, Nojima S, Terashima K, Misawa S, Fukazawa T, Ueno T, Sato H, Shintani M, Kiso Y, Hayashi H. Structure-activity relationship of small-sized HIV protease inhibitors containing allophenylnorstatine. J Med Chem 1999;42:1789–1802.
70. Kempf DJ, Marsh KC, Kumar G, Rodrigues AD, Denissen JF, McDonald E, Kukulka MJ, Hsu A, Granneman GR, Baroldi PA, Sun E, Pizzuti D, Plattner JJ, Norbeck DW,

Leonard JM. Pharmacokinetic enhancement of inhibitors of the human immunodeficiency virus protease by coadministration with ritonavir. Antimicrob Agents Chemother 1997;41:654–660.

71. Sham HL, Kempf DJ, Molla A, Marsh KC, Kumar GN, Chen C-M, Kati W, Stewart K, Lal R, Hsu A, Betebenner D, Korneyeva M, Vasavanonda S, McDonald E, Saldivar A, Wideburg N, Chen X, Niu P, Park C, Jayanti V, Grabowski B, Granneman GR, Sun E, Japour AJ, Leonard JM, Plattner JJ, Norbeck DW. ABT-378, a highly potent inhibitor of the human immunodeficiency virus protease. Antimicrob Agents Chemother 1998;42:3218–3224.

72. Ohno Y, Kiso Y, Kobayashi Y. Solution conformations of KNI-272, a tripeptide HIV protease inhibitor designed on the basis of substrate transition state: determined by NMR spectroscopy and simulated annealing calculations. Bioorg Med Chem 1996;4:1565–1572.

73. Kiso Y, Yamaguchi S, Matsumoto H, Mimoto T, Kato R, Nojima S, Takaku H, Fukazawa T, Kimura T, Akaji K. KNI-577, a potent small-sized HIV protease inhibitor based on the dipeptide containing the hydroxymethylcarbonyl isostere as an ideal transition-state mimic. Arch Pharm Pharm Med Chem 1998;331:87–89.

74. Pai VB, Nahata MC. Nelfinavir mesylate: a protease inhibitor. Ann Pharmacother 1999;33:325–339.

75. Yoshimura K, Kato R, Yusa K, Kavlick MF, Maroun V, Nguyen A, Mimoto T, Ueno T, Shintani M, Falloon J, Masur H, Hayashi H, Erickson J, Mitsuya H. JE-2147: A dipeptide protease inhibitor (PI) that potently inhibits multi-PI-resistant HIV-1. Proc Natl Acad Sci U S A 1999;96:8675–8680.

76. Turner SR, Strohlbach JW, Tommasi RA, Aristoff PA, Johnson PD, Skulnick HI, Dolak LA, Seest EP, Tomich PK, Bohanon MJ, Horng M-M, Lynn JC, Chong K-T, Hinshaw RR, Watenpaugh KD, Janakiraman MN, Thaisrivongs S. Tipranavir (PNU-140690): a potent, orally bioavailable nonpeptidic HIV protease inhibitor of the 5,6-dihydro-4-hydroxy-2-pyrone sulfonamide class. J Med Chem 1998;41:3467–3476.

77. (a) Yamashima T, Kohda Y, Tsuchiya K, Ueno T, Yamashita J, Yoshioka T, Kominami E. Inhibition of ischaemic hippocampal neuronal death in primates with cathepsin B inhibitor CA-074: a novel strategy for neuroprotection based on 'calpaincathepsin hypothesis'. Eur J Neurosci 1998;10:1723–1733; (b) Hill PA, Buttle DJ, Jones SJ, Boyde A, Murata M, Reynolds JJ, Meikle MC. Inhibition of bone resorption by selective inactivators of cysteine proteinases. J Cell Biochem 1994;56:118–130.

78. Gauthier JY, Chauret N, Cromlish W, Desmarais S, Duong LT, Falgueyret JP, Kimmel DB, Lamontagne S, Leger S, LeRiche T, Li CS, Masse F, McKay DJ, Nicoll-Griffith DA, Oballa RM, Palmer JT, Percival MD, Riendeau D, Robichaud J, Rodan GA, Rodan SB, Seto C, Therien M, Truong V-L, Venuti MC, Wesolowski G, Young RN, Zamboni R, Black WC. The discovery of odanacatib (MK-0822), a selective inhibitor of cathepsin K. Bioorg Med Chem Lett 2008;18:923–928.

79. (a) Yasuma T, Oi S, Choh N, Nomura T, Furuyama N, Nishimura A, Fujisawa Y, Sohda T. Synthesis of peptide aldehyde derivatives as selective inhibitors of human cathepsin L and their inhibitory effect on bone resorption. J Med Chem 1998;41:4301–4308; (b) Navab R, Mort JS, Brodt P. Inhibition of carcinoma cell invasion and liver metastases formation by the cysteine proteinase inhibitor E-64. Clin Exp Metast 1997;15:121–129.

80. Iqbal M, Messina PA, Freed B, Das M, Chatterjee S, Tripathy R, Tao M, Josef KA, Dembofsky B, Dunn D, Griffith E, Siman R, Senadhi S, Biazzo W, Bozyczko-Coyne D,

Meyer SL, Ator MA, Bihovsky R. Subsite requirements for peptide aldehyde inhibitors of human calpain I. Bioorg Med Chem Lett 1997;7:539–544.

81. (a) Ku G, Faust T, Lauffer LL, Livingston DJ, Harding MW. Interleukin-1 beta converting enzyme inhibition blocks progression of type II collagen-induced arthritis in mice. Cytokine 1996;8:377–386; (b) Norman J, Yang J, Fink G, Carter G, Ku G, Denham W, Livingston D. Severity and mortality of experimental pancreatitis are dependent on interleukin-1 converting enzyme (ICE). J Interferon Cytokine Res 1997;17:113–118.

82. Karanewsky DS, Bai X, Linton SD, Krebs JF, Wu J, Pham B, Tomaselli KJ. Conformationally constrained inhibitors of caspase-1 (interleukin-1 beta converting enzyme) and of the human CED-3 homologue caspase-3 (CPP32, apopain). Bioorg Med Chem Lett 1998;8:2757–2762.

83. Nguyen J-T, Yamani A, Kiso Y. Views on amyloid hypothesis and secretase inhibitors for treating Alzheimer's disease: progress and problems. Curr Pharm Design 2006;12:4295–4312.

84. (a) Kimura T, Shuto D, Hamada Y, Igawa N, Kasai S, Liu P, Hidaka K, Hamada T, Hayashi Y, Kiso Y. Design and synthesis of highly active Alzheimer's beta-secretase (BACE1) inhibitors, KMI-420 and KMI-429, with enhanced chemical stability. Bioorg Med Chem Lett 2005;15:211–215; (b) Asai M, Hattori C, Iwata N, Saido TC, Sasagawa N, Szabo B, Hashimoto Y, Maruyama K, Tamura S-I, Kiso Y, Ishiura S. J Neurochem. The novel beta-secretase inhibitor KMI-429 reduces amyloid beta peptide production in amyloid precursor protein transgenic and wild-type mice. 2006;96:533–540.

85. Ebina M, Futai E, Tanabe C, Sasagaw N, Kiso Y, Ishiura S. Inhibition of KMI-574 leads to dislocalization of BACE1 from lipid rafts. J Neurosci Res 2009;87:360–368.

86. Hamada Y, Ohta H, Miyamoto N, Yamaguchi R, Yamani A, Hidaka K, Kimura T, Saito K, Hayashi Y, Ishiura S, Kiso Y. BACE1 Inhibitor Peptides: Can an infinitely small $k_{cat}$ value turn the substrate of an enzyme into its inhibitor? Bioorg Med Chem Lett 2008;18:1654–1658.

87. Siemers ER, Quinn JF, Kaye J, Farlow MR, Porsteinsson A, Tariot P, Zoulnouni P, Galvin JE, Holtzman DM, Knopman DS, Satterwhite J, Gonzales C, Dean RA, May PC. Effects of a gamma-secretase inhibitor in a randomized study of patients with Alzheimer disease. Neurology 2006;66:602–604.

88. (a) Luisetti M, Sturani C, Sella D, Madonini E, Galavotti V, Bruno G, Peona V, Kucich U, Dagnino G, Rosenbloom J, Starcher B, Grassi C. MR889, a neutrophil elastase inhibitor, in patients with chronic obstructive pulmonary disease: a double-blind, randomized, placebo-controlled clinical trial. Eur Respir J 1996;9:1482–1486; (b) Cregge RJ, Durham SL, Farr RA, Gallion SL, Hare CM, Hoffman RV, Janusz MJ, Kim HO, Koehl JR, Mehdi S, Metz WA, Peet NP, Pelton JT, Schreuder HA, Sunder S, Tardif C. Inhibition of human neutrophil elastase. 4. Design, synthesis, X-ray crystallographic analysis, and structure-activity relationships for a series of $P_2$-modified, orally active peptidyl pentafluoroethyl ketones. J Med Chem 1998;41:2461–2480; (c) Macdonald SJ, Belton DJ, Buckley DM, Spooner JE, Anson MS, Harrison LA, Mills K, Upton RJ, Dowle MD, Smith RA, Molloy CR, Risley C. Syntheses of trans-5-oxo-hexahydro-pyrrolo[3,2-b]pyrroles and trans-5-oxo-hexahydrofuro[3,2-b]pyrroles (pyrrolidine trans-lactams and trans-lactones): New pharmacophores for elastase inhibition. J Med Chem 1998;41:3919–3922; (d) Anderson GP, Shinagawa K. Neutrophil elastase inhibitors as treatments for emphysema and chronic bronchitis. Curr Opin Anti-inflam Immun Invest Drugs 1999;1:29–38.

89. Seife C. Blunting nature's Swiss army knife [news]. Science 1997;277:1602–1603.

90. Whittaker M, Brown P. Recent advances in matrix metalloproteinase inhibitor research and development. Curr Opin Drug Discov Dev 1998;1:157–164.

91. (a) O'Connell JE, Jardine AG, Davidson G, Connell JM. Candoxatril, an orally active neutral endopeptidase inhibitor, raises plasma atrial natriuretic factor and is natriuretic in essential hypertension. J Hypertens 1992;10:271–277; (b) Elsner D, Muntze A, Kromer EP, Riegger GA. Effectiveness of endopeptidase inhibition (candoxatril) in congestive heart failure. Am J Cardiol 1992;70:494–498.

92. (a) Marie C, Mossiat C, Lecomte JM, Bcralet J. Acute natriuretic effect of fasidotrilat, a mixed inhibitor of neutral endopeptidase and angiotensin I-converting enzyme, in rats with heart failure. Pharmacology 1998;56:291–296; (b) Fournié-Zaluski MC, Gonzalez W, Turcaud S, Pham I, Roques BP, Michel JB. Dual inhibition of angiotensin-converting enzyme and netural endopeptidase by the orally active inhibitor mixanpril: a potential therapeutic approach in hypertension. Proc Natl Acad Sci U S A 1994;91:4072–4076; (c) Wallis EJ, Ramsay LE, Hettiarachchi J. Combined inhibition of neutral endopeptidase and angiotensin-converting enzyme by sampatrilat in essential hypertension. Clin Pharmacol Ther 1998;64:439–449.

93. Venugopal J. Pharmacological modulation of the natriuretic peptide system. Expert Opin Ther Pat 2003;13:1389–1409.

94. Newton RC, Decicco CP. Therapeutic potential and strategies for inhibiting tumor necrosis factor-alpha. J Med Chem 1999;42:2295–2314.

95. Gruber A, Speth C, Lukasser-Vogl E, Zangerle R, Borg-von Zepelin M, Dierich MP, Wurzner R. Human immunodeficiency virus type 1 protease inhibitor attenuates *Candida albicans* virulence properties in vitro. Immunopharmacology 1999;41:227–234.

96. Abad-Zapatero C, Goldman R, Muchmore SW, Hutchins C, Stewart K, Navaza J, Payne CD, Ray TL. Structure of a secreted aspartic protease from *C. albicans* complexed with a potent inhibitor: Implications for the design of antifungal agents. Protein Sci 1996;5:640–652.

97. (a) Njoroge FG, Chen KX, Shih NY, Piwinski JJ. Challenges in modern drug discovery: A case study of boceprevir, an HCV protease inhibitor for the treatment of hepatitis C virus infection. Acc Chem Res 2008;41:50–59; (b) Revill P, Serradell N, Bolos J, Rosa E. Telaprevir. Drug Future 2007;32:788.

98. Nezami A, Kimura T, Hidaka K, Kiso A, Liu J, Kiso Y, Goldberg DE, Freire E. High-affinity inhibition of a family of *Plasmodium falciparum* proteases by a designed adaptive inhibitor. Biochemistry 2003;42:8459–8464.

99. Patick AK, Binford SL, Brothers MA, Jackson RL, Ford CE, Diem MD, Maldonado F, Dragovich PS, Zhou R, Prins TJ, Fuhrman SA, Meador JW, Zalman LS, Matthews DA, Worland ST. In vitro antiviral activity of AG7088, a potent inhibitor of human rhinovirus 3C protease. Antimicrob Agents Chemother 1999;43:2444–2450.

100. Sydnes MO, Hayashi Y, Sharma VK, Hamada T, Bacha U, Barrila J, Freire E, Kiso Y. Synthesis of glutamic acid and glutamine peptides possessing a trifluoromethyl ketone group as SARS-CoV 3CL protease inhibitors. Tetrahedron 2006;62:8601–8609.

# 6

# DISCOVERY OF PEPTIDE DRUGS FROM NATURAL SOURCES

SÓNIA T. HENRIQUES

*Institute for Molecular Bioscience The University of Queensland, Brisbane, Queensland, Australia*

DAVID J. CRAIK

*Institute for Molecular Bioscience The University of Queensland, Brisbane, Queensland, Australia*

## 6.1 INTRODUCTION

Drug discovery strategies are remarkably guided by nature. All organisms screen their environment and produce compounds that provide them with an evolutionary advantage [1, 2]. Thus, natural compounds are often the starting point for the design of drugs with high selectivity and potency [1–4]. Peptides are expressed in all living species and display a large diversity of structures and biological effects. Therefore it is not surprising that natural occurring peptides are attractive drug leads and are making their way into clinical applications [5–15]. A vast number of active peptides have been isolated and characterized from a broad variety of biological sources. Peptides involved in host defense and prey capture are among the best drug candidates, due to their fast-acting protection/capture mechanism. Organisms that produce host-defense peptides with potential applications in drug development include prokaryotes, plants, and animals, and we begin this article with brief descriptions of a few examples.

Bacteria are a rich source of peptides with potential pharmaceutical applications. Both Gram-negative and Gram-positive bacteria produce antimicrobial

*Peptide Chemistry and Drug Design*, First Edition. Edited by Ben M. Dunn.

peptides (AMPs), with one particularly important class being bacteriocins [16], which target closely related bacteria and are nontoxic to mammals. Bacteriocins are attractive candidates both as antimicrobial agents for the treatment of human and animal infections, and as food preservatives [16–18]. Nisin is the only bacteriocin approved by the FDA, and with activity against food-borne pathogens, has been applied in a broad range of commercial products, including dairy products, bakery products, vegetables, meat, and fish [19]. Nisin also has promising applications for the treatment of *Helicobacter* infections, ulcers [20] and intestinal colonization by *Enterococci* [21] and more recently has been suggested to have potential therapeutic anti-tumorigenic properties [22]. The emergence of vancomycin-resistant *Enterococci* has led to the need for alternative therapies to traditional antibiotics, and nisin has entered preclinical trials [23]. Other bacteriocins have attracted attention for the treatment of diarrheagenic bacterial contamination [24] and as spermicidal agents [25].

Plants also produce peptides to defend themselves against pathogen attack [26–28]. We focus here on a group of plant peptides that, as well as having defense properties [29, 30], have topological properties that make them particularly stable and hence suitable as framework in drug design [31–34]. These peptides have a head-to-tail cyclised peptidic backbone and are referred to as cyclotides [35]. Cyclotide-containing plants were first used for medicinal purposes in the Congo region of Africa [36, 37]. It was later determined that they incorporate a unique knotted macrocyclic structure that confers them with great stability relative to conventional linear proteins [35]. Cyclotides have various biological activities with pharmacological relevance, including toxicity against cancer cells [38] and anti-HIV properties [39–41]. Overall, cyclotides are fascinating peptides that have the chemical constitution of proteins but the stability properties of organic chemicals, thus making them useful drug leads. Cyclotides will be explored in more detail later in this chapter, revealing what plants have to offer to the drug design field.

Venomous organisms are spread throughout the animal kingdom and include reptiles, fishes, amphibians, mammals, mollusks, arachnids, and insects. In any niche there is a competition for resources, and the use of venom for prey capturing, or as a defense mechanism, represents a successful adaptative trait [42]. Venoms are typically produced as deadly cocktails, comprising mixtures of peptides adapted by natural selection. These toxins disrupt cardiovascular and neuromuscular systems by disturbing the activity of critical enzymes, receptors, and ion channels. Venom toxins have a high degree of target specificity and they have been used increasingly as pharmacological tools and leads in drug development [42–45].

Amphibians secrete peptides with antimicrobial properties from their skin as part of their defense system [46–48]. The magainins are of particular interest as they have potent antimicrobial activity, with little or no hemolytic activity [49], and they represent early examples of peptides that were considered to have great potential as drugs due to their specificity and broad antibacterial spectrum. Indeed, pexiganan (MSI-78), a 22-residues magainin analog, entered clinical trials as a topical agent for the treatment of foot ulcers [50]. Phase III trials revealed that it was efficacious, but the judgment was made that it was no more effective than existing treatments for diabetic foot ulcers and thus the FDA required more tests [51]. Nevertheless, the interest in magainin stimulated searches for other antibiotics, and peptides with a range of antimicrobial [46, 52, 53], anticancer [54], and antiviral activities [55–57] have now been isolated from amphibian skin.

Peptides with potential therapeutic applications have been found in the venom of a range of other animals, including cone snails, spiders, scorpions, and snakes [58]. Such peptides are particularly abundant in cone snails and due to their small size and suitability for synthesis these peptides, called conotoxins, are valuable drug leads [45, 58]. The genus *Conus* is a large group of carnivorous predators found in tropical marine habitats, and although each *Conus* species is a highly specialized predator, collectively, cone snails have a remarkably broad spectrum of prey. All members of this *genus* use their venom, which contains numerous (100–1000) toxic peptides, for prey capture [59]. They bind to a diverse range of sodium, calcium and potassium channels, membrane receptors and transporters, leading to efficient immobilization of the prey [60, 61].

Conotoxins have great diversity and specificity, and each peptide targets a specific receptor protein. With their ability to discriminate between different isoforms of the same receptor, these peptides are valuable pharmacological probes as well as potential leads in drug design [62]. In fact, a conotoxin extracted from *Conus magus*, is an example of the development of a toxin into an approved drug. Conotoxin MVIIA, which has the generic name ziconotide [63], and the trade name of Prialt®, was approved by the FDA in 2004 and is used as a treatment for chronic pain [64, 65]. This is a relatively rare example of a peptide used without further modification. Several other conotoxins are currently being evaluated in clinical and preclinical trials (e.g., ω-conotoxin CVID (AM336) and a derivative of the χ-conotoxin MrIA (Xen 2174)) [66]. Overall, cone snail venoms contain a huge reservoir of compounds that can be regarded as a combinatorial library of drug leads.

Having introduced a few examples of peptide drug leads, we briefly overview the drug development process before examining particular classes of peptides in more detail. There are a number of key steps in the development of a drug, as shown in Figure 6.1. The first consideration is to satisfy an unmet medical need [3] and so selection of a drug target is usually the starting point for drug development [67]. Other steps include the choice of natural sources containing promising active compounds; the screening of large numbers of compounds [68]; identification and isolation of the most active peptides; characterization of primary structure using sequencing techniques or genomics for gene determination; and three-dimensional (3D) structure determination [9].

Knowledge obtained from peptide structure characterization allows leads to be optimized via medicinal chemistry. Substitution analysis and chemical modification are used to improve stability and activity. Cost of production, stability, selectivity, delivery, and mechanism of action need to be considered [58]. Peptides are particularly amenable to modifications to confer improved selectivity, potency, and stability [51], while maintaining bioactivity [69]. In addition, for a drug candidate to be suitable for clinical testing, it must have sufficient bioavailability and distribution within

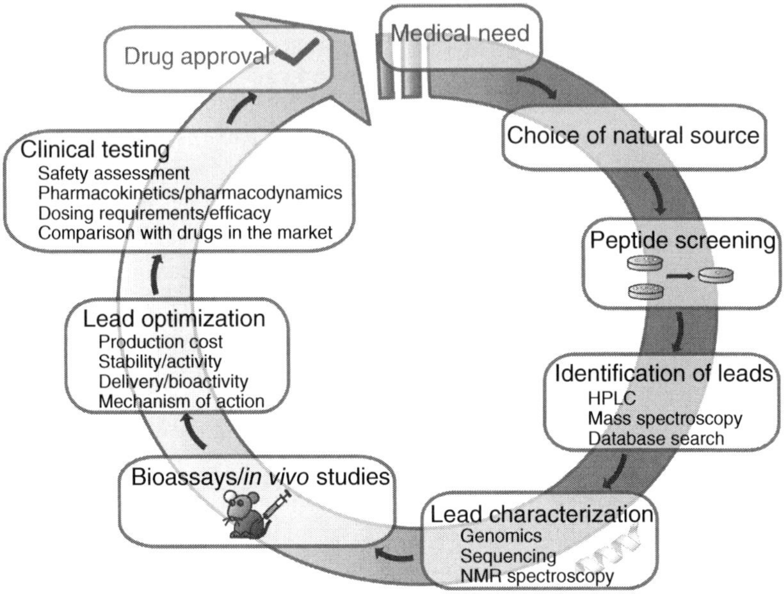


**Figure 6.1** Overview of the drug-development pathway for peptide-based drug design. (*See insert for color representation of this figure.*)

the body to reach the target [67]. Drug candidates must also pass toxicity evaluation in animal models [67].

The aim of this chapter is to illustrate that peptides isolated from natural sources have exciting potential as drugs. Examples from various bacterial, plant and animal sources will be described to highlight the large diversity of chemical structures, modes of action and biological applications of peptide. Strategies to overcome possible drawbacks associated with the application of peptides as drugs will be discussed.

## 6.2 PEPTIDES ARE INVOLVED IN THE HOST DEFENSE MECHANISM OF LIVING ORGANISMS

Living organisms are constantly exposed to multiple harmful microbes and their capacity to overcome infection is essential for survival [70]. AMPs are products of the long-term evolution of host defense mechanisms [71], and virtually all organisms, including plants [27], animals [46, 72], and bacteria [73] produce AMPs as a component of their innate immune system [74–77]. They act as endogenous antibiotics, inducing the direct destruction of microorganisms. Owing to their ability to attack different microorganisms, including bacteria, fungi [78], viruses [10, 79], and even tumor cells [8], together with the growing problem of resistance to conventional antibiotics, AMPs are of much interest for the development of novel human therapeutics.

On the basis of structural homology, two main families of eukaryotic AMPs can be described: cationic AMPs and anionic AMPs [80]. The cationic group is the largest and is discussed in more detail in this chapter, as the anionic group is considered to have lower activity [81]. Prokaryotic AMPs, commonly referred to as bacteriocins, can be broadly divided into two groups: bacteriocins produced by Gram-negative bacteria and bacteriocins produced by Gram-positive bacteria [16]. To highlight the diversity of these peptides, we have chosen examples based not only on their therapeutic value, but also on the novelty of their structures and modes of action.

### 6.2.1 Cationic AMPs from Eukaryotes, Peptides that Target the Membrane

The diversity and number of cationic AMPs discovered is very large. Online databases dedicated to the rapidly increasing number of peptides of class are available; for instance, the Antimicrobial Peptide Database (APD) (http://aps.unmc.edu/AP/main.php), which categorizes AMPs according to their biological activities (e.g., anticancer, antiviral, antifungal, and antibacterial) [82] and the Antimicrobial Sequences Database (AMSDb) (http://www.bbcm.univ.trieste.it/~tossi/amsdb.html), which categorizes AMPs on basis of their secondary structure, are both excellent resources.

***6.2.1.1 Diversity of Structure*** Cationic AMPs have both highly heterogeneous amino acid sequences and substantial variation in their secondary structure [83]. The group includes peptides having linear (noncross-linked) structures and forming α-helical structures as well as Cys-rich peptides with disulfide bonds and β-sheet structures [23]. The α-helical peptides are particularly abundant in the extracellular fluids of insects and amphibians and frequently exist as unstructured monomers in solution, becoming helical upon interaction with phospholipid membranes [84–86]. The β-sheet peptides have a diverse range of primary structures and often possess Cys residues in disulfide bonds, as is demonstrated for example by the defensin family [87].

Cationic AMPs have many potential clinical applications and some examples are given in Table 6.1, which shows the diversity of sequences, structures, and source organisms. For instance, LL-37, an α-helical AMP belonging to cathelicin family and expressed in humans [88], has applications for the treatment of lung microbial infections [89], which are common in cystic fibrosis patients. With high stability and activity under a variety of conditions [90–92], this peptide is more resistant to proteolytic degradation than other α-helical AMPs [93]. Conventional antibiotics are largely ineffective against cystic fibrosis due to elevated salt concentrations that inhibit the usual antibacterial defenses in the lung [89].

As mentioned earlier, magainin 2 is among the most extensively studied AMPs. However, the use of amphibian skin as a source of antibiotics is not restricted to this example, and several other peptide families with antimicrobial activity have been identified, including temporins [94], bombinins [95], and bombinins H [96]. In contrast to the majority of natural AMPs, which contain high numbers of positively charged amino acids, temporins, and bombinins are short and mildly cationic peptides. Because of their small size and potent activities, they are of particular interest for drug design.

**TABLE 6.1 Sources, Sequences, Structures, and Possible Applications of a Selection of Cationic Antimicrobial Peptides Involved in the Eukaryotic Innate Immune System.**

| Peptide (Source) | Sequence[a] | Structure[b,c] | Potential Applications |
|---|---|---|---|
| LL-37 (*Homo sapiens*) | 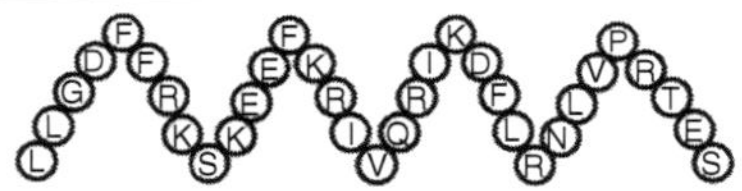 |  | Treatment of lung microbial infection |
| Bombinin H4 (*Bombina variegata*) | 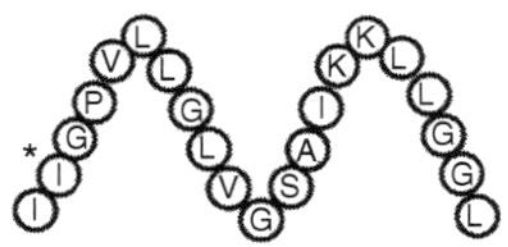 | 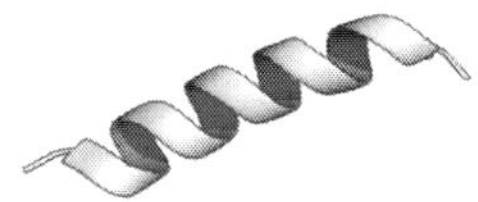 | Anti-*Leishmanial* agent |
| Temporin A (*Rana Temporaria*) |   |  | General microbicide |
| θ-Defensin RTD1 (*Macaca mulatta*) | 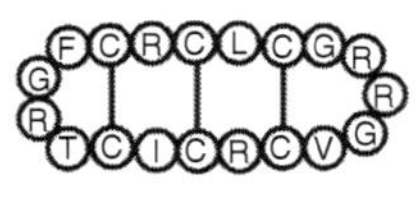 |  | Anti-HIV-1 activity |

[a] *Denotes D-amino acid in position 2 in the bombinin H4 sequence, the disulfide connectivity between cysteine residues is shown by the solid black lines in the q-Defensin RTD-1 sequence.

[b] The disulfide bonds are shown in black.

[c] Protein data base (PDB) ID codes: LL-37 (2k6o); bombinin H4 (2ap8); temporin A (no PDB code available, modeled structure shown); θ-Defensin RTD-1 (1hvz).

Bombinins and bombinins H are isolated from the skin secretions of frogs from the genus *Bombina*. The bombinin family comprises 20–27 residues peptides with activity against both Gram-negative bacteria and Gram-positive bacteria [95, 97]. These peptides generally adopt a random coil structure an aqueous environments whereas in an apolar environment they have an amphipathic α-helix structure [97]. Although these peptides are antimicrobial, they show no cytolytic activity against mammalian cells [95, 97]. In contrast, bombinins H have both hemolytic and antimicrobial activities. They are termed bombinins H as they are more hydrophobic than bombinins. Peptides belonging to this family contain 17–20 residues, and D-amino acids are found in some peptides in this class, increasing their stability [96]. Like bombinins, these peptides typically adopt a random coil structure in an aqueous environment and an amphipathic α-helix in apolar environments [98, 99]. The presence of D-amino acids results from a post-translational modification involving a L–D isomerization [100], with these peptides displaying better antimicrobial activity against some bacterial strains than the pure L-isomer [98]. D-amino acid substitution thus represents an approach developed by nature to modulate not only the solubility of a peptide [101], but also biostability by protection from proteolytic degradation [98]. These peptides have potent activity against *Leishmania* by rapid perturbation of the plasma membrane, and are of interest for the development of new drugs against this global infectious disease [102].

Temporins were first identified in the frog *Rana temporaria* [94] and are the shortest α-helical peptides isolated from amphibians (10–14 residues). They tend to form an amphipathic α-helical structure in hydrophobic environments, and have a net charge of 0 to +3 [103]. They are active against a wide range of pathogens (bacteria, viruses, fungi, yeasts, and protozoa) [94, 104–106] and are not toxic to mammalian cells at concentrations that kill microbes [94]. An exception is temporin L, which is highly active on bacteria, erythrocytes, and cancer cells [107]. Their mode of action seems to involve perturbation of the cytoplasmic membrane. However, this occurs in a different way to that proposed for the majority of cationic α-helical AMPs, as temporins bind and permeate both *zwitterionic* and *anionic* phospholipid bilayers [108]. Peptides belonging to temporin family have attractive properties, such as high activity in physiological conditions, high stability in serum [106], and low cost synthesis due to their short amino acid sequence [108], making them exciting peptides for drug design applications.

Defensins are cysteine-rich peptides that participate in the host defense of mammals [109], insects [110], and plants [111]. They are characterized by intramolecular disulfide bonds that stabilize the structures, and frequently contain small β-sheet structures [87]. Some members have an N-terminal α-helical structure [109]. Although defensins have been isolated from many species, the α- and β-defensins of human origin are the best studied. In the α-defensins, the cysteines are paired with a 1–6, 2–4, and 3–5 configuration, whereas in the β-defensins the pairing is 1–5, 2–4, and 3–6 [72]. Many potential therapeutic applications have been suggested for defensins due to their activity against Gram-positive and Gram-negative bacteria, fungi, viruses, and cancer cells [109]. Recently, another group of defensins, namely the θ-defensins, which has an unusual intramolecular head-to-tail ligation has been discovered [112, 113], and their potent activity against HIV-1 has attracted much interest [114, 115].

***6.2.1.2 Cell Membranes, the Main Target of AMPs*** Positively charged amino acids, hydrophobic residues [83], and the potential to adopt amphipathic structures (e.g., structures with separate hydrophobic and hydrophilic faces) are properties commonly found in cationic AMPs [116]. Cationic and amphiphilic features of AMPs are crucial for their insertion into the membrane [116], as their ability to inactivate microbes is intimately related to membrane targeting [117–119]. Fundamental differences exist between microbial and mammalian cells, including membrane composition and architecture, transmembrane potential and polarization, and structural features, including the presence of a cell wall [83]. Bacterial membranes contain substantial amounts of negatively charged phospholipids, such as phosphatidylgercerol and cardiolipin, on the external leaflet. Furthermore, the microbial wall is enriched with either the anionic lipopolyschacaride (LPS) or the anionic peptidoglycan in Gram-negative bacteria or Gram-positive bacteria, respectively. In contrast, the outer membrane layer of eukaryotic cells is composed mainly of phosphatidylcholine, sphingomyelin and cholesterol, all of which are neutral at physiological pH [83].

Because of their cationic nature, AMPs favor negatively charged bacterial membranes over neutral eukaryotic membranes [120], explaining the varying degrees of selective toxicity among distinct AMPs [83]. Generally, a net positive charge enables the accumulation of peptides at anionic microbial surfaces, enriched with anionic LPS in Gram-negative bacteria or with peptidoglycan in Gram-positive bacteria. These peptides make contact with the anionic outer layer of bacterial cytoplasmic membranes, while the hydrophobic domain favors insertion in the membrane [121]. AMPs exert their activity by either membrane permeabilization through pore formation, or translocation across the membrane to gain access to cell's interior to attack internal targets [118, 122]. Several disruption models have been proposed, including pore formation by a barrel-stave pore [123], a toroidal pore [124], or a carpet model [125]. Each of these mechanisms depends on the electrostatic and hydrophobic properties of both the peptide and membrane [80, 126]. When the activity is associated with a cytoplasmic target, the peptide might translocate without membrane permeabilization [127, 128].

***6.2.1.3 Applications of Cationic AMPs as Alternatives to Conventional Antibiotics*** The effectiveness of antibiotics has become limited due to an increase in bacterial resistance [129], and the lack of discovery of new molecules with low host toxicity and a broad spectrum of pathogen activity [68, 130]. Thus, AMPs or their derivatives are being increasingly evaluated as a new generation of antibiotics. They are considered in some cases to have the possibility of synergistic effects with conventional antibiotics [131]. Cationic AMPs mainly target bacterial cell membranes leading to cell lysis and death without affecting host cells, whereas traditional antibiotics specifically target protein receptors. Because AMPs can disrupt membranes in a nonspecific way, they are potentially less susceptible to the development of bacterial resistance than conventional antibiotics [83].

Some cationic AMPs are potent against cancer cells as well as bacteria [132], but not toward normal mammalian cells [133]. The altered membrane composition of cancer cells appears to be a major factor in this selectivity [133]. Typically, cancer cells have a greater negative charge due to higher expression of anionic phospholipids such as phosphatidylgercerol [134], and have a more negative transmembrane potential [135] and greater membrane fluidity [136]. These factors increase the potential applications of AMPs and emphasize the importance of membrane composition for their activity.

Several structural features have been identified as important for the antimicrobial activity of peptides. These include size, amino acid sequence, net charge, hydrophobicity, and amphipathicity [137, 138]. However, an unambiguous identification of how AMPs distinguish between pathogen and host cells and indeed between Gram-positive and Gram-negative bacteria, and among different bacterial strains, holds the promise of designing novel anti-infective agents with greater selective toxicity [83]. Although electrostatic attraction increases peptide concentrations at the membrane surface, disruption of the membrane depends on hydrophobic interactions between peptide and membrane [137–139]. Therefore, subtle properties of peptides are crucial in determining the extent of insertion and disruption of membrane integrity [126, 138, 140]. Neither the role of membrane composition nor the structural features of peptides required for specificity are, as yet, fully understood [107] and predicting antimicrobial or cytotoxic activity from a given amino acid sequence is not an easy task.

Interestingly, many plants and animals synthesize a large number of peptides with relatively minor differences in sequence and structure between them. Combinations of structurally related peptides can increase the spectrum of antimicrobial activity by inducing changes in the biophysical properties of the peptides [141]. A single mutation can dramatically alter the biological activity of a peptide, and the large diversity of AMPs reflects the adaptation of each species to the unique microbial environments that characterize their niches [51].

With diverse modes of action, structures, sequences and sources (see Table 6.1), AMPs have significant advantages over traditional antibiotics and bacteria find it hard to circumvent their action [142]. Therefore, natural AMPs are likely to contribute to the future development of new peptide-based anti-infective therapeutics to overcome resistance of microbes to commonly used drugs [141, 143]. They provide good templates for rational drug improvement, putting into practice the information on protein structure and protein–lipid interactions gathered over natural evolution and laboratory research [83].

### 6.2.2 Peptides and the Host defense in bacteria – Bacteriocins

The production of antimicrobial compounds is a strategy employed by bacteria to survive competitive conditions. Bacteriocins thus typically target closely related bacteria found in the same nutritional niche as the producer organism [144]. Bacteriocins of Gram-negative bacteria are either smaller than 10 kDa (microcins) or larger than 20 kDa (colicins). They are usually released through cell lysis and are often dependent on host regulatory pathways [16, 145]. Gram-positive bacteria usually produce bacteriocins smaller than 8 kDa, which resemble many of the AMPs produced by eukaryotes, such as defensins. They are generally cationic, amphiphilic, and membrane-permeabilizing peptides [146]. Due to the large number and diversity in structure and function of Gram-positive bacteriocins, they have been further subdivided and although different classification systems have been proposed [5, 16, 147–151], a consensus has not yet been reached. Two bacteriocins, microcin J25 and subtilosin A, from Gram-negative bacteria and Gram-positive bacteria, respectively, were chosen here to illustrate applications (Table 6.2).

***6.2.2.1 Microcin J25, a Lariat Peptide Found in Bacteria*** Microcins are small peptides produced by diverse strains of *Enterobacteriaceae*, mostly *Escherichia coli* [152]. Their production is stimulated in nutritionally limited media and they are actively secreted into the extracellular medium [145]. They are thermostable, resistant to extreme pH and some proteases, and are relatively hydrophobic [145]. Their structures are diverse and range from linear, unmodified peptides, to structures having extensive post-translational modifications [153]. Like their chemical structures, their biological applications vary widely [154] and this diversity has encouraged their use in the design of new-generation drugs for cancer [155, 156] and for infectious diseases [156, 157].

Microcin J25 (MccJ25) is plasmid-encoded, ribosomally synthesized and was first isolated from *E. coli* AY25 [158]. It is active against some pathogenic *E. coli*, *Salmonella* and *Shigella* species [158] and has been studied extensively due to its unusual structural features. MccJ25 is active at extremes of pH (from pH 2 to 12) and also after exposure to temperatures as high as 120 °C [158]. Initially MccJ25 was thought to be a macrocyclic peptide with a head-to-tail cyclization [159]. However, further inspection showed that it instead incorporates a sidechain-to-backbone cycle that sequesters the N terminus, but also protects the C-terminus via a threading mechanism. It contains an eight-residues cyclic segment, resulting from the formation of an internal lactam bond between the α-amino group of Gly1 and the γ-carboxyl group of Glu8, followed by a 13-residues linear segment that loops back and threads through the cyclic segment [160–162]. The tail is sterically entrapped within the ring due to the bulky side chains of Phe19 and Tyr20 (see Table 6.2), making this peptide resistant to denaturing conditions [162, 163]. Despite this unusual structure, only a small number of residues are essential for MccJ25 function and many residues can be substituted [164]. MccJ25 production and release increases when cells reach stationary phase and nutrients become limiting [158, 165] and occurs both under aerobic and anaerobic conditions [158], independently of pH [165], giving MccJ25-producing cells an advantage over non-producers.

MccJ25 appears to have two intracellular targets: (i) RNA polymerase (RNAP) [166, 167] and (ii) the respiratory chain [168, 169]. Bacterial RNAP is the central enzyme of gene expression [170] and is an attractive target for drug design [171]. The observation that RNA synthesis is inhibited by MccJ25 in a dose-dependent manner first prompted the suggestion that RNAP was the cellular target of MccJ25 [166]. Yuzenkova et al. [170] hypothesized that MccJ25 inhibits transcription by binding to

**TABLE 6.2 Sources, Sequences, Structures, and Possible Applications of the Bacteriocins Microcin J25 and Subtilosin A.**

| Peptide (Source) | Sequence[a] | Structure[b,c] | Potential Applications |
|---|---|---|---|
| Microcin J25 (*Escherichia coli*) | 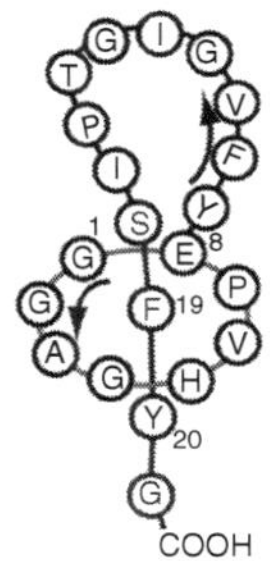 | 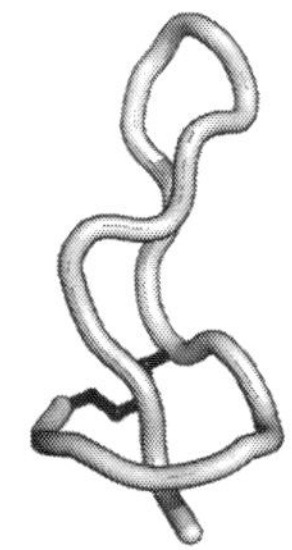 | Antibacterial food preservative |
| Subtilosin A (*Bacillus subtillus*) | 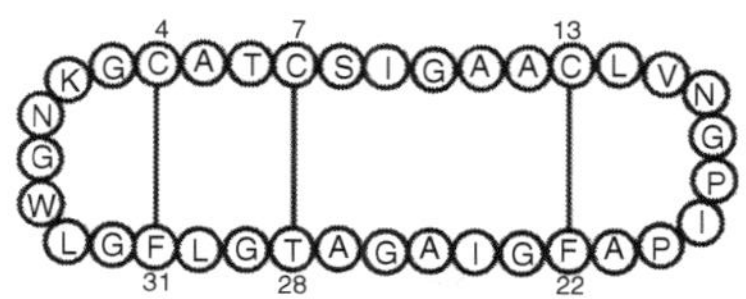 | 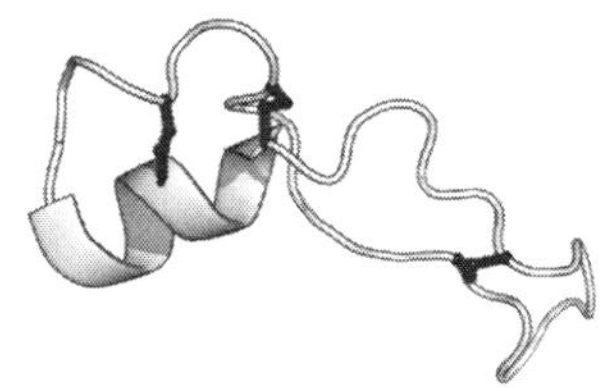 | Antibacterial spermicidal agent |

[a]The thioether linkages in the subtilosin A sequence are indicated by the solid black lines.
[b]The internal lactam bond in microcin J25 and the thioether linkages in subtilosin A are shown in black.
[c]PDB ID codes: microcin J25 (1q71); subtilosin A (1px9).

RNAP, thus blocking the access to RNAP catalytic center. This hypothesis was supported by molecular modeling [172] and kinetic analysis of the transcription process in the presence of MccJ25 [173].

MccJ25 also apparently targets the respiratory chain, as shown from a study of *E. coli* MccJ25-resistant-RNAP strains [168]. The growth of this *E. coli* strain is inhibited in the presence of MccJ25 in aerobic conditions, but not in anaerobic conditions. A significant inhibition of oxygen consumption and increase in reactive oxygen species when MccJ25 is present seems to be the reason, while in anaerobic conditions MccJ25 lost the antibiotic effect [168]. In this alternative mechanism MccJ25 uptake is required to attack intracellular targets affecting oxygen consumption, suggesting that peptide-membrane interactions and MccJ25 uptake are determinants for the mechanism [168]. This suggestion is supported by the fact that MccJ25 can interact with artificial model membranes, leading to permeabilization of the bilayer structure [174]. The ability to modulate cytoplasmic membrane permeability, and subsequent depolarization was further confirmed *in vivo* with *Salmonella newport* with a consequent inhibition of oxygen consumption [175] and also on rat heart mitochondria [176]. Peptide insertion, permeability, electrical potential dissipation, and inhibition of the respiratory chain was reported [176]. In addition, the outer-membrane receptor FhuA-dependent TonB-pathway and the inner-membrane SbmA transporter seem to be responsible for the uptake of the MccJ25 into the cytoplasm [177]. Overall, the dual independent mechanisms of action of MccJ25 help explain the successful action of the intriguing antibiotic peptide.

MccJ25 has several advantageous properties over other peptides from a drug design perspective. It is resistant to extreme pH and to high temperatures [158]; it is resistant to most endoproteases [159]; it is active against *E. coli*, *Salmonella* and *Shigella* strains with clinical relevance [158]; it is active in blood, even after 24 h of incubation [157]; it has no hemolytic activity [157]; and it displays a prolonged systemic antimicrobial activity [157]. Together, these properties suggest that MccJ25 has potential applications not only as a food preservative [24], but also as a human therapeutic agent [171] and further encourages the potential application of this molecule for systemic administration and treatment of otherwise antibiotic-resistant infections [157].

The fact that MccJ25 has a relatively narrow antimicrobial spectrum, affecting only Gram-negative bacteria, might limit broader application. Its inability to attack a broad range of strains seems to be related to an inability to cross membranes in a nonspecific way [171]. Therefore, modulation of its molecular properties to overcome bacterial membrane impermeability could improve its antimicrobial spectrum. The observation that only a few residues of MccJ25 are essential for its structure and activity [164] suggests that it should be possible to construct MccJ25 derivatives with a higher potency and/or broader specificities.

#### *6.2.2.2 Subtilosin A, a Lantibiotic with Unusual Post-translational Modifications*

Lantibiotics are small, gene-encoded and post-translationally modified antibiotic peptides that possess thioether cross-links called lanthionine and/or β-methyllanthionine linkages [178]. They are generally active against most Gram-positive bacteria and often target lipid II, the precursor of peptidoglycan in the bacterial cell wall. Nisin is the most widely used example, and its mode of action and applications have been reviewed [179–181], so we focus here on another lantibiotic that has attracted much attention recently.

Subtilosin A is a cyclic lantibiotic, originally identified in the soil bacterium *Bacillus subtillus*, which shows activity against a diverse range of Gram-negative [182] and Gram-positive bacteria [183]. It is ribosomally generated and highly post-translationally modified [182, 184]. Subtilosin A was first reported to have a cyclic structure [182] but it soon became evident that it also had several other post-translational modifications [185]. The initially proposed sequence [182] and structure [184] were revised and three unusual linkages involving Cys4, Cys7, and Cys13 with Phe31, Thr28, and Phe22, respectively, were identified [185]. Specifically, thioether linkages were found between the sulfur of each Cys and the α-carbon of Phe or Thr [186] (Table 6.2). Presumably, these features and the head-to-tail cyclic backbone give subtilosin A significant rigidity, which could restrict significant conformational changes upon target binding.

Although originally identified in *B. subtillus* [182] subtilosin A production has been confirmed in a range of related subspecies as well as the closely related species *Bacillus atrophaeus* [187] and *Bacillus amyloliquefaciens* [188]. It is expressed at the end of exponential growth, particularly under conditions of stress [182]. The sulfide bridges are necessary for the antimicrobial activity of subtilosin A [189]. In contrast with many antibacterial peptides, subtilosin A is nearly insoluble in aqueous solution, due to a high content of hydrophobic amino acids and a net anionic charge. This suggests that a surface receptor is probably the main target [189], as the anionic charge does not favor electrostatic interaction with bacterial membranes, whereas the sulfide bridges might have a role in holding the conformation to target the receptor [189]. Heat stress increases the effectiveness of subtilosin A against Gram-negative bacteria [183] and its unusual bridged and cyclic structure makes it stable up to 100 °C [185].

Thennarasu et al. [190] reported that subtilosin A is capable of binding to lipid bilayers, adopting an orientation where the Trp-containing domain becomes buried in the hydrophobic core of the membrane. However, NMR studies suggested that, neither nonlamellar lipid phase formation, or micellization of lipid bilayers were evident at concentrations at which other antimicrobial permeabilizer peptides showed membrane destabilization [190]. Thus, interaction with membranes might only be a supplementary role to enhance the interaction with a receptor, rather than a mechanism of action *per se* [190]. These findings support a receptor-dependent mechanism for subtilosin A.

Subtilosin A has potent activity against *Listeria monocytogenes* [184, 187], making it an attractive preservative in the food industry. It also has activity against vaginal pathogens such as *Gardnerella vaginalis* and *Sterptococcus agalactiae* [188], and also has spermicidal activity against boar, bovine, horse, rat [191], and human spermatozoa [25], with no cytotoxicity to human tissues or healthy human microbiota [25]. These results suggest that subtilosin A has potential application as a general spermicidal agent [25].

***6.2.2.3 Applications of Bacteriocins*** Bacteriocins are considered ideal candidates for food preservation and personal care applications because the range of their activity is limited only to species closely related to the producing species. Therefore, while they might target a specific pathogen, they theoretically have no harmful effects on humans. The fact that bacteriocins are active against food-borne pathogens (e.g., *Listeria monocytigenes*, *Staphylococcus aureus*, and *Bacillus cereus*), together with the approval of nisin as food preservative [19], has stimulated the use of bacteriocins for various applications in animal and human health [192, 193]. However, it is necessary to fully understand the biology of bacteriocins to elucidate their structure-function relationships, production, immunity, regulation and mode of action [5]. Attempts to create improved bioengineered derivatives of bacteriocins have been successful and increased solubility, stability [194, 195] and/or improvement of antimicrobial activity against some bacterial strains [196, 197] have been reported.

### 6.2.3 Cyclotides, Ultra-Stable Peptides that are Part of Plant Defense Mechanism

Cyclotides were first isolated from the African plant *Oldenlandia affinis* after they were recognized as an active component of an indigenous medicinal tea [198] used to accelerate childbirth. This use highlighted the fact that cyclotides are resistant to high temperatures and are apparently orally available [198]. Subsequently, it was found that cyclotides possess a knotted macrocyclic structure that confers exceptional stability on them [35], and additional members of the cyclotide family were identified. A complete list of cyclotides reported so far can be found in CyBase (www.cybase.org.au), an online database dedicated to circular peptides [199]. Cyclotides are approximately 30 residues in size, and occur in plants from the Rubiaceae (coffee), Violaceae (violet) Cucurbitaceae (squash), Solanaceae (nightshade), and Fabaceae (legume) families [32].

***6.2.3.1 Structure*** The structure of the prototypic cyclotide, kalata B1 (kB1) was first reported in 1995 [200] and is shown in Table 6.3. Three-dimensional structures have been determined using NMR spectroscopy for approximately 2 dozen cyclotides and reveal a consensus structural motif termed the cyclic cystine knot (CCK) [35]. More than 300 cyclotides are now known [199] and it has been estimated that the family could grow to 50,000 members [201]. Although the six Cys residues that make up the cystine knot are completely conserved, there is a significant degree of variation in the rest of the sequence. The cyclotides have six loops between the six Cys residues, with loops 1 and 4 having more sequence conservation than the others [202]. They typically contain a surface-exposed patch of hydrophobic residues which is believed to promote their ability to bind to membranes [203, 204].

Originally, cyclotides were subdivided in two major subfamilies, namely the Möbius and bracelet subfamilies (see Table 6.3), so called because of the presence or absence of a 180° twist in the circular backbone, respectively. Cyclotides belonging to the Möbius subfamily have a Pro residue in loop 5 that is preceded by a *cis*-peptide bond, which is responsible for the conceptual backbone twist.

**TABLE 6.3 Sources, Sequences, Structures, and Possible applications of Cyclotides Belonging to Different Subfamilies.**

| Peptide (Source) | Sequence[a] | Structure[b,c] | Potential Applications |
|---|---|---|---|
| *Möbius subfamily*<br>Kalata B1 (*Oldelandia affinis*) | 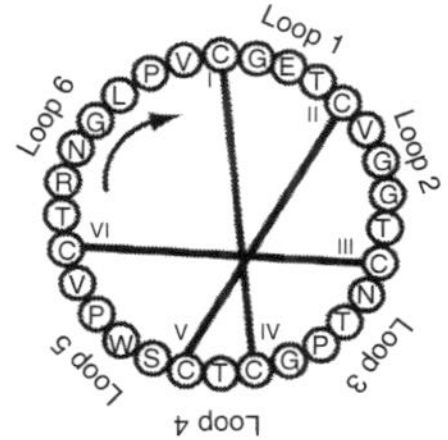  | 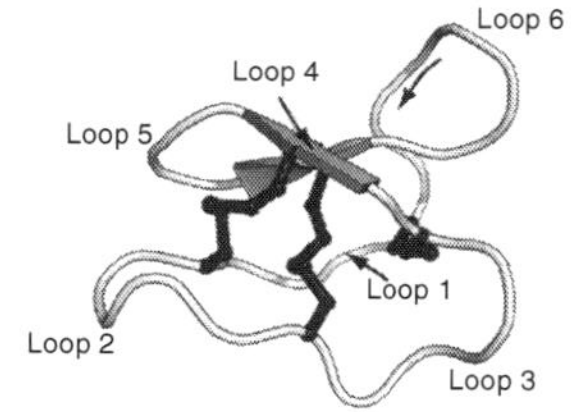  | Insecticidal<br>Grafting framework for membrane and intracellular targets |
| *Bracelet subfamily*<br>Cycloviolacin O1 (*Viola odorata*) | 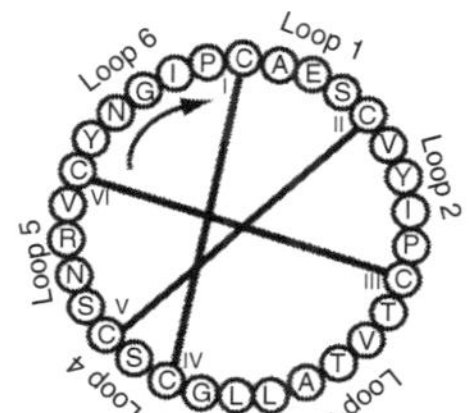  | 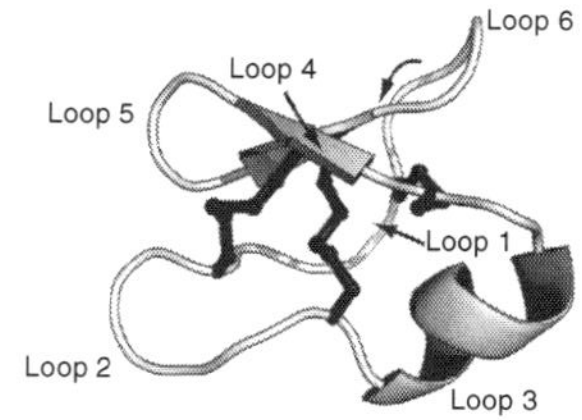  | Pesticide<br>Grafting framework for extracellular targets |
| *Trypsin inhibitors*<br>MCoTI-II (*Mormodica cochinchinensis*) | 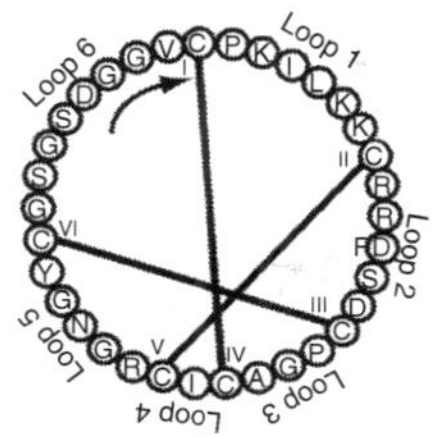  | 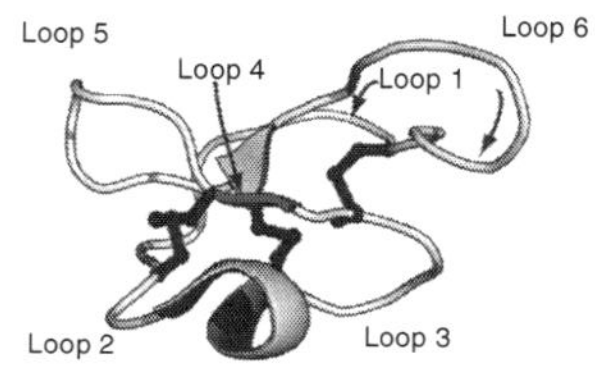  | Grafting framework for intracellular targets |

[a]The disulfide connectivity between cysteine residues is indicated by the solid black lines.
[b]The disulfide bonds are shown in black.
[c]PDB ID codes: kalata B1 (1nb1); cycloviolacin O1 (1nbj); MCoTI-II (1ib9).

In the bracelet subfamily, all backbone peptide bonds are in the *trans* configuration [35]. More recently, a third subfamily, the trypsin inhibitor cyclotides, has been described [205], which currently contains only a handful of members found in seeds of *Momordica cochinchinensis* from the Cucurbitaceae family [206, 207]. Related, but acyclic cystine knot peptides are also found in *Momordica charantia* [208]. Apart from a similar 3D structure and the six Cys residues, the *Momordica* peptides show very little similarity to the Möbius and bracelet subfamilies [206, 209] (e.g., see MCoTI-II sequence and structure in Table 6.3).

***6.2.3.2 Biological Activity and Mode of Action*** The natural biological role of cyclotides has not been unequivocably elucidated, but their high level of expression in leaves [37], together with the presence of a large number of different cyclotides in any given plant [210, 211], suggest that they probably have a role in plant defense. This is supported by their insecticidal activity [29]. Kalata peptides dramatically affect development of larvae from Lepidopteran species, *Helicoverpa punctigera* [29] and *Helicoverpa armigera* [212] and damage to microvilli cells in their midguts by membrane disruption was observed after cyclotide ingestion [213]. In addition to the defense function, many other activities have been identified for cyclotides, including uterotonic [198], anticancer [38], hemolytic [214, 215], anti-fouling [216], anti-HIV [39–41], and molluscicidal [217] activities.

Subtle changes in sequence have been shown to have a significant influence on biological activity and on the ability of cyclotides to bind to model membranes. For instance, Ala- [218] and Lys-scans [219] of kB1 have shown that specific point mutations abolish insecticidal, anti-HIV and hemolytic activities. Interestingly, the residues that seem to be important for these activities are surface-exposed and colocalize on one side of the molecule [218, 219]. Recently, these residues have also been shown to be important to specifically target phosphatidylethanolamine (PE)-phospholipids and/or facilitate insertion into the hydrophobic core of the lipid bilayer. A strong correlation between bioactivity and affinity for cell membranes with PE-phospholipids was found [220]. A mechanism in which kB1 targets cell membranes through specific interaction with PE-headgroups, followed by insertion into the membrane hydrophobic core and disruption of the cell membrane was proposed [203, 220].

Similar membrane binding properties have also been observed for other native cyclotides belonging to the Möbius and bracelet subfamilies [204]. In summary, despite considerable sequence diversity, all the tested native cyclotides have a conserved bioactive face and hydrophobic patch. The bioactive patch is believed to be involved in specifically recognizing PE-phospholipids, whereas the hydrophobic patch is required for insertion into the hydrophobic region of the cell membrane and cell permeabilization [203, 204].

***6.2.3.3 Applications in Drug Design*** In general, the utility of conventional peptides in pharmaceutical and biotechnological applications is limited by their poor stability and bioavailability but cyclotides have chemical and structural properties that make them capable of overcoming these disadvantages. In particular, cyclotides are exceptionally resistant to thermal and chemical denaturation and to enzymatic degradation [221], making them a valuable scaffold for drug design applications [31, 32, 34]. The discovery and characterization of an increasing number of cyclotide sequences has highlighted the plasticity of the CCK framework (i.e., its tolerance to sequence substitutions) [202], and thus its amenability to molecular grafting studies. More than a dozen different peptide sequences have been successfully grafted into the CCK scaffold and shown to retain full biological activity [32]. Some examples include angiogenic epitopes with potential cardiovascular and wound healing applications [222], a melanocortin antagonist sequence with the potential to treat obesity [223], and a bradykinin receptor antagonist for the treatment of inflammatory pain [224]. In these examples the engineered cyclotides target extracellular/membrane proteins. Significantly, in the case of the bradykinin antagonist, the grafted peptide was orally active, marking a major milestone in the development of cyclotide-based drug leads.

Importantly from a drug design perspective, MCoTI-I/II [225, 226], and kB1 [227] have recently been shown to be able to internalize into cells. The amenability of cyclotides toward grafting, together with their remarkable stability and cell-penetrating properties make them a very attractive framework for the delivery of peptide epitopes to intracellular targets. Indeed, the potential of engineered cyclotides to specifically inhibit an intracellular target was recently demonstrated by Camarero and coworkers [228]. In that study, PMI, a 12-amino acid sequence known to block the interaction of p53 with the oncoprotein Hdm2, was engineered into MCoTI-I and shown to modulate the p53 tumor suppressor pathway *in vivo*, while retaining the stability of the scaffold [228].

In conclusion, cyclotides have a range of drug design applications, including their use as a template for the insertion (e.g., "grafting") of biologically active epitopes into their stable molecular framework to prolong the biological half-life of the epitopes that target either extracellular or intracellular targets.

## 6.3 ANIMAL VENOMS, A RICH SOURCE OF PEPTIDES WITH THERAPEUTIC POTENTIAL

Venomous animals, including snakes, scorpions, spiders, and cone snails contain a wealth of highly selective and potent toxins in their venoms. These venoms have evolved for rapid prey immobilization and target vital physiological processes. Of the venom peptides characterized to date, a significant number are also highly selective for mammalian receptors associated with pain [58]. Peptides extracted from venoms are thus excellent sources of lead compounds for the development of therapeutic agents. Cone snail toxins have been particularly extensively studied for pharmaceutical applications [7].

### 6.3.1 Conotoxins, a Naturally Occurring Combinatorial Peptide Library

Peptides from cone snail venom, referred to as conopeptides, are divided into two classes: those with one or no disulfide bonds, and those with two or more disulfide bonds, with the latter commonly called conotoxins [43, 61, 66, 229, 230]. Conotoxins are small (10–30 residues) peptides and specifically target various components of neural transmission pathways. Examples are shown in Table 6.4.

Conotoxin genes encode precursor proteins, from which the mature conotoxins are processed. Conotoxins are frequently post-translationally modified [231], with the modifications including C-terminal amidation, proline hydroxylation, *O*-glycosylation, glutamine γ-carboxylation, and N-terminal glutaminyl ring closure to pyroglutamate [232], further enhancing the sequence diversity of these peptides. The precise role of post-translational modifications is not yet known, but the large chemical diversity resulting from these modifications enlarges both the variability of conotoxins and their biological specificity and/or functional efficacy [233].

The nomenclature employed for conotoxin classification, as originally proposed by Cruz et al. [234], categorizes them into superfamilies based on signal sequence and framework homology and then into pharmacological families based on the targets they interact with. With the increasing number of sequences reported, this classification is constantly expanding [61, 235, 236], and the latest update has been published recently [237]. An online database dedicated to conopeptides, ConoServer (www.conoserver.org), is also available [238].

The most extensively studied pharmacological families of conotoxins are the α-conotoxins, which target nicotinic acetylcholine receptors (nAChRs) [59]; the ω-conotoxins, which target presynaptic calcium channels [239, 240]; the μ-conotoxins, which target muscle sodium channels [241]; the δ-conotoxins, which delay inactivation of sodium channels [242] and the κ-conotoxins, which target Shaker potassium channels [243]. Table 6.4 shows examples of these conotoxins and their pharmacological applications.

***6.3.1.1 Structure*** Because of their small size, NMR has been particularly valuable for determination of the 3D structures of conotoxins [244]. NMR not only has the ability to study the structure and dynamics of peptides under varying conditions of temperature and pH, but is also suitable for determining structures of post-translationally modified peptides [245]. So far, more than 120 NMR structures of conotoxins have been solved [238], with most having rigid conformations stabilized by multiple disulfide cross-links [13, 246–250] (see Table 6.4).

The sequence variability of conotoxins is reflected in their 3D diversity, with their structures including a range of well-defined secondary structural elements, such as β-sheets, α-helices, and β-turn motifs. Besides these classical secondary structure motifs, more complex motifs, such as the cystine-stabilized *αβ* (CS*αβ*) motif occur. This motif comprises an α-helix and a β-structure connected by two or three disulfide bridges [251] and is seen for example in the μ-conotoxin GIIIB [252]. We mention it specifically because this motif contains most of the secondary structural elements found in protein structures and it has been proposed as a scaffold for protein engineering [253], not only due to its suitability for chemical synthesis, but also due to its high stability and tolerance to sequence mutations [253]. Another important motif identified in conotoxins is the cystine knot, similar to that observed for cyclotides. This motif occurs in conotoxins belonging to the ω-conotoxin family, such as MVIIA

**TABLE 6.4 Sources, Sequences, Structures, and Possible Applications of Conotoxins Belonging to Different Pharmacological Families.**

| Peptide (Source) | Sequence[a] | Structure[b,c] | Potential Applications |
| --- | --- | --- | --- |
| α-Conotoxin Vc1.1. (*Conus victoriae*) | 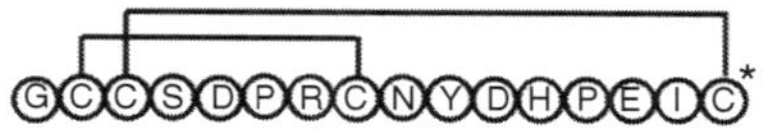 | 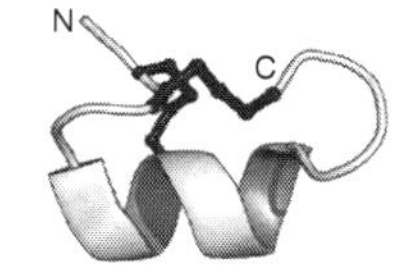 | Analgesic for neuropathic pain |
| m-Conotoxin GIIIB (*Conus geographus*) | 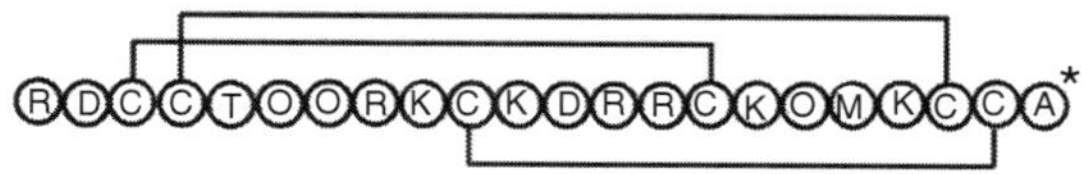 | 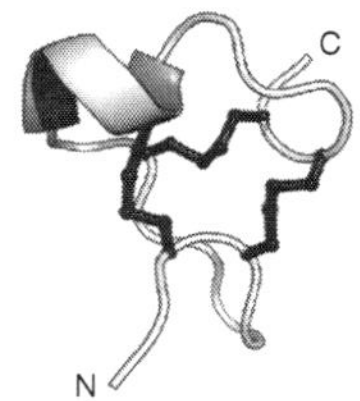 | Scaffold for protein engineering |
| w-Conotoxin MVIA (*Conus magus*) | 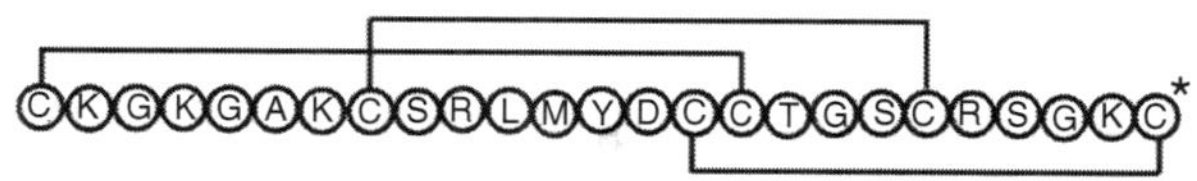 | 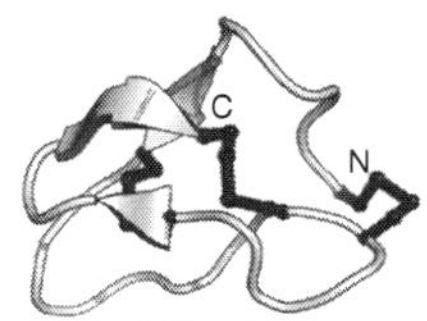 | Analgesic for neuropathic pain |

(*continued*)

**TABLE 6.4 (Continued)**

| Peptide (Source) | Sequence[a] | Structure[b,c] | Potential Applications |
|---|---|---|---|
| k-Conotoxin PVIIA *(Conus purpurascens)* | 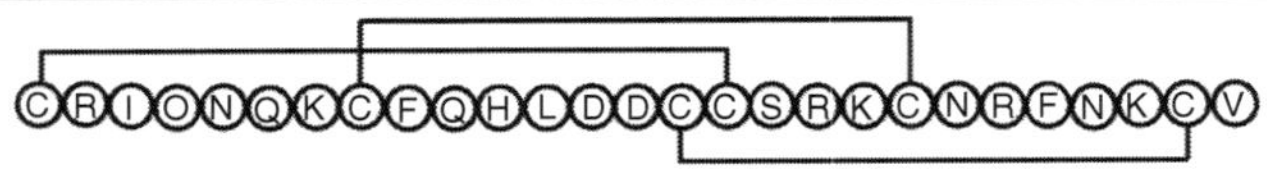 | 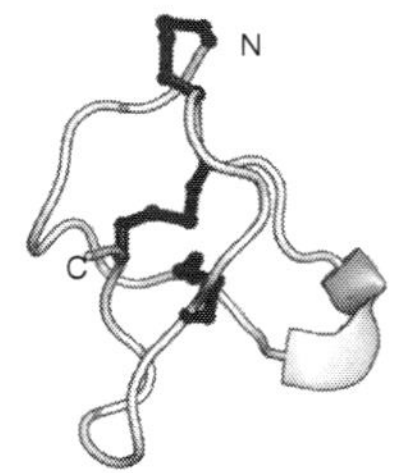 | Cardioprotective activity |
| d-Conotoxin TxVIA *(Conus textile)* | 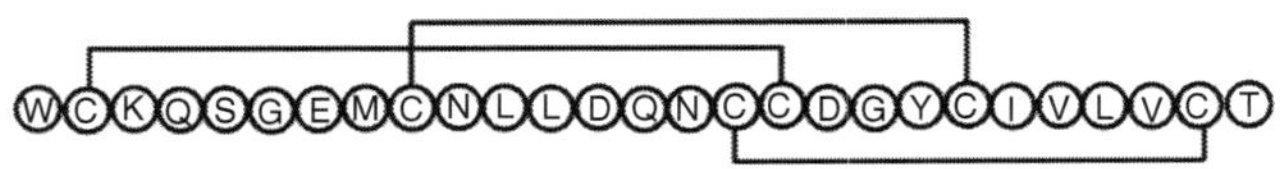 | 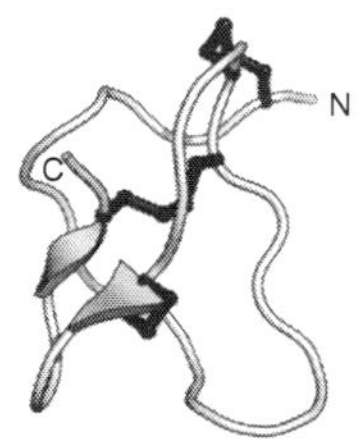 | Pharmacological probe |

[a]The disulfide connectivity between cysteine residues is indicated by the solid black lines in the sequences. * Denotes amidated C terminus and O denotes hydroxyproline.
[b]The N and C termini are indicated in each structure and the disulfide bonds are shown in black.
[c]PDB ID codes: VC1.1 (2h8s); GIIIB (1gib); MVIA (1mvi); PVIIA (1av3); TxVIA (1fu3).

[254, 255], TxVII [256], and GVIA [248, 257–259], conferring great stability to these peptides.

***6.3.1.2 Mode of Action*** The many conotoxins expressed in a single cone snail species are thought to act in a synergistic fashion to immobilize prey [61]. Their targets include voltage-sensitive potassium, calcium and sodium channels and *N*-methyl-D-aspartate, glutamate, vasoperessin, serotonin, and acetylcholine receptors [60]. An assembly of conotoxins acting together to a specific end point has been termed a "toxin cabal" [61]. The lightning strike cabal is responsible for the instantaneous immobilization of the prey, causing a massive depolarization of the axons near the venom injection site and includes peptides that inhibit voltage-gated sodium channels and peptides that block potassium channels. The motor cabal induces a block of neuromuscular transmission, an effect more slowly achieved because it acts at sites far from the venom injection site [236, 260]; this cabal includes peptides that target nAChRs and block skeletal muscle sodium channels and peptides that target presynaptic calcium channels, which control neurotransmitter release [61, 260].

To further illustrate the specificity of conotoxins, the mechanism of action of α-conotoxins is described here in more detail. The α-conotoxins interact with neuronal and muscle type nAChRs [235]. These receptors are pentameric ligand-gated ion channels, which have varying subunit compositions and this combinatorial diversity results in receptor subtypes with distinct pharmacological and physiological properties [261]. The high selectivity of the α-conotoxins for different nAChR subtypes arises from subtle differences in the sequences of α-conotoxins [262, 263]. They can be regarded as essentially rigid frameworks that bind to their receptors without significant variation of their conformations [264], but variations in amino acids displayed on their surface determine their receptor selectivity [262].

The α-conotoxins are divided into different subfamilies: $\alpha 3/5$; $\alpha 4/3$; $\alpha 4/6$; and $\alpha 4/7$, depending on the number of amino acids between the second and third Cys residues (loop 1) and the third and fourth Cys residus (loop 2) (see Table 6.4), respectively [265]. Besides the four Cys residues, the α-conotoxins have a Ser and a Pro conserved in loop 1, which are thought to have a role in maintenance of secondary structure [266]. A significant degree of sequence variation occurs in the remaining residues, particularly in loop 2, which seems to confer the selectivity and potency of different α-conotoxins for different nAChRs subtypes [261, 266].

***6.3.1.3 Applications in Drug Design*** One of the most exciting aspects of conotoxins is their pharmacological potential; every conotoxin can be regarded as a specialist, optimized for a particular target, showing extraordinary potency and an ability to differentiate between related synaptic components [267]. Due to their small size, conotoxins are convenient for chemical synthesis [12, 43], making them attractive leads in drug design programs. Furthermore, the diversity of conotoxins arising from hypermutation can be compared with combinatory libraries used by pharmaceutical companies when searching for new drug leads. The introduction of modified functional groups in conotoxins, accomplished through post-translational modifications, parallels medicinal chemistry approaches used to optimize initial lead compounds [236].

An example that illustrates the pharmacological value of conotoxins is the clinical use of ω-conotoxin MVIIA, Prialt, for the treatment of chronic pain [64, 65]. Besides applications as pain killers, conotoxins have other pharmacological applications [267]. For instance, κ-conotoxin PVIIA, has been evaluated for its cardioprotective effects [268]. The role of nAChRs in the pathophysiology of schizophrenia [269], Alzheimer's disease [270], and Parkinson's disease [271] suggests that α-conotoxins might have potential for the treatment of these neuropsychiatric disorders [60].

Notwithstanding these favorable features, the application of conotoxins as drugs potentially suffers from the generic drawbacks of other peptides *in vivo*, including poor absorption, susceptibility to proteolysis and a short half-life. Therefore, stabilizing conotoxins for therapeutic or diagnostic applications and for improving their route of delivery are of interest [272]. The ω-conotoxin MVIIA is delivered intrathecally, (e.g., direct injection into the spinal cord) for example, and next generation therapeutics would benefit from a less invasive delivery route, ideally via the oral route. The stabilization of peptides to achieve broader therapeutic value is addressed in the following section.

## 6.4 OPTIMIZATION OF PEPTIDES FOR DRUG DEVELOPMENT

Although purified natural peptides can become the final drug without modification (e.g., ω-conotoxin MVIIA sold as Prialt), the initial lead is often subject to extensive modification via medicinal chemistry before entering clinical trials. Natural product leads often suffer from deficiencies, such as low stability and poor bioavailability, which compromise their broader application. They can potentially be further improved, in terms of efficacy and selectivity for the target, or achieving optimal pharmacokinetic and pharmacodynamic properties [3].

### 6.4.1 Chemical Modifications to Improve Activity

Many factors have to be taken in consideration during the drug development process, including the chemical properties (e.g., the molecule should be stable and easy to synthesize), the pharmacological properties (e.g., selective high-affinity binding and potent functional effects at the target receptor), the pharmacokinetics (adequate bioavailability for the selected route of administration and adequate half-life of biodistribution), and the safety and toxicity of the compound [67]. As we described for natural conotoxins, the post-translational modification of peptides is an efficient strategy for regulating peptide localization, function and turnover, and influences physicochemical properties, solubility, stability, aggregation, propensity to be degraded by protease activity, and specificity of peptides [273]. In a similar way, pharmaceutical companies modify drug leads as a strategy to improve their properties. Some examples of chemical modifications to improve peptide properties and their value as therapeutics are discussed below.

***6.4.1.1 Amino Acid Substitution*** One of the first methods adopted to improve the specificity and stability of peptides is amino acid substitution. Some amino acids are more susceptible to degradation than others and replacement of these amino acids can increase the stability of peptides [272]. For instance, Met is sensitive to oxidation [274], Asn is susceptible to deamination, and Asp is prone to isomerisation [275]. Trypsin and chymotrypsin in the human gastrointestinal tract have the potential to decrease the bioavailability of peptide-based therapeutics by causing proteolysis. Peptide bonds following Lys or Arg are cleaved by trypsin [276, 277], whereas chymotrypsin cleaves at hydrophobic residues such as Phe, Tyr, and Trp [277]. Therefore, modification of the primary structure of peptide drug lead to minimize reactivity is an important consideration in the design of peptide therapeutics.

Alternatively, amino acid substitution is frequently employed to enhance affinity for receptors by alteration of amino acids involved in binding interactions [278]. The introduction of positively charged amino acids in AMPs can lead to more active peptides by improving affinity for membranes [68, 279], whereas substitutions with less hydrophobic residues, frequently decreases their hemolytic activity [279].

The cost of production is important in pharmaceutical development and a residue modification strategy is one way that can be used to reduce the cost of synthesis. For example, substitution of γ-carboxyl glutamic acid, common in conotoxins, with an unmodified glutamic acid, often does not induce a loss of activity but substantially decreases production costs [272]. However, it is important to consider that altering amino acids can sometimes influence the conformation of peptides, which can impact on their stability and binding properties. Thus, substitutions should be done to ensure that no loss of biological activity or undesirable side effects occurs.

***6.4.1.2 Nonnatural Amino Acids Analogs*** The use of nonnatural amino acids is a strategy to provide favorable biophysical properties, such as protease resistance and improved stability. D-amino acids occur naturally in some peptides, including for example the bombinins H. These peptides have better stability [96] and higher antimicrobial activity against some bacterial strains [98] than their all L-analogs. In this case a single D-amino acid substitution was an approach developed by nature to modulate not only the solubility [101] but also the biostability of a peptide [98]. The use of D-amino acids has also been adapted by the pharmaceutical industry and is now common in peptide-based drug design [83, 280]. Another possible strategy is the incorporation of β-amino acids, which also generally increases resistance to enzymatic degradation [281] while maintaining a stable secondary structure [282], and the functional properties of the natural peptide [281].

Modification of peptide termini by *N*-acetylation, *C*-amidation and PEGylation (polyethylene-glycol modification at C-terminus) are other strategies used to stabilize peptides. Capping by *N*-acetylation or *C*-amidation reduces susceptibility to carboxy-peptidases, improving the stability of natural peptides [283, 284]. PEGylation has been used as a modification technology to improve pharmacokinetic and pharmacodynamic properties [285].

***6.4.1.3 Cyclization*** The large diversity of naturally occurring circular peptides suggests that cyclization has evolved independently in a wide range of organisms, from bacteria to plants, to animals. Cyclic peptides are particularly important due to their resistance to enzymatic degradation, pH and temperature [286]. Linear peptides are often less stable, and more flexible, leading to reduced binding affinity and lower biological activity than their cyclic counterparts. Joining the ends of peptides removes susceptibility to exopeptidases and increases thermodynamic stability [286]. The exceptional stability of cyclic structures and their use as a framework in drug design has been explored for conotoxins [202], and a successful proof-of-concept example is the cyclization of α-conotoxin MII [287], a potent inhibitor of the nAChR [288] that has been implicated as a possible treatment in Parkinson's disease therapy [289]. A cyclic version of α-conotoxin MII retained the activity of the native form but had greatly improved resistance to proteolytic degradation [287]. This pathfinder study has been followed up with a number of other examples of the cyclization of conotoxins [290–293], the most successful of which resulted in the development of a cyclic analog of conotoxin Vc1.1 that was orally active in a rat model of neuropathic pain [293]. Thus in this case cyclization not only improved stability but led to oral activity that was not present in the parent linear conotoxin.

Not all examples of cyclization will lead to improved peptides, as seen with cyclic versions of some AMPs. A cyclic melittin analog exhibited increased antibacterial activity, with reduced hemolytic propensity, whereas a cyclic magainin 2 derivative was not so successful and had reduced antibacterial activity and increased hemolytic propensity [294]. The proper design of bioactive cyclic peptides requires detailed knowledge of the role of each amino acid residue, so that for example, cyclization should be designed to not affect residues that are crucial for activity [278]. Another consideration is the selection of a correctly sized linker, which must span the distance between the N and C termini. The adverse effects of removing stabilizing charge-charge interactions between the termini have to be overcome with linkers of correct length [287]. Nevertheless, with due consideration of these potential caveats peptide cyclization is a widely applied technique in the pharmaceutical industry, which decreases proteolytic degradation, prolongs half-life and stability and can improve binding efficiency [278].

***6.4.1.4 Disulfide Bond Engineering*** Disulfide bonds are a prerequisite for the proper biological function of many proteins [295]. They are very important for the folding and stability of proteins, and in peptides they introduce conformational constraints that confer a bioactive and thermodynamically stable conformation [296]. Disulfide-rich peptides can be used as stable scaffolds to graft exogenous peptide epitopes onto their stable structure, giving them new, and desired properties. Such scaffolds include the cyclotides [202], the defensins [297, 298], and the conotoxins [299] already described in this article. Because of their various disulfide connectivities and a wide range of activities, these natural peptides offer a large diversity of stable molecular scaffolds. To supplement this natural set of scaffolds, the engineering of new intramolecular disulfide bonds into peptide structures is a valuable strategy for the design of peptidic compounds with desired structural and active properties [300]. For example, nonnative disulfide bonds have been used to induce a constrained and stable structure in peptides, such as an amphipathic α-helix [301–303] or β-hairpin [279, 300]. Peptides with potential antimicrobial activity were shown to possess better membrane binding, and enhanced antimicrobial potency, when a nonnative bond was introduced [279, 303, 304]. A variety of alternative chemical linkers have also been used recently as more stable surrogates for disulfide bonds [305]. The use of diselenide bonds in place of disulfide bonds has been a particularly popular approach as the surrogate is almost isosteric but is more resistant to reduction [306–308].

## 6.5 CONCLUSIONS

Organisms are exquisitely adapted to their niche and possess an arsenal of compounds that give them evolutionary advantages. The potency and selectivity of these natural compounds, including peptides, has made them of interest in the field of drug design. In some cases, natural peptides have already been approved and are used as drugs or as food preservatives, while many others are in the pipeline of pharmaceutical companies. In this review, some examples of peptides isolated from different organisms with potential as therapeutic compounds have been illustrated. Such applications are facilitated by chemical modifications and peptide engineering to improve drug-like properties of peptides. Although only limited examples have been described, the future appears to be bright for applications of natural peptides as drug leads.

## ACKNOWLEDGMENTS

Work in our laboratory on peptide-based drug design is supported by an Australian Research Council (ARC) linkage grant (LP110200213) and by a National Health and Medical Research Council (NHMRC) project grant (APP1060225). DJC is a NHMRC Professorial Fellow (APP1026501). STH is the recipient of a Discovery Early Career Researcher Award (DE120103152), awarded by ARC.

## REFERENCES

1. Wender PA, Verma VA, Paxton TJ, Pillow TH. Function-oriented synthesis, step economy, and drug design. Acc Chem Res 2008;41:40–49.
2. Ogbourne SM, Parsons PG. The value of Nature's natural product library for the discovery of New Chemical Entities: the discovery of ingenol mebutate. Fitoterapia 2014;98:36–44.
3. Ganesan A. The impact of natural products upon modern drug discovery. Curr Opin Chem Biol 2008;12:306–317.
4. Basmadjian C, Zhao Q, Bentouhami E, Djehal A, Nebigil CG, Johnson RA, Serova M, de Gramont A, Faivre S, Raymond E, Desaubry LG. Cancer wars: natural products strike back. Front Chem 2014;2:20.
5. Cotter PD, Hill C, Ross RP. Bacteriocins: developing innate immunity for food. Nat Rev Microbiol 2005;3:777–788.
6. Molinski TF, Dalisay DS, Lievens SL, Saludes JP. Drug development from marine natural products. Nat Rev Drug Discov 2009;8:69–85.
7. Olivera BM, Hillyard DR, Marsh M, Yoshikami D. Combinatorial peptide libraries in drug design: lessons from venomous cone snails. Trends Biotechnol 1995;13:422–426.

8. Bhutia SK, Maiti TK. Targeting tumors with peptides from natural sources. Trends Biotechnol 2008;26:210–217.
9. Adermann K, John H, Standker L, Forssmann WG. Exploiting natural peptide diversity: novel research tools and drug leads. Curr Opin Biotechnol 2004;15:599–606.
10. Huther A, Dietrich U. The emergence of peptides as therapeutic drugs for the inhibition of HIV-1. AIDS Rev 2007;9:208–217.
11. Craik DJ, Fairlie DP, Liras S, Price D. The future of peptide-based drugs. Chem Biol Drug Des 2013;81:136–147.
12. Gongora-Benitez M, Tulla-Puche J, Albericio F. Multifaceted roles of disulfide bonds. Peptides as therapeutics. Chem Rev 2014;114:901–926.
13. Schroeder CI, Swedberg JE, Craik DJ. Recent progress towards pharmaceutical applications of disulfide-rich cyclic peptides. Curr Protein Pept Sci 2013;14:532–542.
14. Veiseh M, Gabikian P, Bahrami SB, Veiseh O, Zhang M, Hackman RC, Ravanpay AC, Stroud MR, Kusuma Y, Hansen SJ, Kwok D, Munoz NM, Sze RW, Grady WM, Greenberg NM, Ellenbogen RG, Olson JM. Tumor paint: a chlorotoxin:Cy5.5 bioconjugate for intraoperative visualization of cancer foci. Cancer Res 2007;67:6882–6888.
15. Akcan M, Stroud MR, Hansen SJ, Clark RJ, Daly NL, Craik DJ, Olson JM. Chemical re-engineering of chlorotoxin improves bioconjugation properties for tumor imaging and targeted therapy. J Med Chem 2011;54:782–787.
16. Gillor O, Ghazaryan L. Recent advances in bacteriocin application as antimicrobials. Recent Pat Antiinfect Drug Discov 2007;2:115–122.
17. Hammami R, Fernandez B, Lacroix C, Fliss I. Anti-infective properties of bacteriocins: an update. Cell Mol Life Sci: CMLS 2013;70:2947–2967.
18. Snyder AB, Worobo RW. Chemical and genetic characterization of bacteriocins: antimicrobial peptides for food safety. J Sci Food Agric 2014;94:28–44.
19. Delves-Broughton J, Blackburn P, Evans RJ, Hugenholtz J. Applications of the bacteriocin, nisin. Antonie Van Leeuwenhoek 1996;69:193–202.
20. Zhang XJ, Duan G, Zhang R, Fan Q. Optimized expression of Helicobacter pylori ureB gene in the Lactococcus lactis nisin-controlled gene expression (NICE) system and experimental study of its immunoreactivity. Curr Microbiol 2009;58:308–314.
21. Millette M, Cornut G, Dupont C, Shareck F, Archambault D, Lacroix M. Capacity of human nisin- and pediocin-producing lactic acid bacteria to reduce intestinal colonization by vancomycin-resistant enterococci. Appl Environ Microbiol 2008;74:1997–2003.
22. Joo NE, Ritchie K, Kamarajan P, Miao D, Kapila YL. Nisin, an apoptogenic bacteriocin and food preservative, attenuates HNSCC tumorigenesis via CHAC1. Cancer Med 2012;1:295–305.
23. Reddy KV, Yedery RD, Aranha C. Antimicrobial peptides: premises and promises. Int J Antimicrob Agents 2004;24:536–547.
24. Sable S, Pons AM, Gendron-Gaillard S, Cottenceau G. Antibacterial activity evaluation of microcin J25 against diarrheagenic Escherichia coli. Appl Environ Microbiol 2000;66:4595–4597.
25. Sutyak KE, Anderson RA, Dover SE, Feathergill KA, Aroutcheva AA, Faro S, Chikindas ML. Spermicidal activity of the safe natural antimicrobial peptide subtilosin. Infect Dis Obstet Gynecol 2008;2008:540758.
26. Thevissen K, Ferket KK, Francois IE, Cammue BP. Interactions of antifungal plant defensins with fungal membrane components. Peptides 2003;24:1705–1712.

27. Thomma BP, Cammue BP, Thevissen K. Plant defensins. Planta 2002;216:193–202.
28. Marmiroli N, Maestri E. Plant peptides in defense and signaling. Peptides 2014;56: 30–44.
29. Jennings C, West J, Waine C, Craik D, Anderson M. Biosynthesis and insecticidal properties of plant cyclotides: the cyclic knotted proteins from Oldenlandia affinis. Proc Natl Acad Sci U S A 2001;98:10614–10619.
30. Mylne JS, Wang CK, van der Weerden NL, Craik DJ. Cyclotides are a component of the innate defense of Oldenlandia affinis. Biopolymers 2010;94:635–646.
31. Craik DJ, Daly NL, Waine C. The cystine knot motif in toxins and implications for drug design. Toxicon 2001;39:43–60.
32. Poth AG, Chan LY, Craik DJ. Cyclotides as grafting frameworks for protein engineering and drug design applications. Biopolymers 2013;100:480–491.
33. Gould A, Ji Y, Aboye TL, Camarero JA. Cyclotides, a novel ultrastable polypeptide scaffold for drug discovery. Curr Pharm Des 2011;17:4294–4307.
34. Henriques ST, Craik DJ. Cyclotides as templates in drug design. Drug Discov Today 2010;15:57–64.
35. Craik DJ, Daly NL, Bond T, Waine C. Plant cyclotides: a unique family of cyclic and knotted proteins that defines the cyclic cystine knot structural motif. J Mol Biol 1999;294:1327–1336.
36. Gran L. On the isolation of tetramethylputrescine from Oldenlandia affinis. Lloydia 1973;36:209–210.
37. Gran L. Oxytocic principles of Oldenlandia affinis. Lloydia 1973;36:174–178.
38. Lindholm P, Goransson U, Johansson S, Claeson P, Gullbo J, Larsson R, Bohlin L, Backlund A. Cyclotides: a novel type of cytotoxic agents. Mol Cancer Ther 2002;1:365–369.
39. Bokesch HR, Pannell LK, Cochran PK, Sowder RC 2nd, McKee TC, Boyd MR. A novel anti-HIV macrocyclic peptide from Palicourea condensata. J Nat Prod 2001;64:249–250.
40. Hallock YF, Sowder RC 2nd, Pannell LK, Hughes CB, Johnson DG, Gulakowski R, Cardellina JH 2nd, Boyd MR. Cycloviolins A-D, anti-HIV macrocyclic peptides from Leonia cymosa. J Org Chem 2000;65:124–128.
41. Ireland DC, Wang CK, Wilson JA, Gustafson KR, Craik DJ. Cyclotides as natural anti-HIV agents. Biopolymers (Peptide science) 2008;90:51–60.
42. Calvete JJ, Sanz L, Angulo Y, Lomonte B, Gutierrez JM. Venoms, venomics, antivenomics. FEBS Lett 2009;583:1736–1743.
43. Akondi KB, Muttenthaler M, Dutertre S, Kaas Q, Craik DJ, Lewis RJ, Alewood PF. Discovery, synthesis, and structure-activity relationships of conotoxins. Chem Rev 2014;114:5815–5847.
44. Fedosov AE, Moshkovskii SA, Kuznetsova KG, Olivera BM. Conotoxins: from the biodiversity of gastropods to new drugs]. Biomeditsinskaia Khimiia 2013;59:267–294.
45. Vetter I, Lewis RJ. Therapeutic potential of cone snail venom peptides (conopeptides). Curr Top Med Chem 2012;12:1546–1552.
46. Simmaco M, Mignogna G, Barra D. Antimicrobial peptides from amphibian skin: what do they tell us? Biopolymers 1998;47:435–450.
47. Conlon JM, Mechkarska M, King JD. Host-defense peptides in skin secretions of African clawed frogs (Xenopodinae, Pipidae). Gen Comp Endocrinol 2012;176:513–518.

48. Bowie JH, Separovic F, Tyler MJ. Host-defense peptides of Australian anurans. Part 2. Structure, activity, mechanism of action, and evolutionary significance. Peptides 2012;37:174–188.
49. Zasloff M. Magainins, a class of antimicrobial peptides from Xenopus skin: isolation, characterization of two active forms, and partial cDNA sequence of a precursor. Proc Natl Acad Sci U S A 1987;84:5449–5453.
50. Jacob L, Zasloff M. Potential therapeutic applications of magainins and other antimicrobial agents of animal origin. Ciba Found Symp 1994;186:197–216.
51. Zasloff M. Antimicrobial peptides of multicellular organisms. Nature 2002;415: 389–395.
52. Barra D, Simmaco M. Amphibian skin: a promising resource for antimicrobial peptides. Trends Biotechnol 1995;13:205–209.
53. Conlon JM, Kolodziejek J, Nowotny N. Antimicrobial peptides from the skins of North American frogs. Biochim Biophys Acta 2009;1788:1556–1563.
54. Lu CX, Nan KJ, Lei Y. Agents from amphibians with anticancer properties. Anticancer Drugs 2008;19:931–939.
55. Lorin C, Saidi H, Belaid A, Zairi A, Baleux F, Hocini H, Belec L, Hani K, Tangy F. The antimicrobial peptide dermaseptin S4 inhibits HIV-1 infectivity in vitro. Virology 2005;334:264–275.
56. VanCompernolle SE, Taylor RJ, Oswald-Richter K, Jiang J, Youree BE, Bowie JH, Tyler MJ, Conlon JM, Wade D, Aiken C, Dermody TS, KewalRamani VN, Rollins-Smith LA, Unutmaz D. Antimicrobial peptides from amphibian skin potently inhibit human immunodeficiency virus infection and transfer of virus from dendritic cells to T cells. J Virol 2005;79:11598–11606.
57. Bergaoui I, Zairi A, Tangy F, Aouni M, Selmi B, Hani K. In vitro antiviral activity of dermaseptin S(4) and derivatives from amphibian skin against herpes simplex virus type 2. J Med Virol 2013;85:272–281.
58. Lewis RJ, Garcia ML. Therapeutic potential of venom peptides. Nat Rev Drug Discov 2003;2:790–802.
59. Olivera BM, Rivier J, Clark C, Ramilo CA, Corpuz GP, Abogadie FC, Mena EE, Woodward SR, Hillyard DR, Cruz LJ. Diversity of Conus neuropeptides. Science 1990;249:257–263.
60. Adams DJ, Alewood PF, Craik DJ, Drinkwater RD, Lewis RJ. Conotoxins and their potential pharmaceutical applications. Drug Develop Res 1999;46:219–234.
61. Terlau H, Olivera BM. Conus venoms: a rich source of novel ion channel-targeted peptides. Physiol Rev 2004;84:41–68.
62. Essack M, Bajic VB, Archer JA. Conotoxins that confer therapeutic possibilities. Marine Drugs 2012;10:1244–1265.
63. Miljanich GP. Ziconotide: neuronal calcium channel blocker for treating severe chronic pain. Curr Med Chem 2004;11:3029–3040.
64. Klotz U. Ziconotide - a novel neuron-specific calcium channel blocker for the intrathecal treatment of severe chronic pain - a short review. Int J Clin Pharm Ther 2006;44:478–483.
65. Garber K. Peptide leads new class of chronic pain drugs. Nat Biotechnol 2005;23:399.
66. Schroeder CI, Craik DJ. Therapeutic potential of conopeptides. Future Med Chem 2012;4:1243–1255.

67. Hefti FF. Requirements for a lead compound to become a clinical candidate. BMC Neurosci 2008;9:S7.
68. Hilpert K, Volkmer-Engert R, Walter T, Hancock RE. High-throughput generation of small antibacterial peptides with improved activity. Nat Biotechnol 2005;23:1008–1012.
69. Deslouches B, Islam K, Craigo JK, Paranjape SM, Montelaro RC, Mietzner TA. Activity of the de novo engineered antimicrobial peptide WLBU2 against Pseudomonas aeruginosa in human serum and whole blood: implications for systemic applications. Antimicrob Agents Chemother 2005;49:3208–3216.
70. Iriti M, Faoro F. Review of innate and specific immunity in plants and animals. Mycopathologia 2007;164:57–64.
71. Holzl MA, Hofer J, Steinberger P, Pfistershammer K, Zlabinger GJ. Host antimicrobial proteins as endogenous immunomodulators. Immunol Lett 2008;119:4–11.
72. Sorensen OE, Borregaard N, Cole AM. Antimicrobial peptides in innate immune responses. Contrib Microbiol 2008;15:61–77.
73. Cheikhyoussef A, Pogori N, Chen W, Zhang H. Antimicrobial proteinaceous compounds obtained from bifidobacteria: from production to their application. Int J Food Microbiol 2008;125:215–222.
74. Bals R. Epithelial antimicrobial peptides in host defense against infection. Respir Res 2000;1:141–150.
75. Tollin M, Bergman P, Svenberg T, Jornvall H, Gudmundsson GH, Agerberth B. Antimicrobial peptides in the first line defence of human colon mucosa. Peptides 2003;24:523–530.
76. Ganz T, Selsted ME, Szklarek D, Harwig SS, Daher K, Bainton DF, Lehrer RI. Defensins. Natural peptide antibiotics of human neutrophils. J Clin Invest 1985;76:1427–1435.
77. Hoffmann JA, Kafatos FC, Janeway CA, Ezekowitz RA. Phylogenetic perspectives in innate immunity. Science 1999;284:1313–1318.
78. Hilpert K, Elliott M, Jenssen H, Kindrachuk J, Fjell CD, Korner J, Winkler DF, Weaver LL, Henklein P, Ulrich AS, Chiang SH, Farmer SW, Pante N, Volkmer R, Hancock RE. Screening and characterization of surface-tethered cationic peptides for antimicrobial activity. Chem Biol 2009;16:58–69.
79. Murakami T, Niwa M, Tokunaga F, Miyata T, Iwanaga S. Direct virus inactivation of tachyplesin I and its isopeptides from horseshoe crab hemocytes. Chemotherapy 1991;37:327–334.
80. Palffy R, Gardlik R, Behuliak M, Kadasi L, Turna J, Celec P. On the physiology and pathophysiology of antimicrobial peptides. Mol Med 2009;15:51–59.
81. Marshall SH, Arenas G. Antimicrobial peptides: a natural alternative to chemical antibiotics and a potential for applied biotechnology. Electron J Biotechnol 2003;6:271–284.
82. Wang G, Li X, Wang Z. APD2: the updated antimicrobial peptide database and its application in peptide design. Nucleic Acids Res 2009;37:D933–D937.
83. Yeaman MR, Yount NY. Mechanisms of antimicrobial peptide action and resistance. Pharmacol Rev 2003;55:27–55.
84. Ladokhin AS, White SH. Folding of amphipathic alpha-helices on membranes: energetics of helix formation by melittin. J Mol Biol 1999;285:1363–1369.
85. Wieprecht T, Apostolov O, Beyermann M, Seelig J. Thermodynamics of the alpha-helix-coil transition of amphipathic peptides in a membrane environment: implications for the peptide-membrane binding equilibrium. J Mol Biol 1999;294:785–794.

86. Ali MF, Soto A, Knoop FC, Conlon JM. Antimicrobial peptides isolated from skin secretions of the diploid frog, Xenopus tropicalis (Pipidae). Biochim Biophys Acta 2001;1550:81–89.
87. Martin E, Ganz T, Lehrer RI. Defensins and other endogenous peptide antibiotics of vertebrates. J Leukoc Biol 1995;58:128–136.
88. Zanetti M, Gennaro R, Romeo D. Cathelicidins: a novel protein family with a common proregion and a variable C-terminal antimicrobial domain. FEBS Lett 1995;374:1–5.
89. Bals R, Wang X, Zasloff M, Wilson JM. The peptide antibiotic LL-37/hCAP-18 is expressed in epithelia of the human lung where it has broad antimicrobial activity at the airway surface. Proc Natl Acad Sci U S A 1998;95:9541–9546.
90. Travis SM, Anderson NN, Forsyth WR, Espiritu C, Conway BD, Greenberg EP, McCray PB Jr, Lehrer RI, Welsh MJ, Tack BF. Bactericidal activity of mammalian cathelicidin-derived peptides. Infect Immun 2000;68:2748–2755.
91. Turner J, Cho Y, Dinh NN, Waring AJ, Lehrer RI. Activities of LL-37, a cathelin-associated antimicrobial peptide of human neutrophils. Antimicrob Agents Chemother 1998;42:2206–2214.
92. Nagaoka I, Hirota S, Yomogida S, Ohwada A, Hirata M. Synergistic actions of antibacterial neutrophil defensins and cathelicidins. Inflamm Res 2000;49:73–79.
93. Oren Z, Lerman JC, Gudmundsson GH, Agerberth B, Shai Y. Structure and organization of the human antimicrobial peptide LL-37 in phospholipid membranes: relevance to the molecular basis for its non-cell-selective activity. Biochem J 1999;341:501–513.
94. Simmaco M, Mignogna G, Canofeni S, Miele R, Mangoni ML, Barra D. Temporins, antimicrobial peptides from the European red frog Rana temporaria. Eur J Biochem 1996;242:788–792.
95. Simmaco M, Barra D, Chiarini F, Noviello L, Melchiorri P, Kreil G, Richter K. A family of bombinin-related peptides from the skin of Bombina variegata. Eur J Biochem 1991;199:217–222.
96. Mignogna G, Simmaco M, Kreil G, Barra D. Antibacterial and haemolytic peptides containing D-alloisoleucine from the skin of Bombina variegata. EMBO J 1993;12:4829–4832.
97. Gibson BW, Tang DZ, Mandrell R, Kelly M, Spindel ER. Bombinin-like peptides with antimicrobial activity from skin secretions of the Asian toad, Bombina orientalis. J Biol Chem 1991;266:23103–23111.
98. Mangoni ML, Grovale N, Giorgi A, Mignogna G, Simmaco M, Barra D. Structure-function relationships in bombinins H, antimicrobial peptides from Bombina skin secretions. Peptides 2000;21:1673–1679.
99. Zangger K, Gossler R, Khatai L, Lohner K, Jilek A. Structures of the glycine-rich diastereomeric peptides bombinin H2 and H4. Toxicon 2008;52:246–254.
100. Kreil G. D-amino acids in animal peptides. Annu Rev Biochem 1997;66:337–345.
101. Bozzi A, Mangoni ML, Rinaldi AC, Mignogna G, Aschi M. Folding propensity and biological activity of peptides: the effect of a single stereochemical isomerization on the conformational properties of bombinins in aqueous solution. Biopolymers 2008;89:769–778.
102. Mangoni ML, Papo N, Saugar JM, Barra D, Shai Y, Simmaco M, Rivas L. Effect of natural L- to D-amino acid conversion on the organization, membrane binding, and biological function of the antimicrobial peptides bombinins H. Biochemistry 2006;45:4266–4276.

103. Mangoni ML, Rinaldi AC, Di Giulio A, Mignogna G, Bozzi A, Barra D, Simmaco M. Structure-function relationships of temporins, small antimicrobial peptides from amphibian skin. Eur J Biochem 2000;267:1447–1454.

104. Rollins-Smith LA, Carey C, Conlon JM, Reinert LK, Doersam JK, Bergman T, Silberring J, Lankinen H, Wade D. Activities of temporin family peptides against the chytrid fungus (Batrachochytrium dendrobatidis) associated with global amphibian declines. Antimicrob Agents Chemother 2003;47:1157–1160.

105. Chinchar VG, Bryan L, Silphadaung U, Noga E, Wade D, Rollins-Smith L. Inactivation of viruses infecting ectothermic animals by amphibian and piscine antimicrobial peptides. Virology 2004;323:268–275.

106. Mangoni ML, Saugar JM, Dellisanti M, Barra D, Simmaco M, Rivas L. Temporins, small antimicrobial peptides with leishmanicidal activity. J Biol Chem 2005;280:984–990.

107. Rinaldi AC, Mangoni ML, Rufo A, Luzi C, Barra D, Zhao H, Kinnunen PK, Bozzi A, Di Giulio A, Simmaco M. Temporin L: antimicrobial, haemolytic and cytotoxic activities, and effects on membrane permeabilization in lipid vesicles. Biochem J 2002;368:91–100.

108. Mangoni ML. Temporins, anti-infective peptides with expanding properties. Cell Mol Life Sci: CMLS 2006;63:1060–1069.

109. Lehrer RI, Lichtenstein AK, Ganz T. Defensins: antimicrobial and cytotoxic peptides of mammalian cells. Annu Rev Immunol 1993;11:105–128.

110. Lehrer RI, Ganz T. Antimicrobial peptides in mammalian and insect host defence. Curr Opin Immunol 1999;11:23–27.

111. Lay FT, Anderson MA. Defensins--components of the innate immune system in plants. Curr Protein Pept Sci 2005;6:85–101.

112. Tang YQ, Yuan J, Osapay G, Osapay K, Tran D, Miller CJ, Ouellette AJ, Selsted ME. A cyclic antimicrobial peptide produced in primate leukocytes by the ligation of two truncated alpha-defensins. Science 1999;286:498–502.

113. Leonova L, Kokryakov VN, Aleshina G, Hong T, Nguyen T, Zhao C, Waring AJ, Lehrer RI. Circular minidefensins and posttranslational generation of molecular diversity. J Leukoc Biol 2001;70:461–464.

114. Cole AM, Hong T, Boo LM, Nguyen T, Zhao C, Bristol G, Zack JA, Waring AJ, Yang OO, Lehrer RI. Retrocyclin: a primate peptide that protects cells from infection by T- and M-tropic strains of HIV-1. Proc Natl Acad Sci U S A 2002;99:1813–1818.

115. Venkataraman N, Cole AL, Ruchala P, Waring AJ, Lehrer RI, Stuchlik O, Pohl J, Cole AM. Reawakening retrocyclins: ancestral human defensins active against HIV-1. PLoS Biol 2009;7:e95.

116. Shai Y. Mode of action of membrane active antimicrobial peptides. Biopolymers 2002;66:236–248.

117. Shai Y. Mechanism of the binding, insertion and destabilization of phospholipid bilayer membranes by alpha-helical antimicrobial and cell non-selective membrane-lytic peptides. Biochim Biophys Acta 1999;1462:55–70.

118. Nicolas P. Multifunctional host defense peptides: intracellular-targeting antimicrobial peptides. FEBS J 2009;276:6483–6496.

119. Bush K. Antimicrobial agents targeting bacterial cell walls and cell membranes. Rev Sci Tech 2012;31:43–56.

120. Matsuzaki K. Why and how are peptide-lipid interactions utilized for self-defense? Magainins and tachyplesins as archetypes. Biochim Biophys Acta 1999;1462:1–10.

121. Hancock RE, Sahl HG. Antimicrobial and host-defense peptides as new anti-infective therapeutic strategies. Nat Biotechnol 2006;24:1551–1557.

122. Cudic M, Otvos L Jr. Intracellular targets of antibacterial peptides. Curr Drug Targets 2002;3:101–106.

123. Sansom MS. Alamethicin and related peptaibols--model ion channels. Eur Biophys J: EBJ 1993;22:105–124.

124. Ludtke SJ, He K, Heller WT, Harroun TA, Yang L, Huang HW. Membrane pores induced by magainin. Biochemistry 1996;35:13723–13728.

125. Lohner K, Prenner EJ. Differential scanning calorimetry and X-ray diffraction studies of the specificity of the interaction of antimicrobial peptides with membrane-mimetic systems. Biochim Biophys Acta 1999;1462:141–156.

126. Melo MN, Ferre R, Castanho MA. Antimicrobial peptides: linking partition, activity and high membrane-bound concentrations. Nat Rev Microbiol 2009;7:245–250.

127. Henriques ST, Melo MN, Castanho MA. Cell-penetrating peptides and antimicrobial peptides: how different are they? Biochem J 2006;399:1–7.

128. Torcato IM, Huang YH, Franquelim HG, Gaspar DD, Craik DJ, Castanho MA, Henriques ST. The antimicrobial activity of Sub3 is dependent on membrane binding and cell-penetrating ability. ChemBioChem 2013;14:2013–2022.

129. Bush K, Courvalin P, Dantas G, Davies J, Eisenstein B, Huovinen P, Jacoby GA, Kishony R, Kreiswirth BN, Kutter E, Lerner SA, Levy S, Lewis K, Lomovskaya O, Miller JH, Mobashery S, Piddock LJ, Projan S, Thomas CM, Tomasz A, Tulkens PM, Walsh TR, Watson JD, Witkowski J, Witte W, Wright G, Yeh P, Zgurskaya HI. Tackling antibiotic resistance. Nat Rev Microbiol 2011;9:894–896.

130. Piddock LJ. The crisis of no new antibiotics--what is the way forward? Lancet Infect Dis 2012;12:249–253.

131. Hancock RE. Cationic peptides: effectors in innate immunity and novel antimicrobials. Lancet Infect Dis 2001;1:156–164.

132. Gaspar D, Veiga AS, Castanho MA. From antimicrobial to anticancer peptides. A review. Front Microbiol 2013;4:294.

133. Hoskin DW, Ramamoorthy A. Studies on anticancer activities of antimicrobial peptides. Biochim Biophys Acta 2008;1778:357–375.

134. Utsugi T, Schroit AJ, Connor J, Bucana CD, Fidler IJ. Elevated expression of phosphatidylserine in the outer membrane leaflet of human tumor cells and recognition by activated human blood monocytes. Cancer Res 1991;51:3062–3066.

135. Dobrzynska I, Szachowicz-Petelska B, Sulkowski S, Figaszewski Z. Changes in electric charge and phospholipids composition in human colorectal cancer cells. Mol Cell Biochem 2005;276:113–119.

136. Sok M, Sentjurc M, Schara M. Membrane fluidity characteristics of human lung cancer. Cancer Lett 1999;139:215–220.

137. Dathe M, Wieprecht T. Structural features of helical antimicrobial peptides: their potential to modulate activity on model membranes and biological cells. Biochim Biophys Acta 1999;1462:71–87.

138. Dathe M, Wieprecht T, Nikolenko H, Handel L, Maloy WL, MacDonald DL, Beyermann M, Bienert M. Hydrophobicity, hydrophobic moment and angle subtended by charged residues modulate antibacterial and haemolytic activity of amphipathic helical peptides. FEBS Lett 1997;403:208–212.

139. Dathe M, Schumann M, Wieprecht T, Winkler A, Beyermann M, Krause E, Matsuzaki K, Murase O, Bienert M. Peptide helicity and membrane surface charge modulate the balance of electrostatic and hydrophobic interactions with lipid bilayers and biological membranes. Biochemistry 1996;35:12612–12622.

140. Melo MN, Ferre R, Feliu L, Bardaji E, Planas M, Castanho MA. Prediction of antibacterial activity from physicochemical properties of antimicrobial peptides. PLoS ONE 2011;6:e28549.

141. Rosenfeld Y, Barra D, Simmaco M, Shai Y, Mangoni ML. A synergism between temporins toward Gram-negative bacteria overcomes resistance imposed by the lipopolysaccharide protective layer. J Biol Chem 2006;281:28565–28574.

142. Perron GG, Zasloff M, Bell G. Experimental evolution of resistance to an antimicrobial peptide. Proc Biol Sci 2006;273:251–256.

143. Nigam A, Gupta D, Sharma A. Treatment of infectious disease: beyond antibiotics. Microbiol Res 2014;169:643–651.

144. Holtsmark I, Eijsink VG, Brurberg MB. Bacteriocins from plant pathogenic bacteria. FEMS Microbiol Lett 2008;280:1–7.

145. Pons AM, Lanneluc I, Cottenceau G, Sable S. New developments in non-post translationally modified microcins. Biochimie 2002;84:531–537.

146. Maqueda M, Sanchez-Hidalgo M, Fernandez M, Montalban-Lopez M, Valdivia E, Martinez-Bueno M. Genetic features of circular bacteriocins produced by Gram-positive bacteria. FEMS Microbiol Rev 2008;32:2–22.

147. Franz CM, van Belkum MJ, Holzapfel WH, Abriouel H, Galvez A. Diversity of enterococcal bacteriocins and their grouping in a new classification scheme. FEMS Microbiol Rev 2007;31:293–310.

148. Klaenhammer TR. Genetics of bacteriocins produced by lactic acid bacteria. FEMS Microbiol Rev 1993;12:39–85.

149. Nes IF, Diep DB, Havarstein LS, Brurberg MB, Eijsink V, Holo H. Biosynthesis of bacteriocins in lactic acid bacteria. Antonie Van Leeuwenhoek 1996;70:113–128.

150. van Belkum MJ, Stiles ME. Molecular characterization of genes involved in the production of the bacteriocin leucocin A from Leuconostoc gelidum. Appl Environ Microbiol 1995;61:3573–3579.

151. Hassan M, Kjos M, Nes IF, Diep DB, Lotfipour F. Natural antimicrobial peptides from bacteria: characteristics and potential applications to fight against antibiotic resistance. J Appl Microbiol 2012;113:723–736.

152. Asensio C, Perez-Diaz JC. A new family of low molecular weight antibiotics from enterobacteria. Biochem Biophys Res Commun 1976;69:7–14.

153. Duquesne S, Petit V, Peduzzi J, Rebuffat S. Structural and functional diversity of microcins, gene-encoded antibacterial peptides from enterobacteria. J Mol Microbiol Biotechnol 2007;13:200–209.

154. Cornut G, Fortin C, Soulieres D. Antineoplastic properties of bacteriocins: revisiting potential active agents. Am J Clin Oncol 2008;31:399–404.

155. Hetz C, Bono MR, Barros LF, Lagos R. Microcin E492, a channel-forming bacteriocin from Klebsiella pneumoniae, induces apoptosis in some human cell lines. Proc Natl Acad Sci U S A 2002;99:2696–2701.

156. Lagos R, Tello M, Mercado G, Garcia V, Monasterio O. Antibacterial and antitumorigenic properties of microcin E492, a pore-forming bacteriocin. Curr Pharm Biotechnol 2009;10:74–85.

157. Lopez FE, Vincent PA, Zenoff AM, Salomon RA, Farias RN. Efficacy of microcin J25 in biomatrices and in a mouse model of Salmonella infection. J Antimicrob Chemother 2007;59:676–680.

158. Salomon RA, Farias RN. Microcin 25, a novel antimicrobial peptide produced by Escherichia coli. J Bacteriol 1992;174:7428–7435.

159. Blond A, Peduzzi J, Goulard C, Chiuchiolo MJ, Barthelemy M, Prigent Y, Salomon RA, Farias RN, Moreno F, Rebuffat S. The cyclic structure of microcin J25, a 21-residue peptide antibiotic from Escherichia coli. Eur J Biochem 1999;259:747–755.

160. Bayro MJ, Mukhopadhyay J, Swapna GV, Huang JY, Ma LC, Sineva E, Dawson PE, Montelione GT, Ebright RH. Structure of antibacterial peptide microcin J25: a 21-residue lariat protoknot. J Am Chem Soc 2003;125:12382–12383.

161. Rosengren KJ, Clark RJ, Daly NL, Goransson U, Jones A, Craik DJ. Microcin J25 has a threaded sidechain-to-backbone ring structure and not a head-to-tail cyclized backbone. J Am Chem Soc 2003;125:12464–12474.

162. Wilson KA, Kalkum M, Ottesen J, Yuzenkova J, Chait BT, Landick R, Muir T, Severinov K, Darst SA. Structure of microcin J25, a peptide inhibitor of bacterial RNA polymerase, is a lassoed tail. J Am Chem Soc 2003;125:12475–12483.

163. Rosengren KJ, Blond A, Afonso C, Tabet JC, Rebuffat S, Craik DJ. Structure of thermolysin cleaved microcin J25: extreme stability of a two-chain antimicrobial peptide devoid of covalent links. Biochemistry 2004;43:4696–4702.

164. Pavlova O, Mukhopadhyay J, Sineva E, Ebright RH, Severinov K. Systematic structure-activity analysis of microcin J25. J Biol Chem 2008;283:25589–25595.

165. Chiuchiolo MJ, Delgado MA, Farias RN, Salomon RA. Growth-phase-dependent expression of the cyclopeptide antibiotic microcin J25. J Bacteriol 2001;183:1755–1764.

166. Delgado MA, Rintoul MR, Farias RN, Salomon RA. Escherichia coli RNA polymerase is the target of the cyclopeptide antibiotic microcin J25. J Bacteriol 2001;183:4543–4550.

167. Kuznedelov K, Semenova E, Knappe TA, Mukhamedyarov D, Srivastava A, Chatterjee S, Ebright RH, Marahiel MA, Severinov K. The antibacterial threaded-lasso peptide capistruin inhibits bacterial RNA polymerase. J Mol Biol 2011;412:842–848.

168. Bellomio A, Vincent PA, de Arcuri BF, Farias RN, Morero RD. Microcin J25 has dual and independent mechanisms of action in Escherichia coli: RNA polymerase inhibition and increased superoxide production. J Bacteriol 2007;189:4180–4186.

169. Niklison-Chirou MV, Dupuy F, Pena LB, Gallego SM, Barreiro-Arcos ML, Avila C, Torres-Bugeau C, Arcuri BE, Bellomio A, Minahk C, Morero RD. Microcin J25 triggers cytochrome c release through irreversible damage of mitochondrial proteins and lipids. Int J Biochem Cell Biol 2010;42:273–281.

170. Yuzenkova J, Delgado M, Nechaev S, Savalia D, Epshtein V, Artsimovitch I, Mooney RA, Landick R, Farias RN, Salomon R, Severinov K. Mutations of bacterial RNA polymerase leading to resistance to microcin j25. J Biol Chem 2002;277:50867–50875.

171. Vincent PA, Morero RD. The Structure and Biological Aspects of Peptide Antibiotic Microcin J25. Curr Med Chem 2009;16:538–549.
172. Mukhopadhyay J, Sineva E, Knight J, Levy RM, Ebright RH. Antibacterial peptide microcin J25 inhibits transcription by binding within and obstructing the RNA polymerase secondary channel. Mol Cell 2004;14:739–751.
173. Adelman K, Yuzenkova J, La Porta A, Zenkin N, Lee J, Lis JT, Borukhov S, Wang MD, Severinov K. Molecular mechanism of transcription inhibition by peptide antibiotic Microcin J25. Mol Cell 2004;14:753–762.
174. Rintoul MR, de Arcuri BF, Morero RD. Effects of the antibiotic peptide microcin J25 on liposomes: role of acyl chain length and negatively charged phospholipid. Biochim Biophys Acta 2000;1509:65–72.
175. Rintoul MR, de Arcuri BF, Salomon RA, Farias RN, Morero RD. The antibacterial action of microcin J25: evidence for disruption of cytoplasmic membrane energization in Salmonella newport. FEMS Microbiol Lett 2001;204:265–270.
176. Niklison Chirou MV, Minahk CJ, Morero RD. Antimitochondrial activity displayed by the antimicrobial peptide microcin J25. Biochem Biophys Res Commun 2004;317:882–886.
177. Mathavan I, Beis K. The role of bacterial membrane proteins in the internalization of microcin MccJ25 and MccB17. Biochem Soc Trans 2012;40:1539–1543.
178. Xie L, van der Donk WA. Post-translational modifications during lantibiotic biosynthesis. Curr Opin Chem Biol 2004;8:498–507.
179. Lubelski J, Rink R, Khusainov R, Moll GN, Kuipers OP. Biosynthesis, immunity, regulation, mode of action and engineering of the model lantibiotic nisin. Cell Mol Life Sci: CMLS 2008;65:455–476.
180. Bauer R, Dicks LM. Mode of action of lipid II-targeting lantibiotics. Int J Food Microbiol 2005;101:201–216.
181. Zhou XX, Li WF, Ma GX, Pan YJ. The nisin-controlled gene expression system: construction, application and improvements. Biotechnol Adv 2006;24:285–295.
182. Babasaki K, Takao T, Shimonishi Y, Kurahashi K. Subtilosin A, a new antibiotic peptide produced by Bacillus subtilis 168: isolation, structural analysis, and biogenesis. J Biochem 1985;98:585–603.
183. Shelburne CE, An FY, Dholpe V, Ramamoorthy A, Lopatin DE, Lantz MS. The spectrum of antimicrobial activity of the bacteriocin subtilosin A. J Antimicrob Chemother 2007;59:297–300.
184. Zheng G, Yan LZ, Vederas JC, Zuber P. Genes of the sbo-alb locus of Bacillus subtilis are required for production of the antilisterial bacteriocin subtilosin. J Bacteriol 1999;181:7346–7355.
185. Marx R, Stein T, Entian KD, Glaser SJ. Structure of the Bacillus subtilis peptide antibiotic subtilosin A determined by 1H-NMR and matrix assisted laser desorption/ionization time-of-flight mass spectrometry. J Protein Chem 2001;20:501–506.
186. Kawulka K, Sprules T, McKay RT, Mercier P, Diaper CM, Zuber P, Vederas JC. Structure of subtilosin A, an antimicrobial peptide from Bacillus subtilis with unusual posttranslational modifications linking cysteine sulfurs to alpha-carbons of phenylalanine and threonine. J Am Chem Soc 2003;125:4726–4727.
187. Stein T, Dusterhus S, Stroh A, Entian KD. Subtilosin production by two Bacillus subtilis subspecies and variance of the sbo-alb cluster. Appl Environ Microbiol 2004;70:2349–2353.

188. Sutyak KE, Wirawan RE, Aroutcheva AA, Chikindas ML. Isolation of the Bacillus subtilis antimicrobial peptide subtilosin from the dairy product-derived Bacillus amyloliquefaciens. J Appl Microbiol 2008;104:1067–1074.

189. Kawulka KE, Sprules T, Diaper CM, Whittal RM, McKay RT, Mercier P, Zuber P, Vederas JC. Structure of subtilosin A, a cyclic antimicrobial peptide from Bacillus subtilis with unusual sulfur to alpha-carbon cross-links: formation and reduction of alpha-thio-alpha-amino acid derivatives. Biochemistry 2004;43:3385–3395.

190. Thennarasu S, Lee DK, Poon A, Kawulka KE, Vederas JC, Ramamoorthy A. Membrane permeabilization, orientation, and antimicrobial mechanism of subtilosin A. Chem Phys Lipids 2005;137:38–51.

191. Silkin L, Hamza S, Kaufman S, Cobb SL, Vederas JC. Spermicidal bacteriocins: lacticin 3147 and subtilosin A. Bioorg Med Chem Lett 2008;18:3103–3106.

192. van Kraaij C, de Vos WM, Siezen RJ, Kuipers OP. Lantibiotics: biosynthesis, mode of action and applications. Nat Prod Rep 1999;16:575–587.

193. Yang SC, Lin CH, Sung CT, Fang JY. Antibacterial activities of bacteriocins: application in foods and pharmaceuticals. Front Microbiol 2014;5:241.

194. Johnsen L, Fimland G, Eijsink V, Nissen-Meyer J. Engineering increased stability in the antimicrobial peptide pediocin PA-1. Appl Environ Microbiol 2000;66:4798–4802.

195. Rollema HS, Kuipers OP, Both P, de Vos WM, Siezen RJ. Improvement of solubility and stability of the antimicrobial peptide nisin by protein engineering. Appl Environ Microbiol 1995;61:2873–2878.

196. Liu W, Hansen JN. Enhancement of the chemical and antimicrobial properties of subtilin by site-directed mutagenesis. J Biol Chem 1992;267:25078–25085.

197. Yuan J, Zhang ZZ, Chen XZ, Yang W, Huan LD. Site-directed mutagenesis of the hinge region of nisinZ and properties of nisinZ mutants. Appl Microbiol Biotechnol 2004;64:806–815.

198. Gran L. On the effect of a polypeptide isolated from "Kalata-Kalata" (Oldenlandia affinis DC) on the oestrogen dominated uterus. Acta Pharmacol Toxicol (Copenh) 1973;33:400–408.

199. Wang CK, Kaas Q, Chiche L, Craik DJ. CyBase: a database of cyclic protein sequences and structures, with applications in protein discovery and engineering. Nucleic Acids Res 2008;36:D206–D210.

200. Saether O, Craik DJ, Campbell ID, Sletten K, Juul J, Norman DG. Elucidation of the primary and three-dimensional structure of the uterotonic polypeptide kalata B1. Biochemistry 1995;34:4147–4158.

201. Gruber CW, Elliott AG, Ireland DC, Delprete PG, Dessein S, Goransson U, Trabi M, Wang CK, Kinghorn AB, Robbrecht E, Craik DJ. Distribution and evolution of circular miniproteins in flowering plants. Plant cell 2008;20:2471–2483.

202. Clark RJ, Daly NL, Craik DJ. Structural plasticity of the cyclic-cystine-knot framework: implications for biological activity and drug design. Biochem J 2006;394:85–93.

203. Henriques ST, Craik DJ. Importance of the cell membrane on the mechanism of action of cyclotides. ACS Chem Biol 2012;7:626–636.

204. Henriques ST, Huang YH, Castanho MA, Bagatolli LA, Sonza S, Tachedjian G, Daly NL, Craik DJ. Phosphatidylethanolamine binding is a conserved feature of cyclotide-membrane interactions. J Biol Chem 2012;287:33629–33643.

205. Craik DJ, Daly NL, Mulvenna J, Plan MR, Trabi M. Discovery, structure and biological activities of the cyclotides. Curr Protein Pept Sci 2004;5:297–315.

206. Hernandez JF, Gagnon J, Chiche L, Nguyen TM, Andrieu JP, Heitz A, Trinh Hong T, Pham TT, Le Nguyen D. Squash trypsin inhibitors from Momordica cochinchinensis exhibit an atypical macrocyclic structure. Biochemistry 2000;39:5722–5730.

207. Mylne JS, Chan LY, Chanson AH, Daly NL, Schaefer H, Bailey TL, Nguyencong P, Cascales L, Craik DJ. Cyclic peptides arising by evolutionary parallelism via asparaginyl-endopeptidase-mediated biosynthesis. Plant cell 2012;24:2765–2778.

208. He WJ, Chan LY, Clark RJ, Tang J, Zeng GZ, Franco OL, Cantacessi C, Craik DJ, Daly NL, Tan NH. Novel inhibitor cystine knot peptides from Momordica charantia. PLoS ONE 2013;8:e75334.

209. Heitz A, Hernandez JF, Gagnon J, Hong TT, Pham TT, Nguyen TM, Le-Nguyen D, Chiche L. Solution structure of the squash trypsin inhibitor MCoTI-II.A new family for cyclic knottins. Biochemistry 2001;40:7973–7983.

210. Ireland DC, Colgrave ML, Craik DJ. A novel suite of cyclotides from Viola odorata: sequence variation and the implications for structure, function and stability. Biochem J 2006;400:1–12.

211. Plan MR, Goransson U, Clark RJ, Daly NL, Colgrave ML, Craik DJ. The cyclotide fingerprint in oldenlandia affinis: elucidation of chemically modified, linear and novel macrocyclic peptides. ChemBioChem 2007;8:1001–1011.

212. Jennings CV, Rosengren KJ, Daly NL, Plan M, Stevens J, Scanlon MJ, Waine C, Norman DG, Anderson MA, Craik DJ. Isolation, solution structure, and insecticidal activity of kalata B2, a circular protein with a twist: do Mobius strips exist in nature? Biochemistry 2005;44:851–860.

213. Barbeta BL, Marshall AT, Gillon AD, Craik DJ, Anderson MA. Plant cyclotides disrupt epithelial cells in the midgut of lepidopteran larvae. Proc Natl Acad Sci U S A 2008;105:1221–1225.

214. Daly NL, Love S, Alewood PF, Craik DJ. Chemical synthesis and folding pathways of large cyclic polypeptides: studies of the cystine knot polypeptide kalata B1. Biochemistry 1999;38:10606–10614.

215. Tam JP, Lu YA, Yang JL, Chiu KW. An unusual structural motif of antimicrobial peptides containing end-to-end macrocycle and cystine-knot disulfides. Proc Natl Acad Sci U S A 1999;96:8913–8918.

216. Goransson U, Sjogren M, Svangard E, Claeson P, Bohlin L. Reversible antifouling effect of the cyclotide cycloviolacin O2 against barnacles. J Nat Prod 2004;67:1287–1290.

217. Plan MR, Saska I, Cagauan AG, Craik DJ. Backbone cyclised peptides from plants show molluscicidal activity against the rice pest Pomacea canaliculata (golden apple snail). J Agric Food Chem 2008;56:5237–5241.

218. Simonsen SM, Sando L, Rosengren KJ, Wang CK, Colgrave ML, Daly NL, Craik DJ. Alanine scanning mutagenesis of the prototypic cyclotide reveals a cluster of residues essential for bioactivity. J Biol Chem 2008;283:9805–9813.

219. Huang YH, Colgrave ML, Clark RJ, Kotze AC, Craik DJ. Lysine-scanning mutagenesis reveals an amendable face of the cyclotide kalata B1 for the optimization of nematocidal activity. J Biol Chem 2010;285:10797–10805.

220. Henriques ST, Huang YH, Rosengren KJ, Franquelim HG, Carvalho FA, Johnson A, Sonza S, Tachedjian G, Castanho MA, Daly NL, Craik DJ. Decoding the membrane

activity of the cyclotide kalata B1: the importance of phosphatidylethanolamine phospholipids and lipid organization on hemolytic and anti-HIV activities. J Biol Chem 2011;286:24231–24241.

221. Colgrave ML, Craik DJ. Thermal, chemical, and enzymatic stability of the cyclotide kalata B1: the importance of the cyclic cystine knot. Biochemistry 2004;43:5965–5975.

222. Chan LY, Gunasekera S, Henriques ST, Worth NF, Le SJ, Clark RJ, Campbell JH, Craik DJ, Daly NL. Engineering pro-angiogenic peptides using stable, disulfide-rich cyclic scaffolds. Blood 2011;118:6709–6717.

223. Eliasen R, Daly NL, Wulff BS, Andresen TL, Conde-Frieboes KW, Craik DJ. Design, synthesis, structural and functional characterization of novel melanocortin agonists based on the cyclotide kalata B1. J Biol Chem 2012;287:40493–40501.

224. Wong CT, Rowlands DK, Wong CH, Lo TW, Nguyen GK, Li HY, Tam JP. Orally active peptidic bradykinin B1 receptor antagonists engineered from a cyclotide scaffold for inflammatory pain treatment. Angew Chem 2012;51:5620–5624.

225. Contreras J, Elnagar AY, Hamm-Alvarez SF, Camarero JA. Cellular uptake of cyclotide MCoTI-I follows multiple endocytic pathways. J Controlled Release: Off J Controlled Release Soc 2011;155:134–143.

226. Greenwood KP, Daly NL, Brown DL, Stow JL, Craik DJ. The cyclic cystine knot miniprotein MCoTI-II is internalized into cells by macropinocytosis. Int J Biochem Cell Biol 2007;39:2252–2264.

227. Cascales L, Henriques ST, Kerr MC, Huang YH, Sweet MJ, Daly NL, Craik DJ. Identification and characterization of a new family of cell-penetrating peptides: cyclic cell-penetrating peptides. J Biol Chem 2011;286:36932–36943.

228. Ji Y, Majumder S, Millard M, Borra R, Bi T, Elnagar AY, Neamati N, Shekhtman A, Camarero JA. In vivo activation of the p53 tumor suppressor pathway by an engineered cyclotide. J Am Chem Soc 2013;135:11623–11633.

229. Lewis RJ, Dutertre S, Vetter I, Christie MJ. Conus Venom Peptide Pharmacology. Pharmacol Rev 2012;64:259–298.

230. Kaas Q, Craik DJ. Conotoxins and other conopeptides. In: La Barre S, Kornprobst J-M, editors. *Outstanding Marine Molecules: Chemistry, Biology, Analysis*. Germany: Wiley-VCH Verlag GmbH & Co KGaA; 2014. p 319–332.

231. Craig AG, Bandyopadhyay P, Olivera BM. Post-translationally modified neuropeptides from Conus venoms. Eur J Biochem 1999;264:271–275.

232. Myers RA, Cruz LJ, Rivier JE, Olivera BM. Conus peptides as chemical probes for receptors and ion channels. Chem Rev 1993;93:1923–1936.

233. Buczek O, Bulaj G, Olivera BM. Conotoxins and the posttranslational modification of secreted gene products. Cell Mol Life Sci: CMLS 2005;62:3067–3079.

234. Cruz LJ, Gray WR, Yoshikami D, Olivera BM. *Conus* venoms: a rich source of neuroactive peptides. Toxin Rev 1985;4:107–132.

235. McIntosh JM, Santos AD, Olivera BM. Conus peptides targeted to specific nicotinic acetylcholine receptor subtypes. Annu Rev Biochem 1999;68:59–88.

236. Olivera BM. *Conus* venom peptides: reflecations from the biology of clades and species. Annu Rev Ecol Syst 2002;33:25–47.

237. Halai R, Craik DJ. Conotoxins: natural product drug leads. Nat Prod Rep 2009;26:526–536.

238. Kaas Q, Westermann JC, Halai R, Wang CK, Craik DJ. ConoServer, a database for conopeptide sequences and structures. Bioinformatics 2008;24:445–446.

239. Olivera BM, McIntosh JM, Cruz LJ, Luque FA, Gray WR. Purification and sequence of a presynaptic peptide toxin from Conus geographus venom. Biochemistry 1984;23:5087–5090.

240. Cruz LJ, Johnson DS, Olivera BM. Characterization of the omega-conotoxin target. Evidence for tissue-specific heterogeneity in calcium channel types. Biochemistry 1987;26:820–824.

241. Cruz LJ, Gray WR, Olivera BM, Zeikus RD, Kerr L, Yoshikami D, Moczydlowski E. Conus geographus toxins that discriminate between neuronal and muscle sodium channels. J Biol Chem 1985;260:9280–9288.

242. Fainzilber M, Kofman O, Zlotkin E, Gordon D. A new neurotoxin receptor site on sodium channels is identified by a conotoxin that affects sodium channel inactivation in molluscs and acts as an antagonist in rat brain. J Biol Chem 1994;269:2574–2580.

243. Shon KJ, Stocker M, Terlau H, Stuhmer W, Jacobsen R, Walker C, Grilley M, Watkins M, Hillyard DR, Gray WR, Olivera BM. kappa-Conotoxin PVIIA is a peptide inhibiting the shaker K + channel. J Biol Chem 1998;273:33–38.

244. Daly NL, Rosengren KJ, Henriques ST, Craik DJ. NMR and protein structure in drug design: application to cyclotides and conotoxins. Euro Biophys J: EBJ 2011;40:359–370.

245. Marx UC, Daly NL, Craik DJ. NMR of conotoxins: structural features and an analysis of chemical shifts of post-translationally modified amino acids. Mag Reson Chem 2006;44:S41–S50.

246. Kobayashi Y, Ohkubo T, Kyogoku Y, Nishiuchi Y, Sakakibara S, Braun W, Go N. Solution conformation of conotoxin GI determined by 1H nuclear magnetic resonance spectroscopy and distance geometry calculations. Biochemistry 1989;28:4853–4860.

247. Lancelin JM, Kohda D, Tate S, Yanagawa Y, Abe T, Satake M, Inagaki F. Tertiary structure of conotoxin GIIIA in aqueous solution. Biochemistry 1991;30:6908–6916.

248. Davis JH, Bradley EK, Miljanich GP, Nadasdi L, Ramachandran J, Basus VJ. Solution structure of omega-conotoxin GVIA using 2-D NMR spectroscopy and relaxation matrix analysis. Biochemistry 1993;32:7396–7405.

249. Kohno T, Sasaki T, Kobayashi K, Fainzilber M, Sato K. Three-dimensional solution structure of the sodium channel agonist/antagonist delta-conotoxin TxVIA. J Biol Chem 2002;277:36387–36391.

250. Savarin P, Guenneugues M, Gilquin B, Lamthanh H, Gasparini S, Zinn-Justin S, Menez A. Three-dimensional structure of kappa-conotoxin PVIIA, a novel potassium channel-blocking toxin from cone snails. Biochemistry 1998;37:5407–5416.

251. Cornet B, Bonmatin JM, Hetru C, Hoffmann JA, Ptak M, Vovelle F. Refined three-dimensional solution structure of insect defensin A. Structure 1995;3:435–448.

252. Hill JM, Alewood PF, Craik DJ. Three-dimensional solution structure of mu-conotoxin GIIIB, a specific blocker of skeletal muscle sodium channels. Biochemistry 1996;35:8824–8835.

253. Vita C, Roumestand C, Toma F, Menez A. Scorpion toxins as natural scaffolds for protein engineering. Proc Natl Acad Sci U S A 1995;92:6404–6408.

254. Basus VJ, Nadasdi L, Ramachandran J, Miljanich GP. Solution structure of omega-conotoxin MVIIA using 2D NMR spectroscopy. FEBS Lett 1995;370:163–169.

255. Kohno T, Kim JI, Kobayashi K, Kodera Y, Maeda T, Sato K. Three-dimensional structure in solution of the calcium channel blocker omega-conotoxin MVIIA. Biochemistry 1995;34:10256–10265.
256. Kobayashi K, Sasaki T, Sato K, Kohno T. Three-dimensional solution structure of omega-conotoxin TxVII, an L-type calcium channel blocker. Biochemistry 2000;39:14761–14767.
257. Pallaghy PK, Duggan BM, Pennington MW, Norton RS. Three-dimensional structure in solution of the calcium channel blocker omega-conotoxin. J Mol Biol 1993;234:405–420.
258. Sevilla P, Bruix M, Santoro J, Gago F, Garcia AG, Rico M. Three-dimensional structure of omega-conotoxin GVIA determined by 1H NMR. Biochem Biophys Res Commun 1993;192:1238–1244.
259. Skalicky JJ, Metzler WJ, Ciesla DJ, Galdes A, Pardi A. Solution structure of the calcium channel antagonist omega-conotoxin GVIA. Protein Sci 1993;2:1591–1603.
260. Terlau H, Shon KJ, Grilley M, Stocker M, Stuhmer W, Olivera BM. Strategy for rapid immobilization of prey by a fish-hunting marine snail. Nature 1996;381:148–151.
261. Nicke A, Wonnacott S, Lewis RJ. Alpha-conotoxins as tools for the elucidation of structure and function of neuronal nicotinic acetylcholine receptor subtypes. Eur J Biochem 2004;271:2305–2319.
262. Ulens C, Hogg RC, Celie PH, Bertrand D, Tsetlin V, Smit AB, Sixma TK. Structural determinants of selective alpha-conotoxin binding to a nicotinic acetylcholine receptor homolog AChBP. Proc Natl Acad Sci U S A 2006;103:3615–3620.
263. Dutertre S, Ulens C, Buttner R, Fish A, van Elk R, Kendel Y, Hopping G, Alewood PF, Schroeder C, Nicke A, Smit AB, Sixma TK, Lewis RJ. AChBP-targeted alpha-conotoxin correlates distinct binding orientations with nAChR subtype selectivity. Embo J 2007;26:3858–3867.
264. Celie PH, Kasheverov IE, Mordvintsev DY, Hogg RC, van Nierop P, van Elk R, van Rossum-Fikkert SE, Zhmak MN, Bertrand D, Tsetlin V, Sixma TK, Smit AB. Crystal structure of nicotinic acetylcholine receptor homolog AChBP in complex with an alpha-conotoxin PnIA variant. Nat Struct Mol Biol 2005;12:582–588.
265. Dutton JL, Craik DJ. alpha-Conotoxins: nicotinic acetylcholine receptor antagonists as pharmacological tools and potential drug leads. Curr Med Chem 2001;8:327–344.
266. Millard EL, Daly NL, Craik DJ. Structure-activity relationships of alpha-conotoxins targeting neuronal nicotinic acetylcholine receptors. Eur J Biochem 2004;271:2320–2326.
267. Becker S, Terlau H. Toxins from cone snails: properties, applications and biotechnological production. Appl Microbiol Biotechnol 2008;79:1–9.
268. Zhang SJ, Yang XM, Liu GS, Cohen MV, Pemberton K, Downey JM. CGX-1051, a peptide from Conus snail venom, attenuates infarction in rabbit hearts when administered at reperfusion. J Cardiovasc Pharmacol 2003;42:764–771.
269. Freedman R, Coon H, Myles-Worsley M, Orr-Urtreger A, Olincy A, Davis A, Polymeropoulos M, Holik J, Hopkins J, Hoff M, Rosenthal J, Waldo MC, Reimherr F, Wender P, Yaw J, Young DA, Breese CR, Adams C, Patterson D, Adler LE, Kruglyak L, Leonard S, Byerley W. Linkage of a neurophysiological deficit in schizophrenia to a chromosome 15 locus. Proc Natl Acad Sci U S A 1997;94:587–592.
270. Nordberg A, Nilsson-Hakansson L, Adem A, Hardy J, Alafuzoff I, Lai Z, Herrera-Marschitz M, Winblad B. The role of nicotinic receptors in the pathophysiology of Alzheimer's disease. Prog Brain Res 1989;79:353–362.

271. Janson AM, Fuxe K, Sundstrom E, Agnati LF, Goldstein M. Chronic nicotine treatment partly protects against the 1-methyl-4-phenyl-2,3,6-tetrahydropyridine-induced degeneration of nigrostriatal dopamine neurons in the black mouse. Acta Physiol Scand 1988;132:589–591.
272. Craik DJ, Adams DJ. Chemical modification of conotoxins to improve stability and activity. ACS Chem Biol 2007;2:457–468.
273. Bobst CE, Abzalimov RR, Houde D, Kloczewiak M, Mhatre R, Berkowitz SA, Kaltashov IA. Detection and characterization of altered conformations of protein pharmaceuticals using complementary mass spectrometry-based approaches. Anal Chem 2008;80:7473–7481.
274. Boschi-Muller S, Gand A, Branlant G. The methionine sulfoxide reductases: Catalysis and substrate specificities. Arch Biochem Biophys 2008;474:266–273.
275. Wakankar AA, Borchardt RT. Formulation considerations for proteins susceptible to asparagine deamidation and aspartate isomerization. J Pharm Sci 2006;95:2321–2336.
276. Olsen JV, Ong SE, Mann M. Trypsin cleaves exclusively C-terminal to arginine and lysine residues. Mol Cell Proteomics 2004;3:608–614.
277. Vajda T, Szabo T. Specificity of trypsin and alpha-chymotrypsin towards neutral substrates. Acta Biochim Biophys Acad Sci Hung 1976;11:287–294.
278. Kanodia S, Kast WM. Peptide-based vaccines for cancer: realizing their potential. Expert Rev Vaccines 2008;7:1533–1545.
279. Frecer V, Ho B, Ding JL. De novo design of potent antimicrobial peptides. Antimicrob Agents Chemother 2004;48:3349–3357.
280. Martinez-Rodriguez S, Martinez-Gomez AI, Rodriguez-Vico F, Clemente-Jimenez JM, Las Heras-Vazquez FJ. Natural occurrence and industrial applications of D-amino acids: an overview. Chem Biodiv 2010;7:1531–1548.
281. Steer DL, Lew RA, Perlmutter P, Smith AI, Aguilar MI. Beta-amino acids: versatile peptidomimetics. Curr Med Chem 2002;9:811–822.
282. Martinek TA, Fulop F. Side-chain control of beta-peptide secondary structures. Eur J Biochem 2003;270:3657–3666.
283. Marschutz MK, Zauner W, Mattner F, Otava A, Buschle M, Bernkop-Schnurch A. Improvement of the enzymatic stability of a cytotoxic T-lymphocyte-epitope model peptide for its oral administration. Peptides 2002;23:1727–1733.
284. Brinckerhoff LH, Kalashnikov VV, Thompson LW, Yamshchikov GV, Pierce RA, Galavotti HS, Engelhard VH, Slingluff CL Jr. Terminal modifications inhibit proteolytic degradation of an immunogenic MART-1(27–35) peptide: implications for peptide vaccines. Int J Cancer 1999;83:326–334.
285. Bailon P, Won CY. PEG-modified biopharmaceuticals. Expert Opin Drug Delivery 2009;6:1–16.
286. Craik DJ. Chemistry. Seamless proteins tie up their loose ends. Science 2006;311:1563–1564.
287. Clark RJ, Fischer H, Dempster L, Daly NL, Rosengren KJ, Nevin ST, Meunier FA, Adams DJ, Craik DJ. Engineering stable peptide toxins by means of backbone cyclization: stabilization of the alpha-conotoxin MII. Proc Natl Acad Sci U S A 2005;102:13767–13772.
288. Cartier GE, Yoshikami D, Gray WR, Luo S, Olivera BM, McIntosh JM. A new alpha-conotoxin which targets alpha3beta2 nicotinic acetylcholine receptors. J Biol Chem 1996;271:7522–7528.

289. Quik M, Polonskaya Y, Kulak JM, McIntosh JM. Vulnerability of 125I-alpha-conotoxin MII binding sites to nigrostriatal damage in monkey. J Neurosci 2001;21:5494–5500.
290. Lovelace ES, Armishaw CJ, Colgrave ML, Wahlstrom ME, Alewood PF, Daly NL, Craik DJ. Cyclic MrIA: a stable and potent cyclic conotoxin with a novel topological fold that targets the norepinephrine transporter. J Med Chem 2006;49:6561–6568.
291. Armishaw CJ, Dutton JL, Craik DJ, Alewood PF. Establishing regiocontrol of disulfide bond isomers of alpha-conotoxin ImI via the synthesis of N-to-C cyclic analogs. Biopolymers 2010;94:307–313.
292. Halai R, Callaghan B, Daly NL, Clark RJ, Adams DJ, Craik DJ. Effects of cyclization on stability, structure, and activity of alpha-conotoxin RgIA at the alpha9alpha10 nicotinic acetylcholine receptor and GABA(B) receptor. J Med Chem 2011;54:6984–6992.
293. Clark RJ, Jensen J, Nevin ST, Callaghan BP, Adams DJ, Craik DJ. The engineering of an orally active conotoxin for the treatment of neuropathic pain. Angew Chem 2010;49:6545–6548.
294. Unger T, Oren Z, Shai Y. The effect of cyclization of magainin 2 and melittin analogues on structure, function, and model membrane interactions: implication to their mode of action. Biochemistry 2001;40:6388–6397.
295. Hogg PJ. Disulfide bonds as switches for protein function. Trends Biochem Sci 2003;28:210–214.
296. Bulaj G. Formation of disulfide bonds in proteins and peptides. Biotechnol Adv 2005;23:87–92.
297. Landon C, Barbault F, Legrain M, Guenneugues M, Vovelle F. Rational design of peptides active against the gram positive bacteria Staphylococcus aureus. Proteins 2008;72:229–239.
298. Ahn HS, Cho W, Kim JM, Joshi BP, Park JW, Lohani CR, Cho H, Lee KH. Design and synthesis of cyclic disulfide-bonded antibacterial peptides on the basis of the alpha helical domain of Tenecin 1, an insect defensin. Bioorg Med Chem 2008;16:4127–4137.
299. Hu SH, Gehrmann J, Alewood PF, Craik DJ, Martin JL. Crystal structure at 1.1 A resolution of alpha-conotoxin PnIB: comparison with alpha-conotoxins PnIA and GI. Biochemistry 1997;36:11323–11330.
300. Santiveri CM, Leon E, Rico M, Jimenez MA. Context-dependence of the contribution of disulfide bonds to beta-hairpin stability. Chemistry 2008;14:488–499.
301. Jackson DY, King DS, Chmielewsky J, Singh S, Schultz PG. General Approach to the synthesis of short α-helical peptides. J Am Chem Soc 1991;113:9391–9392.
302. Pease JH, Storrs RW, Wemmer DE. Folding and activity of hybrid sequence, disulfide-stabilized peptides. Proc Natl Acad Sci U S A 1990;87:5643–5647.
303. Vila-Perello M, Tognon S, Sanchez-Vallet A, Garcia-Olmedo F, Molina A, Andreu D. A minimalist design approach to antimicrobial agents based on a thionin template. J Med Chem 2006;49:448–451.
304. Oren Z, Shai Y. Cyclization of a cytolytic amphipathic alpha-helical peptide and its diastereomer: effect on structure, interaction with model membranes, and biological function. Biochemistry 2000;39:6103–6114.
305. Craik DJ. Protein folding: turbo-charged crosslinking. Nat Chem 2012;4:600–602.
306. Walewska A, Han TS, Zhang MM, Yoshikami D, Bulaj G, Rolka K. Expanding chemical diversity of conotoxins: peptoid-peptide chimeras of the sodium channel blocker mu-KIIIA and its selenopeptide analogues. Eur J Med Chem 2013;65:144–150.

307. Muttenthaler M, Nevin ST, Grishin AA, Ngo ST, Choy PT, Daly NL, Hu SH, Armishaw CJ, Wang CI, Lewis RJ, Martin JL, Noakes PG, Craik DJ, Adams DJ, Alewood PF. Solving the alpha-conotoxin folding problem: efficient selenium-directed on-resin generation of more potent and stable nicotinic acetylcholine receptor antagonists. J Am Chem Soc 2010;132:3514–3522.

308. de Araujo AD, Mobli M, Castro J, Harrington AM, Vetter I, Dekan Z, Muttenthaler M, Wan J, Lewis RJ, King GF, Brierley SM, Alewood PF. Selenoether oxytocin analogues have analgesic properties in a mouse model of chronic abdominal pain. Nat Commun 2014;5:3165.

# 7

# MODIFICATION OF PEPTIDES TO LIMIT METABOLISM

Isuru R. Kumarasinghe and Victor J. Hruby

*Department of Chemistry and Biochemistry, The University of Arizona, Tucson, AZ, USA*

## 7.1 INTRODUCTION

A common property associated with many naturally occurring peptides is that they have a short effective half-life *in vivo* due to proteolytic degradation. The effective dose of peptide required to maintain *in vivo* concentrations often requires administration of large amounts of peptide which can lead to side effects [1]. On the other hand, in the body large arrays of peptides are constantly synthesized and used as hormones, neurotransmitters, and for other functions which are targeted toward receptors, acceptors, enzymes, and other activities needed to maintain life and homeostasis. They exist either folded, or posttranslationally modified, or attached to carrier transport molecules which often mask them from being recognized by the proteolytic enzymes. Furthermore, proteolytic processing comes into action and plays a vital role in the body when there is misfolding, for degradation of undesired proteins, or for producing biologically active peptides from larger precursor proteins, or for responding to a foreign host, or for decreasing the amount of neurotransmitters to avoid receptor over-activation, or for performing actions in a particular disease state.

In this chapter, we will focus on approaches that can be used to limit peptide metabolism. For developing peptides as potential drugs for treatment of disease, questions related to the peptides mode of action need to be addressed, so that the

*Peptide Chemistry and Drug Design*, First Edition. Edited by Ben M. Dunn.

designed peptides will have the desired biological activity profile. At the same time, it is possible to design strategic modifications of the peptide structure, which will also address issues of stability to proteolysis and biodistribution. The literature in this area is vast so our discussion will be limited to a few examples, which exemplify the approach(es), and thus many aspects will not be discussed.

The design process first requires identification of the pharmacophore residues important for bioactivity [2]. This will be different depending upon whether agonists or antagonists are needed. Structure–activity studies with truncations, deletions of amino acid residues, alanine scans, D-amino acid scans, and so on, are often needed at this stage. In addition it may be possible to identify secondary structure elements important for the desired bioactivity profile using a combination of various cyclic scans, molecular modeling using different force fields and solvent models, and biophysical studies including CD, NMR, and X-ray crystallography [2]. All of these steps provide opportunities to improve peptide stability and bioavailability.

The key considerations in all of these approaches are related to various aspects of structure. In this regard, though we will not explicitly discuss it here (because it is not the topic of this chapter), structural consideration that lead to improved stability of peptides against biodegradation also often improve the chemical–physical properties that render peptides and peptidomimetics more bioavailable and more likely to cross membrane barriers such as the blood–brain-barrier (BBB). This is particularly true when certain secondary structural features are enhanced such as stabilization of α-helical and β-turn structures. Additionally, when certain inherent chemical features such as enhanced lipophilicity, amphipathic properties, pegylation, and so on, are "built into" the structures by design, properties such as interactions with bloodborne proteins or bypassing the liver or kidney elimination routes are obtained. The point to be made is that there are many known approaches that work, and because these generally are only recently exploited in drug design and development, there is much to learn and much to investigate that will provide not only important new scientific principles and knowledge, but also will open up new chemical space for intellectual property enhancement.

We now will systematically, though not comprehensively, examine a number of strategies that can be used to enhance peptide stability *in vivo*.

## 7.2 INTRODUCTION OF UNNATURAL AMINO ACIDS

Naturally occurring amino acids in the native peptide can be replaced by nonnatural amino acids in order to provide a peptide with a side-chain not as easily recognized by proteolytic enzymes, but still compatible with bioactivity. For example, valine can be replaced by the cyclohexylglycine to mimic the side chain of the valine. Likewise, for basic amino acids ornithine can replace lysine. Other possibilities such as 4-hydroxyproline for proline, norleucine instead of the leucine or methionine, cyclohexylalanine or phenylglycine(Phg), 3-pyridylalanine(3Pal), homophenylalanine(HomoPhe), or 2′-naphthylalanine (2′-Nal) instead of phenylalanine [3] (Figure 7.1), or thienylalanine for histidine, and so on.

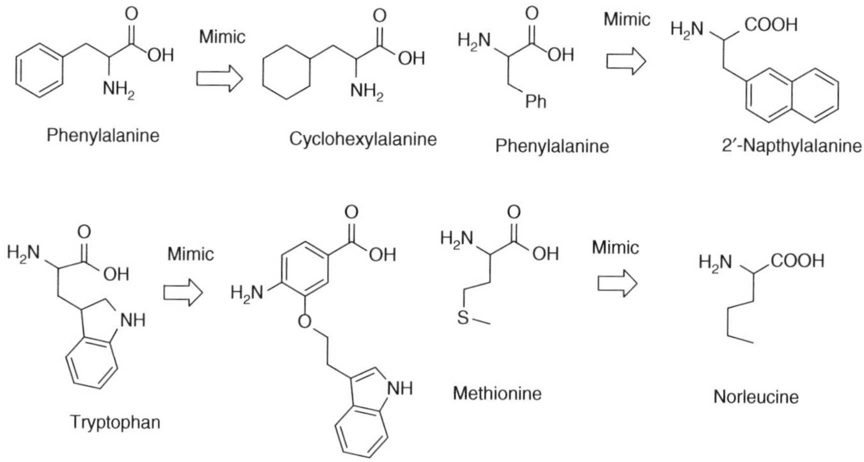


**Figure 7.1** Some examples of nonnatural amino used as substitutes to minimize protease degradation of peptides.

A good example of obtaining a proteolytic resistant peptide is the work of Schmiedeberg et al. [4] who found, in the process of developing an antagonist that can disrupt interactions between the endogenous serine protease urokinase type plasminogen activator (uPA) that substitution of a key Lys residue with unnatural amino acids such as Nle, Dab, Orn, and Dap gave potent uPAR antagonists which were highly resistant to proteolytic degradation.

A more recent example includes substitutions such as 4-amino-3-(benzyloxy) benzoic acid, 3-(2-(1*H*-indol-3-yl)ethoxy)-4-aminobenzoic acid, and 4-amino-3-isobutoxybenzoic acid for phenylalanine, tryptophan, and leucine, respectively [5]. Such modifications can be used to fine-tune the native peptides to be lipophilic, make them less prone to enzymatic degradation, and increase their uptake into the brain.

## 7.3 CYCLIZATION OF LINEAR PEPTIDES TO IMPROVE STABILITY TOWARD BLOOD AND BRAIN PROTEASE DEGRADATION

It has been known for many years that cyclization of a linear peptide usually stabilizes the peptide to proteolytic enzymes compared with the linear counterpart. Moreover, cyclization of a peptide often results in greater selectivity and potency of a peptide for its target. For side-chain-to-side-chain cyclization [6], the identification of the tolerant positions for substitution by amino acids, which can be cyclized without significant loss of activity and affinity is a key step in this process. Cyclization can be done (Figure 7.2) through peptide-bond formation between the amino and carboxyl groups of the N- and C-terminal (e.g., head to tail cyclization), or amide bond formation between the amino group of the N-terminal and a side-chain carboxy group, or amide bond formation of between the C-terminal and a carboxyl side-chain amino group

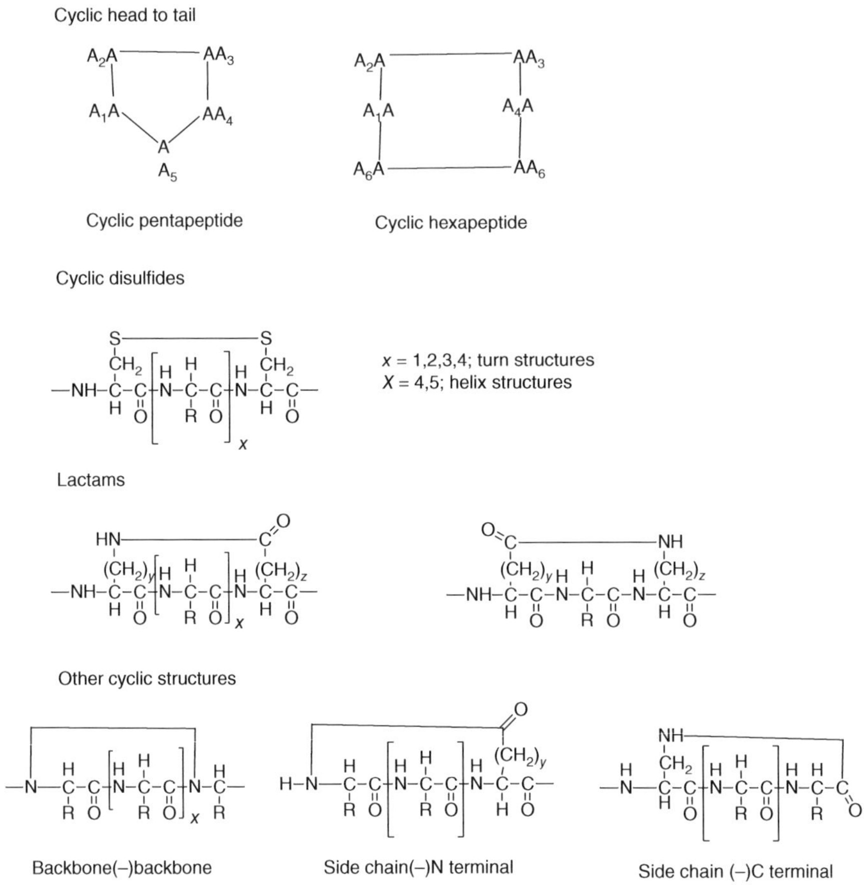


**Figure 7.2** Different types of peptide cyclization [2].

(e.g., C-terminal of any amino acid with the ε-amino group of a lysine), or amide bond formation between the side chain amino and carboxy groups (e.g., carboxylic group of the glutamic acid or aspartic acid side-chain with amino group of the lysine side chain). Of course, disulfide bond formation between thiol groups of cysteine residues is widely used in nature and in peptide design. In addition other types of cyclization have been reported using other functional groups, for example, thioethers, ethers, amides, di-carba bonds, esters, or heterocyclic type cyclizations [2, 7].

Many years ago it was shown that cyclization of linear peptides related to α-MSH around the key pharmacophore His-DPhe-Arg-Trp- via a disulfide structure Ac-Ser-Tyr-Ser-Nle-c[Cys-His-DPhe-Arg-Trp-Cys]-Lys-Pro-Val-$NH_2$, or through a lactam bridge Ac-Nle-c[Asp-His-DPhe-Arg-Trp-Lys]$NH_2$ gave enzymatically stable ligands, the latter of which also crossed the BBB [8–10]. More recently in search for α-MSH derivatives that would be melanocortin-4 receptor selective ligands and that would be orally active [11, 12] made backbone to backbone cyclic peptides around the pharmacophore Phe-DPhe-Arg-Trp-Gly-$NH_2$, which have high metabolic stability to intestinal brush border enzymes compared to linear counterparts (Figure 7.3a and b illustrates the structure and stability of these structures).

In the process of making a selective agonist and antagonist ligands for the neurokinin (NK) receptor subtypes NK1, NK2, and NK3 based on a common C-terminal parent tachykinin peptide Phe-Xaa-Gly-Leu-Met-$NH_2$, Byk et al. [13] found that the linear analogs are less metabolically stable to proteases present in slices or homogenates of liver and of the parotid gland compared to backbone cyclic analogs.

Cyclic peptides are generally metabolically stable comparable to their linear counterparts because they assume conformations that are not compatible with the enzymatic cleave site of proteases. In addition, the bulkiness and rigidity of cyclic peptides may not allow them to enter into the catalytic pocket that is required for proteolysis [14].

Cyclization of linear peptides often results in improved metabolic stability against proteolytic enzymes, and it also can enhance the selectivity and potency of peptides. Leu enkephalin (H-Tyr-Gly-Gly-Phe-Leu-OH) and Met enkephalin (H-Tyr-Gly-Gly-Phe-Met-OH) are endogenous peptides for opioid receptors. These peptides are rapidly degraded *in vivo*. Cyclization of these peptides by substitution the $Gly^2$ and $Leu^5/Met^5$ with cysteine did not improve receptor selectivity significantly. However, with $\beta$,β-dimethylcysteine (penicillamine) the receptor selectivity was enhanced to give delta opioid receptor ligands, and also greatly improved the metabolic stability against the wide variety of enzymes and to the BBB [2, 15–17] (Figure 7.4).

Cyclization can be used even with highly active linear peptides that are somewhat stable to proteases (e.g., biphalin is a mu and delta agonist for the opioid receptor). Cyclization of biphalin with D-cysteine-2 and D-cysteine-2′ improved the potency five times in second messenger assays (Figure 7.5 and Table 7.1) [18].

Cyclization of endogenous neurotransmitters often results in molecules with prolonged activity *in vivo*. Prolong activity *in vivo* presumably occurs, at least in part, when peptides have higher metabolic stability against proteolytic enzymes. For instance cyclic α-MSH neurotransmitter analogs have prolonged activity *in vivo* compared to their linear counterparts [9, 10, 19, 20].

**TABLE 7.1 Binding Affinity, GTPγS Binding Assay, and Bioassay Results [18] of Biphalin and a Cyclic Analog.**

| Drugs | Binding $IC_{50}$(nM) | | GTPγS Binding $IC_{50}$ (nM) | | | | Bioassay $IC_{50}$(nM) | |
|---|---|---|---|---|---|---|---|---|
| | $\delta$ | $\mu$ | $\delta$ | $E_{max}$ (%) | $\mu$ | $E_{max}$ (%) | MVD | GPI |
| Biphalin | 2.6 | 1.4 | 2.5 | 27 | 6 | 25 | 27 | 8.8 |
| Cyclized biphalin | 0.87 | 0.60 | 0.87 | 100 | 0.2 | 47 | 9.9 | 25 |

$E_{max}$ % (net total bound/basal binding × 100) for *in vitro* tissue activities.

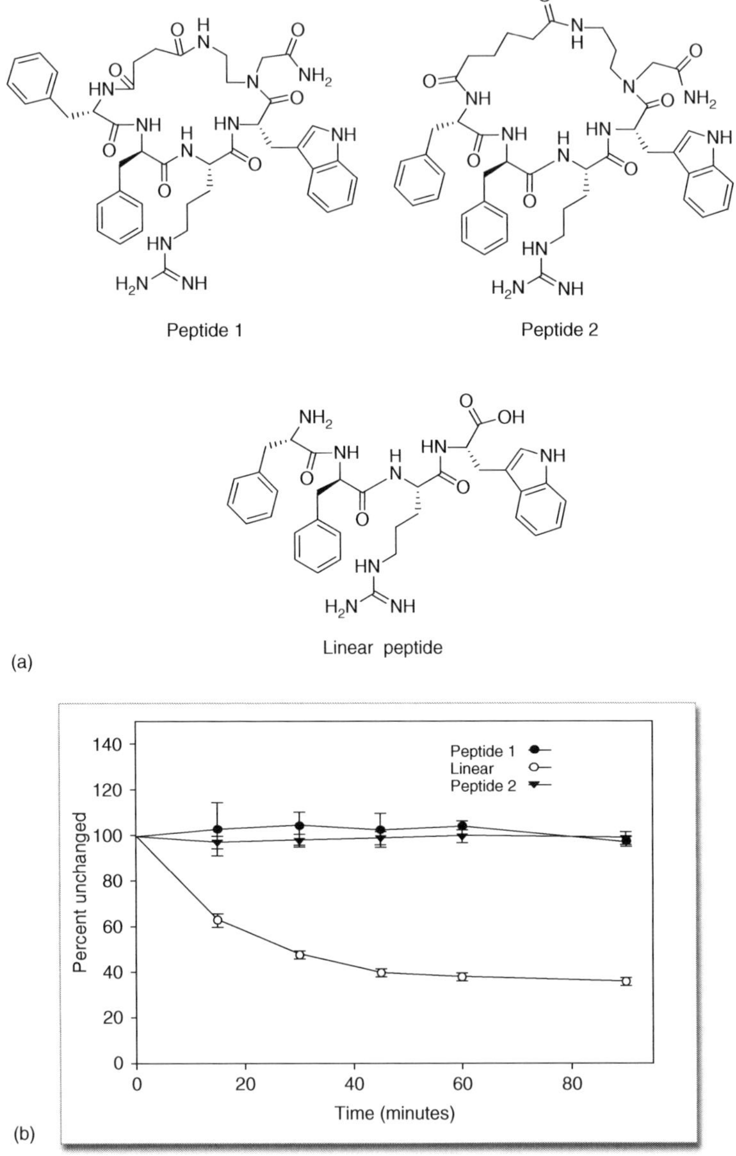


**Figure 7.3** (a) Structure of biostable peptides 1 and 2 and the related linear peptide [11]. (b) Metabolic stabilies of cyclic peptides 1 and 2 and the linear analog [11]. Tested molecules were mixed with purified brush border membrane vesicles (BBM Vs) and incubated at 37 °C for 90 min. Duplicate samples were taken at time 0 and after 15, 30, 45, 60, and 90 min. The samples were diluted 1:1 with ice cooled acetonitrile, centrifuged (7500 *g*, 10 min, 4 °C) and transfereed to analysis, SD < 15% (peptide 1 •, peptide 2 ▾, linear peptide analog ○).

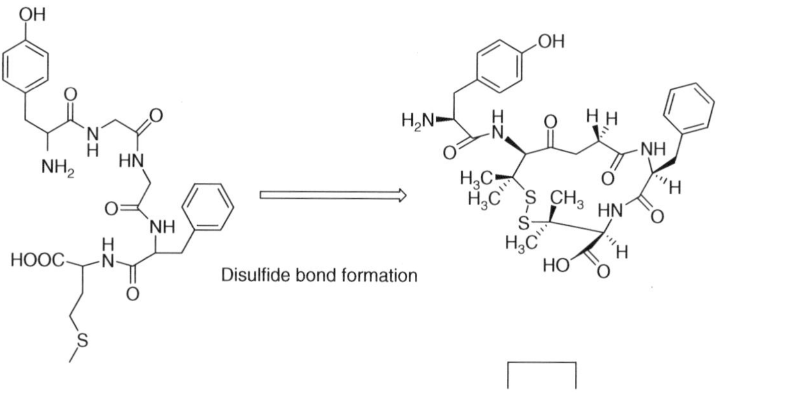


Leu-and met-Enkephalin

*c*[D-Pen$^2$, D-Pen$^5$]-enkephalin (DPDPE)

**Figure 7.4** Cyclization of side chain to side chain to constrain the global conformation [15].

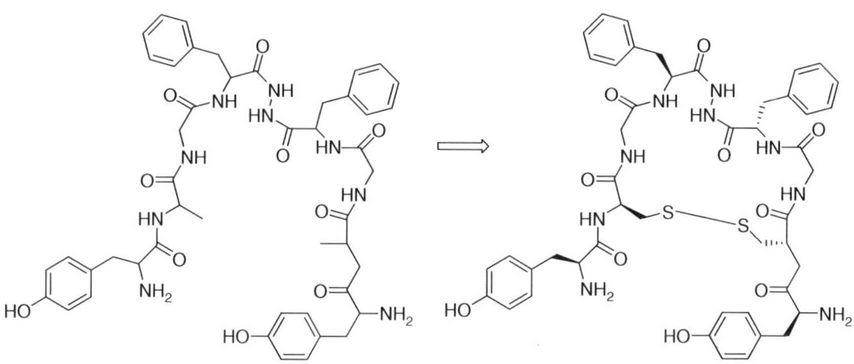

Biphalin

Cyclized biphalin

**Figure 7.5** Cyclization of the biphalin to give most potent cyclized biphalin [18].

## 7.4 INTRODUCTION OF D-AMINO ACIDS INTO PEPTIDES IMPROVES STABILITY TOWARD BLOOD AND BRAIN PROTEASE DEGRADATION

Incorporation of D-amino acids into a peptide sequence often can retain the same or improve bioactivity of the original native peptide, but generally also improves stability against proteolytic enzymes, and has been used often as a general approach in peptide ligand design. For example, Hamamoto et al. [21] have made antimicrobial peptides substituted with D-amino acids. Overall the D-amino substituted peptides had much higher stability toward enzymatic degradation. Sawyer et al. [22] have shown that incorporation of a D-amino acid into a central residue in α-MSH increases the stability of the peptide against serum enzymes.

In the process of developing of an antagonist to the bombesin (Bn) receptor with prolonged *in vivo* inhibitory activity Coy et al. found that modification of the linear peptide encompassing bombesin residues 6 or 7–14 (Asn-Gln-Trp-Ala-Val-Gly-His-Leu-Met-$NH_2$) with D-amino acid substitution at positions 6 and 11, along with C-terminal methyl ester modification resulted in an analog, [D-pentaflouro-$Phe^6$,D-$Ala^{11}$]Bn(6-13)-OMe which had very prolonged activity *in vivo* [23].

Often, D-amino acid substitution requires identification of the tolerant positions of the bioactive peptide, so that it will not lose significant bioactivity. Toke et al. [24] studied the antigenic properties and enzymatic stability of several MUC2 peptides partially substituted with D-amino acids in the flanking regions and found out that partial D-amino acid substitution was well tolerated, and that the peptides showed high resistance against proteolytic degradation in human serum and lysosomal preparations. Moreover, the results indicated that the stability against proteolytic degradation increases proportionally with the number of D-amino acids that were substituted.

## 7.5 INTRODUCTION OF β-AMINO ACIDS INCREASES THE STABILITY TOWARD BLOOD AND BRAIN PROTEASE DEGRADATION

β-Amino acids are alpha peptide analogs, in which a $CH_2$ group is inserted to the backbone of the proteogenic amino acids. Depending on the position of the R groups, β-amino acids can be divided into two types, $\beta^2$ or $\beta^3$. Type $\beta^1$ is when R is a hydrogen (Figure 7.6). A few β-amino acids are naturally occurring, for example, β-alanine [25].

β-Amino acid incorporation into a peptide backbone can induce secondary structural changes, but also can increase the degree of conformational flexibility of the native peptide [26]. In general, alpha peptides are more rapidly degraded by proteases than their β-peptide counterparts.

Incorporation of β-amino acid in normal peptides often can reduce affinity and activity toward their receptors/acceptors. Therefore, it is important to identify the tolerant residues of the native peptide, which can be substituted and still retain the same level of bioactivity.

As an alternative, alpha hydrazine type β-amino acids, where a carbon is replaced with a nitrogen atom in the amino acid have been synthesized, and they provide a different hydrogen bond network when substituted into the peptide to give Aza-$\beta^3$-peptide [31].

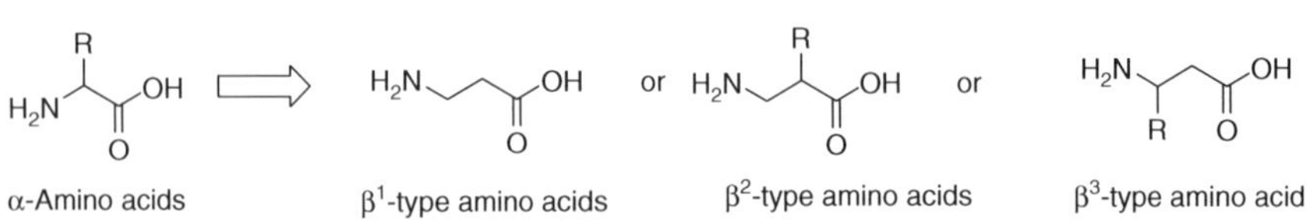


**Figure 7.6** Types of β-amino acids.

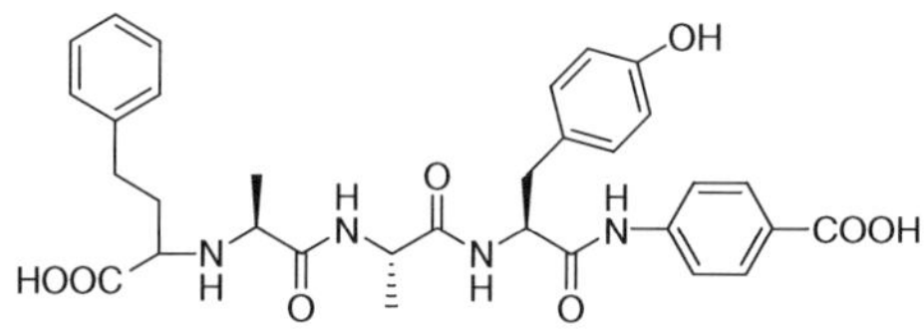

**Figure 7.7** Structure of EP24.15 inhibitor cpp-Ala-Tyr-*p*AB (CFP) [32].

**TABLE 7.2 Amino Acid Sequence and EP24.15 Inhibitory Activity, and Extent of Degradation of CFP – Peptide [32].**

| Code | Sequence | Cal Mr | Obs Mr | $IC_{50}$ (μM) | Extent of Degradation 30 min | 6 h | 24 h |
|---|---|---|---|---|---|---|---|
| CFP-1 | cfp-A-A-Y-*p*AB | 604 | 604.3 | 0.06 | xx | — | — |
| CFP-2 | cfp-A-A-Y-*β*G | 556 | 556.2 | 0.12 | xx | — | — |
| CFP-3 | cfp-A-*β*G-Y-*β*G | 556 | 556.3 | 5.6 | 0 | 0 | 0 |
| CFP-4 | cfp-A-$\beta^2$A-Y-*β*G | 570 | 570.2 | 6.3 | 0 | 0 | 0 |
| CFP-5 | cfp-A-$\beta^3$A-Y-*β*G | 570 | 570.3 | >300 | 0 | 0 | 0 |

0 – No degradation; xx – 80–100% degradation. Cal - Calculated; Obs - Observed; Mr - Molecular weight.

β-Amino acids incorporation increases stability against proteolytic degradation. Until recently, there have been no stable and selective inhibitors for the endopeptidase EC 3.4.24.15 (EP 24.15) to investigate its biological precise role *in vivo*, For example, inhibitors such as *N*-[1,(*R*,*S*)-carboxy-3-phenylpropyl]-Ala-Ala-Tyr-*p*-amino benzoate CFP (Figure 7.7) underwent proteolysis between the alanine and tyrosine residues. In order to stabilize the peptide bond to proteolytic degradation, Steer et al. [32] replaced the Ala adjacent to the scissile bond with a racemic mixture of the *β*, $\beta^2$, and $\beta^3$ types of alanine. Studies revealed that the β-amino acid incorporation increased the stability of the CFP inhibitor (Table 7.2).

Ahamed and Kaur have investigated the stability of four different hexapeptides derived from $\beta^3$ type L-Asp monomers($\beta^3$ hexapeptide **1**), $\beta^2$ L-Dap monomers ($\beta^2$ hexapeptides **2** and **3**), both $\alpha$ and $\beta^3$ acid monomers ($\alpha/\beta^3$ hexapeptide **4**), compared to the $\alpha$ only type hexapeptide **5**. It was found that the peptide containing $\beta$ linkages are totally resistant to proteolytic enzymes in serum. However, both $\alpha/\beta^3$ mixed peptides exhibited some proteolysis [33] (Figure 7.8 and Table 7.3).

Finally peptides with substituted β-amino acids have been used to understand the mechanism of proteolytic enzymatic action. For example, it was found that peptides containing substituted β-amino acids have the same level of stability against proteolytic enzymes no matter what type of substitution it has at the alpha position [25].

## 7.6 INTRODUCTION OF PEPTIDE BOND ISOSTERES

Peptidases range from narrow to broad specificity. Specific peptide bond hydrolysis can be avoided by replacing the peptide bond with isosteres or surrogates [34].

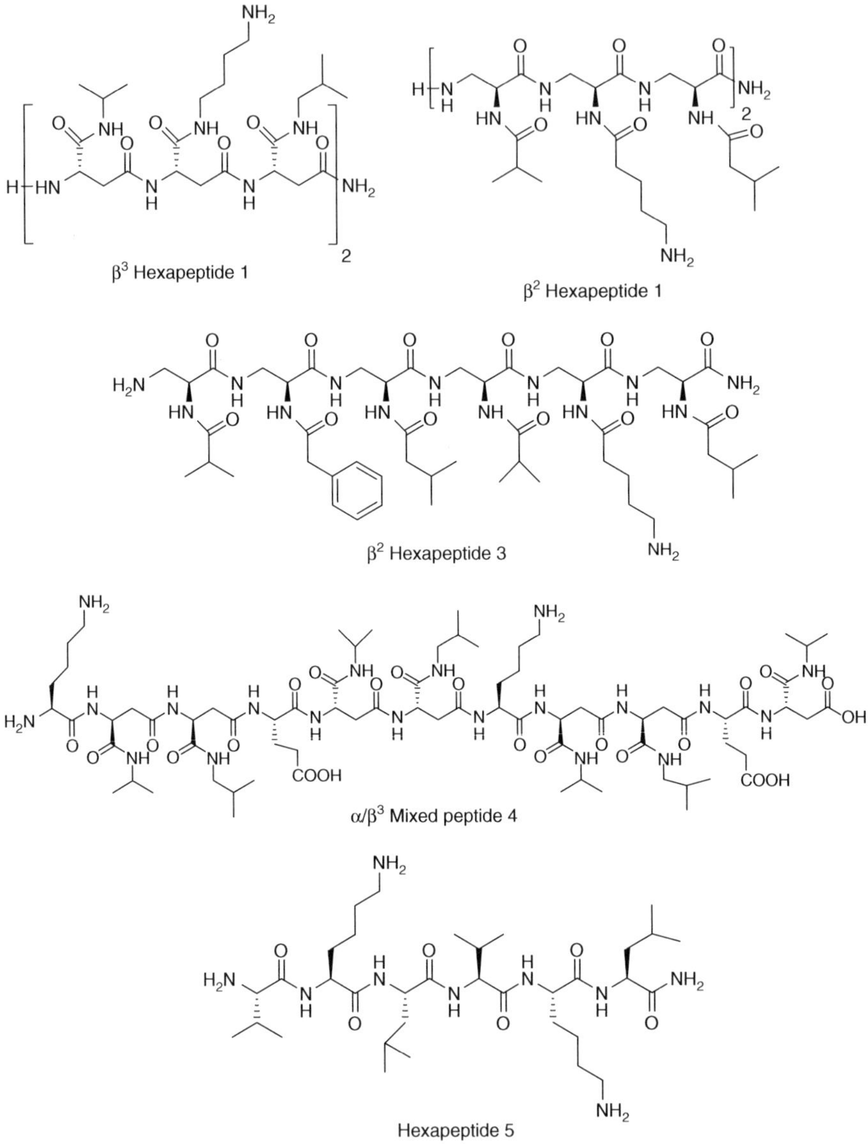


**Figure 7.8** Structure of the hexapepeptides 1–5 [33].

Recognition of the site of peptide bond proteolysis is possible by studying the products after incubation with the known peptidases using mass spectrometry and other bioanalytical techniques.

There is huge literature in this area of research that goes back many years. We will use Kisspeptin, a peptide inhibitor for cancer metathesis, as an example to demonstrate how peptide bond isostere substitution can prevent enzymatic cleavage [35].

Endogenous Kisspeptins can couple to GPR54, which is a Gq protein coupled receptor and signal to suppress cancer metathesis. Kisspeptin stability *in vivo* is

**TABLE 7.3 Treatment of Peptides 1–5 with Three Proteolytic Enzymes or Human Serum [33].**

| Enzyme or Serum | Enzyme Origin | 1 ($\beta^3$) | 2 ($\beta^2$) | 3 ($\beta^2$) | 4 ($\alpha/\beta^3$) | 5 ($\alpha$) |
|---|---|---|---|---|---|---|
| Pronase | *Streptomyces griseus* | — | — | — | + | + |
| Trypsin | Porcine pancreas | — | — | — | + | + |
| Elastase | Hog pancreas | — | — | — | — | + |
| Human serum | | — | — | n.t | n.t | + |

+ Degradation; — no detectable degradation under the experiment conditions; n.t. not tested.

markedly affected due to cleavage of the Gly-Leu bond by metalloproteinases. Peptide-bond-modified isosteres containing GPR54 agonist (Figure 7.9A and B), which are kisspeptin analogs, and wild type GPR54 agonist have been used to analyze stability against the proteolytic enzymes. (*E*)-alkene and its hydroxylated dipeptide isosteres of the GPR45 agonist such as 19 and 25b are more stable

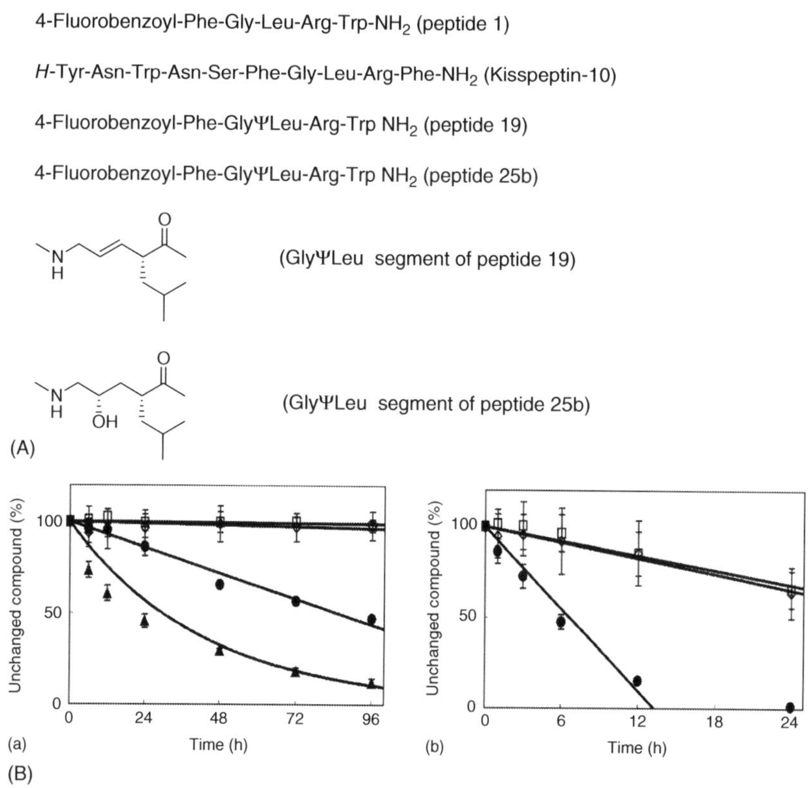


**Figure 7.9** (A) Structure of kisspeptin analogs [35]. (B) Stability evaluation of GPR54 agonists 1, 19, and 25b by treatment with (a) matrix metalloproteinase (MMP)-9 and (b) murine serum: 1 (•), 19 (□), 25b (◇), kisspeptin-10 (▲). Kisspeptin-10 was completely digested in murine serum within 1 h [35].

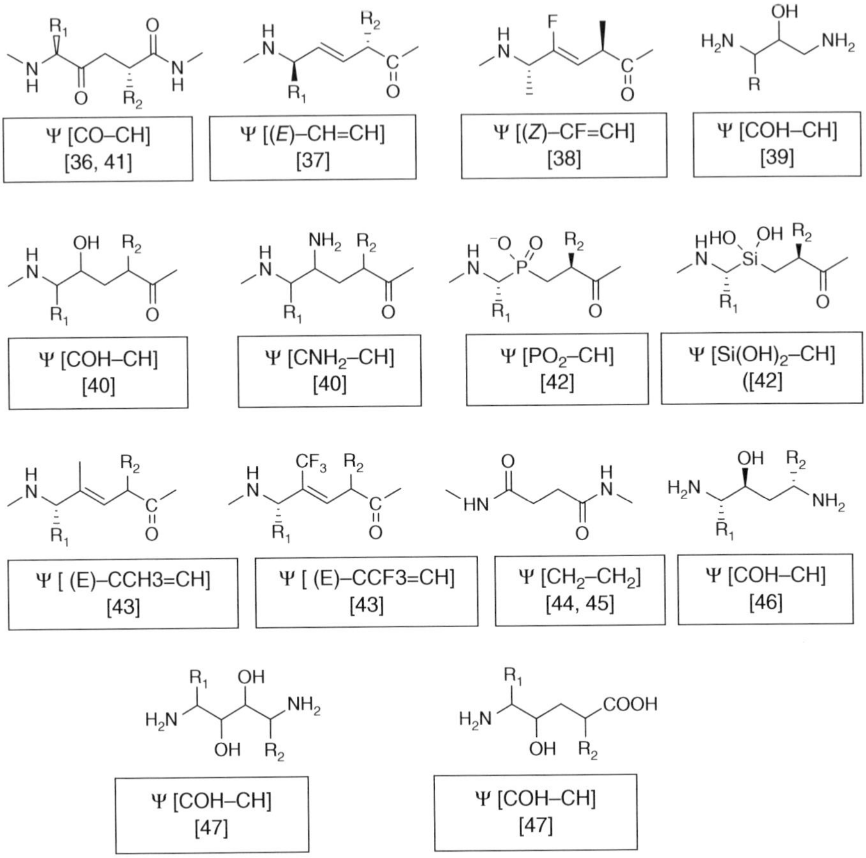


**Figure 7.10** Peptide bond isosters.

compared to the wild type GPR42 agonist 1 or kissapeptine-10 in the presence of the metalloproteinase and murine serum [35]. In addition, many other peptide-bond isosteres have been investigated (e.g., [36–47]) some of which are shown in Figure 7.10.

## 7.7 INTRODUCTION OF A N-METHYLATION OF THE AMIDE BOND OF PEPTIDES CAN IMPROVE THE STABILITY TOWARD BLOOD AND BRAIN PROTEASE DEGRADATION

N-methylation of native peptides increases stability against proteolytic degradation and can result in prolonged activity *in vivo* [48, 49]. Generally N-methylation improves lipophilicity [48], bioavailability [50, 51], and permeability to membranes.

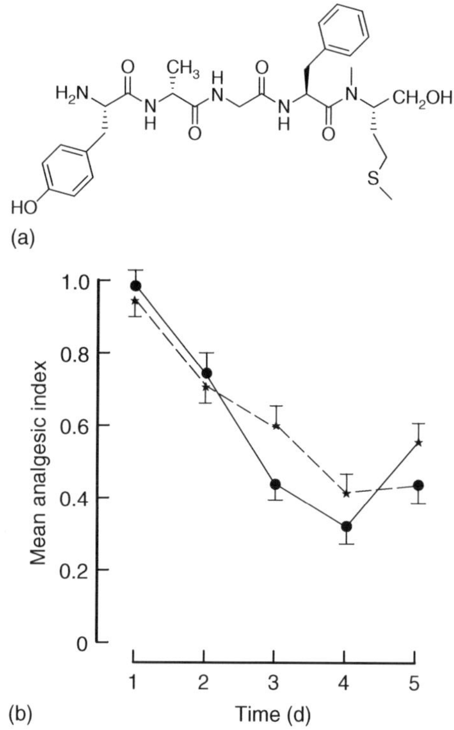


**Figure 7.11** (a) Structure of 33–824 [51]. (b) Mean analgesic index versus time for morphine and compound 33-824 [51]. Mean analgesic index (±s.e.) recorded in the paw pressure test in the rat following s.c. administration of the morphine (7.5 mg $kg^{-1}$, •) and 33–824 (3.2 mg $kg^{-1}$, ★) twice daily for 4 d. On day 5, 33–824 (3.2 mg $kg^{-1}$ s.c.) was given to the morphine pretreated rats and morphine (7.5 mg $kg^{-1}$ s.c.) to the 33–824 pretreated animals. Twenty rats were in each group. The reaction thresholds were recorded immediately before and 30 min after each injection (0815 and 1545 h).

For example, the compound 33–825 (Figure 7.11a and b), a Met-enkephalin analog, was modified by N-alkylation at the methionine-4 residue, D-amino acid substitutions at the second residue, along with a C-terminal alcohol to provide an analog with prolonged activity similar to morphine *in vivo* [52].

For neurotensin (NT) (7–13), Pro-Arg-Arg-Pro-Tyr-Ile-Leu, the peptide bonds at $Arg^8$-$Arg^9$, $Pro^{10}$-$Tyr^{11}$, and $Tyr^{11}$-$Ile^{12}$ are vulnerable to hydrolysis by metalloendopeptidase, neutral endopeptidase and angiotensin converting enzyme, respectively. N-methylation modification at the position 8 has improved stability with a 20-day-half-life in plasma compared to the native radio labeled NT [53]. An excellent review regarding N-methylation and C-methylation of peptides has been published [54].

**TABLE 7.4 Biological Activities of β-Me-Trp Analogs of MTII in Frog Skin Bioassays [55].**

| Compound | Frog Skin Bioassay $IC_{50}$ (nM) | Prolonged[a] Activity | Preferred $\chi 1$[b] Side Chain Conformation (NMR) | |
|---|---|---|---|---|
| α-MSH | 0.1 | — | — | — |
| MTII[c] | 0.1 | +++ | H-*t* | f-*t* |
| | | | R-*t* | W-*t* |
| [(2*S*,3*S*)β-MeTrp$^9$]MTII | 0.44 | — | H-*t* | f-*t* |
| | | | R-*t* | W-*g*(−) |
| [(2*S*,3*R*)β-MeTrp$^9$]MTII | 28.6 | + | H-*t* | f-*t* |
| | | | R-*t* | W-*g*(+) |
| [(2*R*,3*S*)β-MeTrp$^9$]MTII | 0.06 | ++ | H-*t* | f-*g*(+) |
| | | | R-*t* | W-(*g*+) |
| [(2*R*,3*R*)β-MeTrp$^9$]MTII | 0.33 | +++ | H-*g*(+) | f-*t* |
| | | | R-*t* | W-*g*(+) |

[a]Prolong activity: — not prolonged; +++ highly prolonged (irreversible); + modestly prolonged (minutes); ++ quite prolonged (hours).
[b]H = His$^6$; f = DPhe$^7$; R = Arg$^8$; W = Trp$^9$ or β-Me-Trp$^9$ isomer; *t*=trans (180°); g(−) = gauche (−) (−60°);g(+) = gauche(+) (+60°) for an L-amino acid.
[c]MTII = Ac-Nle$^4$-c[Asp$^5$,DPhe$^7$,Lys$^{10}$]α-MSH(4–10)-$NH_2$.

## 7.8 USE OF UNNATURAL AMINO ACIDS – USE OF TOPOGRAPHICALLY CONSTRAINED AMINO ACID

Topography of side chains defined by the $\chi 1$ and $\chi 2$ angles plays a central role in the identification of peptides by peptidases for hydrolysis of peptide bonds. The correct topography of side-chain in chi space is important for peptidase recognition. By maintaining undesired topography of side-chain group to peptidases, we can improve resistance to the peptidase action and enhance peptide biological activity *in vivo* [55]. For example, Ac-Nle$^4$-c[Asp$^5$,DPhe$^7$,Lys$^{10}$]α-MSH-(4–10)-$NH_2$(MTII), which is a potent melanocortin receptor ligand[8] was substituted with topographically constrained 4 diasteromers of β-MeTrp (2*S*,3*S*;2*S*,3*R*;2*R*,3*R*;2*R*,3*S*) to position 9 of MTII (Table 7.4). Biological analysis revealed that the [(2*R*,3*R*) β-MeTrp$^9$]MTII has the highest prolong activity *in vivo* compared the other MTII analogs [55], which provided a topographical explanation for highly prolonged activity of certain DPhe$^7$ analogs of α-MSH.

In addition to the constrained β-position-modified amino acids, alpha carbon dialkyl substitution of amino acid can lead to conformational preference for particular phi and psi angles in the Ramachandran plot and may be useful for increasing the stability of peptides to enzymatic degradation. For example, Aib, Ac-DΦg-NHMe, Ac-Aib-NHMe, Ac-(2S,3S)$c_3$diPhe-NHMe, Ac-(S)$c_3$Dip-NHMe and many other *α*,α-disubstituted amino acids promote specific backbone conformations (Figure 7.12) [56]. Indeed Aib is long known to promote strong helix induce properties due to the reduction of the entropic penalty of helix formation on protein folding [57, 58].

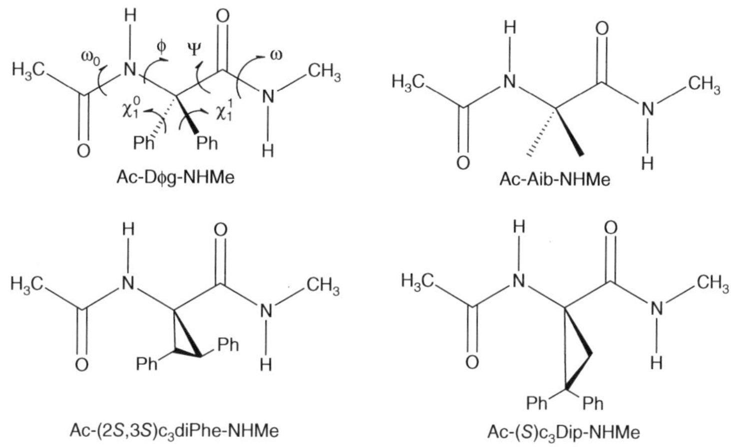


**Figure 7.12** Dialkyl alpha substituted amino acids [56].

## 7.9 USING GLYCOSYLATED AMINO ACIDS TO INCREASE THE RESISTANCE OF THE PROTEOLYTIC DEGRADATION

Modifying proteolytic enzymes' specific sites of cleavage by glycosylation of amino acids near the scissile site can make peptides more stable *in vivo*. For example, Dangoor et al. have reduced peptide-bond degradation between the $Arg^{12}$-$Leu^{13}$ of vasoactive intestinal peptide (VIP) by the enzyme trypsin through glycosylation at the $Thr^{11}$ residue of VIP [59] (Figure 7.13).

Identification of potential sites for glycosylation should be carefully planned, as modification at the wrong site can lead to loss of activity of the peptide. From structure–activity relationship (SAR) studies of mu and delta receptors using opioid ligands such as dermorphin, deltorphin, and enkephalins, we have known for many years that N-terminal modification results in loss of the activity toward those receptors. In addition it is imperative that one retains all the pharmacophoric residues at the C-terminal (message sequence for the opioid receptor) for optimum mu and delta opioid activity (e.g., [60]).

Polt and coworkers [61] have extensively studied opioid glycopeptides. Some of their early work involved the design and synthesis of L-serinyl β-D-glucoside analogs of $[Met^5]$ enkephalins. They found that these glycopeptides can transport across the BBB to bind to the mu and delta receptors of the mouse brain. Moreover, highly desired long lasting analgesia was observed in mice using the tail flick assays and hot plate assays when these glycopeptides were administered intraperitoneally [61].

More recently, Rocchi and coworkers have done similar work using the modified neuropeptides at the C-terminal residues of dermorphins and deltorphins and found that these glycopeptides also retained good activity for mu and delta receptors. Further, they demonstrated that glycosylation increases half-life to the enzymatic breakdown of dermorphin and deltorphin analogs using mouse brain and liver homogenates [62] (Table 7.5).

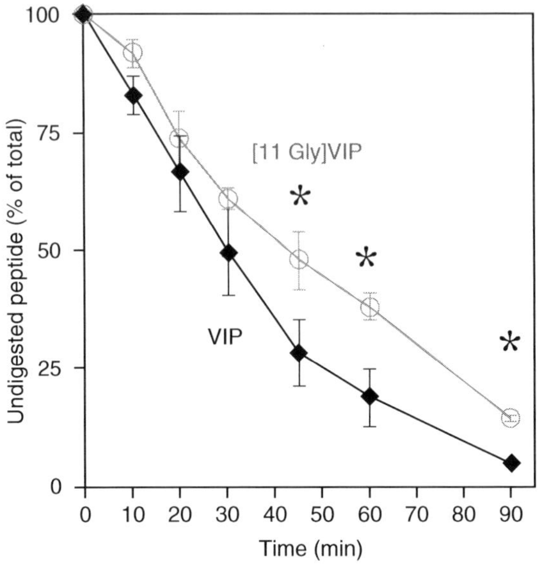


**Figure 7.13** Enzymatic digestion of the native VIP and of the glycosylated VIP analog [11Gly]VIP by trypsin [59]. At the indicated period, the relative quantity of the intact native VIP and of undigested the VIP analog [11Gly]VIP in the reaction mixture was determined by HPLC analysis. Each point represents the percentage (mean ± SD, $n = 4$) of undigested peptide.

**TABLE 7.5 Half Time for Enzymatic Breakdown of Glycosylated Dermorphin and Deltorphin Analogs in Mouse Brain and Liver Homogenates [62].**

| Peptides | Brain $t_{1/2}$ (min) | Liver $t_{1/2}$ (min) |
|---|---|---|
| Dermorphin | 20 ± 5 | 10 ± 4 |
| [($\beta$Glc)Ser$^7$]Dermorphin | 38 ± 6 | 30 ± 5 |
| [$\beta$Glc(Ac)$_4$-Ser$^7$]Dermorphin | 90 ± 10 | 60 ± 8 |
| [Hyp,$^6$Lys$^7$]Dermorphin | 30 ± 5 | 20 ± 4 |
| Deltorphin I | 240 ± 15 | 110 ± 20 |
| [($\beta$Glc)Ser$^7$]Deltorphin | >240 (70%)[a] | 180 ± 25 |
| [$\beta$Glc(Ac)$_4$-Ser$^7$]Deltorphin | >240 (87%)[a] | >240 (70%)[a] |

[a]Numbers in parentheses are the residual biological activity after a 240 min incubation.

## 7.10 CREATION OF PEPTIDES AS MULTIPLE ANTIGEN PEPTIDE (MAP) DENDRIMERIC FORMS INCREASES THE STABILITY TOWARD BLOOD AND BRAIN PROTEASE DEGRADATION

It has been reported that multibranched (e.g., two-branched, tetrabranched, etc.) multiple antigen peptides (MAP) are more stable toward proteolytic degradation

**TABLE 7.6 Proteolytic Stability of Monomeric, Dimeric, and Tetrameric L-Enk as Detected by HPLC.**

| | Plasma | | | Serum | | |
|---|---|---|---|---|---|---|
| | 2 h | 5 h | 24 h | 2 h | 5 h | 24 h |
| YGGFL | + | — | — | — | — | — |
| YGGFL)$_2$ | + | + | + | + | — | — |
| YGGFL)$_4$K$_2$K | + | + | + | + | + | — |

The absence (—) and the presence (+) of peptide substrates after the incubation times [64].

when compared to their monomeric forms. Moreover, this approach can increase peptide affinity and activity toward their respective receptors (e.g., [63]).

Met-enkephalin (YGGFM) and Leu-enkephalin (YGGFL) are endogenous neuropeptides for opioid receptor and are rapidly degraded in the presence of human plasma and serum. However, in its branched form (MAP), (YGGFL)$_2$K is not degraded after 24 h incubation in plasma or after 2 h incubation in serum, and the tetrabranched peptide (MAP), (YGGFL)$_4$K$_2$K is not degraded after 24 h incubation in plasma and 5 h in serum (Table 7.6). In addition, tetrabranched MAP forms of Met and Leu enkephanin shows 10-fold enhancement of their IC$_{50}$ values compared with that of their monomeric counterparts. In the presence of the proteases, monomeric and tetrameric forms show equal IC$_{50}$s [64].

## 7.11 HALOGENATIONS OF AROMATIC RESIDUES IN PEPTIDES CAN REDUCE THE ENZYMATIC RECOGNITION REQUIRED FOR PEPTIDE HYDROLYSIS

Halogenations such as chlorination or fluorination of the aromatic side-chain group in peptides can reduce proteolytic enzyme recognition [65] required for hydrolysis, and increases brain uptake. Some peptides, for example, the enkephalins, are somewhat limited in their access to the CNS due to the presence of peptidases in the blood and at the BBB. Bioavailability is reduced due to the enzymatic hydrolysis [66].

As an example we have focused on biphalin (Figure 7.14a and b), which is a potent analgesic, 257 and 6 times more potent compared to morphine and etorphine after intracerebroventricular administration, respectively [67]. Biphalin uptake to the brain has been observed after interperitoneal (i.p) and subcutaneous (s.c) administration [67]. To increase biphalin BBB uptake and reduce the enzymatic hydrolysis, Biphalin has been halogenated at the 4 and 4′ positions of the phenylalanine residues.

Both biphalin and halogenated biphalin analogs were examined for *in vitro* brain-stability studies. Half time ($t_{1/2}$) disappearance of the enkephalin analogs was calculated by high pressure liquid chromatography (HPLC) analysis. Results revealed that the metabolic half-lives ($t_{1/2}$) of the ρ-[Cl-Phe$^{4,4'}$] biphalin increased two fold compared that of the biphalin [68].

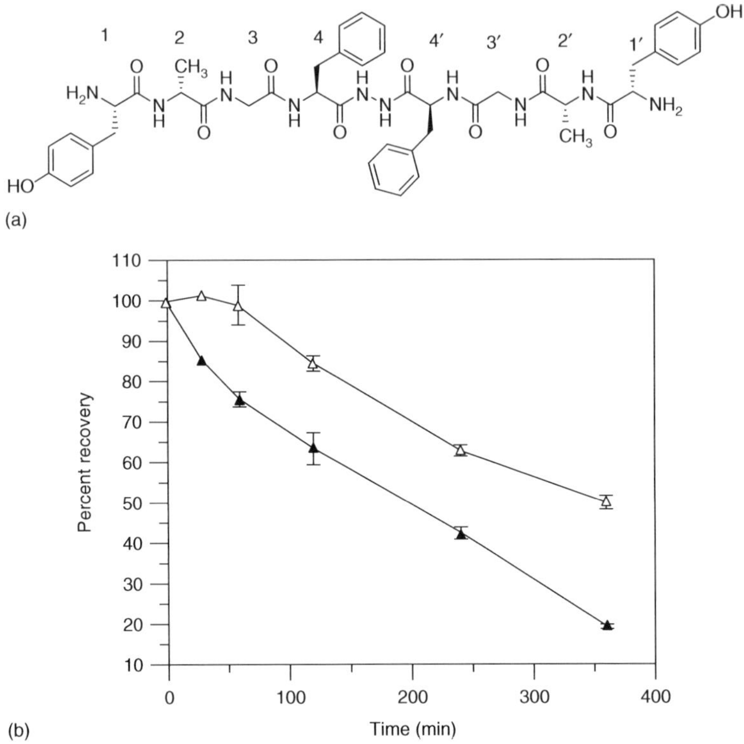


**Figure 7.14** (a) Biphalin [68]. (b) The percentage of recovery of intact biphalin ▲ and ρ-[Cl-Phe$^{4,4'}$] biphalin △ over 360 min time course in brain homogenate [68].

## 7.12 CONCLUDING DISCUSSION

In this chapter, we have briefly discussed several approaches that can be used to stabilize peptides against proteolytic degradation of peptides, which is a major way in which bioactive peptides are metabolized into biologically inactive forms. As we have seen, there are many approaches that have been successfully developed toward this goal. The method or methods chosen for any individual case will depend to a considerable extent on key structural elements of the structure of the bioactive peptide that must be retained for potent biological activity (the pharmacophore). Once that has been established, preferably in three-dimensional space, the utilization of the strategies briefly discussed here can be rapidly implemented, often using molecular modeling and computational chemistry to evaluate the consequence of structural modification on the three-dimensional structure of the modified peptide ligand. From our experience this can be done quite early in SAR studies that are critical for peptide and peptidomimetic design, and thus design for the stability of the peptide against proteolytic degradation in circulation and at membrane barriers can and should be an important aspect of SAR studies from the beginning of peptide and peptidomimetic ligand design. These considerations apply to both smaller

(3–10 residues) and to larger (10–50 residues) peptides. Some knowledge of both primary structure and especially secondary structure can be critical in the design considerations. For example, stable α-helical and β-turn structures generally are not well recognized by proteolytic enzymes and hence one can design stable secondary structures of this kind that are compatible with biological activity, and this can be sufficient to greatly enhance peptide stability to proteolysis *in vivo*. In our experience we have always been able to design bioactive peptides with very significant enhancement of stability against proteolytic enzymes, and retained the desired biological potency and biological activity both *in vitro* and *in vivo*. Of course, biodistribution, membrane barrier permeability, and so on require further considerations, which are not discussed here. Nonetheless, generally enhancing the biostability of peptides is an important component of enhancing biodistribution properties as well.

## REFERENCES

1. Darveau RP, Cunningham MD, Seachord CL, Casiano-Clough L, Cosand WL, Blake J, Watkins CS. β-lactam antibiotics potentiate Magainin II antimicrobial activity *in vitro* and *in vivo*. Antimicrob. Agents Chemother 1991;35:1153–1159.
2. Hruby VJ. Designing peptide receptor agonists and antagonists. Nat Rev Drug Discov 2002;1:847–858.
3. Malakoutikhah M, Teixido M, Giralt E. Toward an optimal blood-brain barrier shuttle by synthesis and evaluation of peptide libraries. J Med Chem 2008;51(16):4881–4889.
4. Schmiedeberg N, Schmitt M, Roelz C, Truffault V, Sukopp M, Buergle M, Wilhelm OG, Schmalix W, Magdolen V, Kessler H. Synthesis, solution structure, and biological evaluation of urokinase type plasminogen activator (uPA)-derived receptor binding domain mimetics. J Med Chem 2002;45:4984–4994.
5. Shaginian A, Whitby LR, Hong S, Hwang I, Farooqi B, Searcey M, Chen J, Vogt PK, Boger DL. Design, synthesis, and evaluation of an .α-helix mimetic library targeting protein-protein interactions. J Am Chem Soc 2009;131:5564–5572.
6. Hruby VJ. Conformational restrictions of biologically active peptides via amino acid side chain groups. Life Sci 1982;31:189–199.
7. Katsara M, Tselios T, Deraos S, Deraos G, Matsoukas MT, Lazoura E, Matsoukas J, Apostolopoulos V. Round and round we go: cyclic peptides in disease. Curr Med Chem 2006;13:2221–2232.
8. Sawyer TK, Hruby VJ, Darman PS, Hadley ME. [4-half-cystine, 10-half-cystine]-α-melanocyte stimulating hormone: a cyclic α-melanotropin exhibiting superagonist biological activity. Proc Natl Acad Sci U S A 1982;79:1751–1755.
9. Al-Obeidi F, Hadley ME, Pettitt BM, Hruby VJ. Design of a new class of superpotent cyclic α-melanotropins based on quenched dynamic simulations. J Am Chem Soc 1989;111:3413–3416.
10. Al-Obeidi F, de Lauro Castrucci AM, Hadley ME, Hruby VJ. Potent and prolonged acting cyclic lactam analogues of α-melanotropin: design based on molecular dynamics. J Med Chem 1989;32:2555–2561.

11. Hess S, Linde Y, Ovadia O, Safrai E, Shalev DE, Swed A, Halbfinger E, Lapidot T, Winkler I, Gabinet Y, Faier A, Yarden D, Xiang Z, Portillo FP, Haskell-Luevano C, Gilon C, Hoffman A. Backbone cyclic peptidomimetic melanocortin-4 receptor agonist as a novel orally administrated drug lead for treating obesity. J Med Chem 2008;51(4):1026–1034.
12. Hess S, Ovadia O, Shalev DE, Senderovich H, Qadri B, Yehezkel T, Salitra Y, Sheynis T, Jelinek R, Gilon C, Hoffman A. Effect of structural and conformation modifications, including backbone cyclization, of hydrophilic hexapeptides on their intestinal permeability and enzymatic stability. J Med Chem 2007;50(24):6201–6211.
13. Byk G, Halle D, Zeltser I, Bitan G, Selinger Z, Gilon C. Synthesis and biological activity of NK-1 selective, N-backbone cyclic analogs of the C-terminal hexapeptide of substance P. J Med Chem 1996;39:3174–3178.
14. Gilon C, Huenges M, Mathae B, Gellerman G, Hornik V, Afargan M, Amitay O, Ziv O, Feller E, Gamliel A, Shohat D, Wanger M, Arad O, Kessler H. A backbone-cyclic, receptor 5-selective somatostatin analog: synthesis, bioactivity, and nuclear magnetic resonance conformational analysis. J Med Chem 1998;41(6):919–929.
15. Mosberg HI, Hurst R, Hruby VJ, Gee K, Yamamura HI, Galligan JJ, Burks TF. Bis-penicillamine enkephalins possess highly improved specificity toward delta opioid receptors. Proc Natl Acad Sci U S A 1983;80(19):5871–5874.
16. Weber SJ, Greene DL, Sharma SD, Yamamura HI, Kramer TH, Burks TF, Hruby VJ, Hersh LB, Davis TP. Distribution and analgesia of [3H][D-Pen2,D-Pen5]enkephalin and two halogenated analogs after intravenous administration. J Pharmacol Exp Ther 1991;259(3):1109–1117.
17. Williams SA, Abbruscato TJ, Hruby VJ, Davis TP. Passage of a δ-opioid receptor selective enkephalin, [D-penicillamine2,5]enkephalin, across the blood-brain and the blood-cerebrospinal fluid barriers. J Neurochem 1996;66(3):1289–1299.
18. Mollica A, Davis P, Ma S, Porreca F, Lai J, Hruby VJ. Synthesis and biological activity of the first cyclic biphalin analogs. Bioorg Med Chem Lett 2006;16:367–372.
19. Cowell SM, Lee YS, Cain JP, Hruby VJ. Exploring ramachandran and chi space: conformationally constrained amino acids and peptides in the design of bioactive polypeptide ligands. Curr Med Chem 2004;11(21):2785–2798.
20. Hadley ME, Dorr RT. Melanocortin peptide therapeutics: historical milestones, clinical studies and commercialization. Peptides 2006;27(4):921–930.
21. Hamamoto K, Kida Y, Zhang Y, Shimizu T, Kuwano K. Antimicrobial activity and stability to proteolysis of small linear cationic peptides with D-amino acid substitutions. Microbiol Immunol 2002;46(11):741–749.
22. Sawyer TK, Sanfilippo PJ, Hruby VJ, Engel MH, Heward CB, Burnett JB, Hadley ME. 4-Norleucine, 7-D-phenylalanine-α-melanocyte-stimulating hormone: a highly potent α-melanotropin with ultralong biological activity. Proc Natl Acad Sci U S A 1980;77(10):5754–5758.
23. Coy DH, Mungan Z, Rossowski WJ, Cheng BL, Lin JT, Mrozinski JE, Jensen RT. Development of a potent bombesin receptor antagonist with prolonged in vivo inhibitory activity on bombesin-stimulated amylase and protein release in the rat. Peptides 1992;13(4):775–781.
24. Toke O, Tugyi R, Uray K, Hudecz F. Solution structure of the all L- and D-amino acid-substituted mucin 2 epitope peptides. Biochem Biophys Res Commun 2007;358(3):739–742.

25. Hook DF, Gessier F, Noti C, Kast P, Seebach D. Probing the proteolytic stability of β-peptides containing α-fluoro- and α-hydroxy-β-amino acids. ChemBioChem 2004;5(5):691–706.

26. Steer D, Lew R, Perlmutter P, Smith AI, Aguilar MI. Inhibitors of metalloendopeptidase EC 3.4.24.15 and EC 3.4.24.16 stabilized against proteolysis by the incorporation of β-amino acids. Biochem 2002;41(35):10819–10826.

27. Frackenpohl J, Arvidsson P, Schreiber JV, Seebach D. The outstanding biological stability of β-and β-peptides toward proteolytic enzymes: an in vitro investigation with fifteen peptidases. ChemBioChem 2001;2:445–455.

28. Hook DF, Bindschaedler P, Mahajan YR, Sebesta R, Kast P, Seebach D. The proteolytic stability of 'designed' β-peptides containing α-peptide-bond mimics and of mixed $\alpha$, β-peptides: application to the construction of MHC-binding peptides. Chem Biodiversity 2005;2(5):591–632.

29. Porter EA, Weisblum B, Gellman SH. Mimicry of host-defense peptides by unnatural oligomers: antimicrobial β-peptides. J Am Chem Soc 2002;124(25):7324–7330.

30. Seebach D, Gardiner J. β-Peptidic peptidomimetics. Acc Chem Res 2008;41(10): 1366–1375.

31. Salauen A, Favre A, Le Grel B, Potel M, Le Grel P. Conformation of N α-substituted hydrazino acetamides in CDCl3, the precious help of the analysis of. $\Delta\delta$. between amidic hydrogens, and correlation to the conformation of Aza-.$\beta$3-peptides. J Org Chem 2006;71:150–158.

32. Steer DL, Lew RA, Perlmutter P, Smith AI, Aguilar MI. Design and synthesis of inhibitors incorporating β-aminoacids of metalloendopeptidase EC 3.4.24.15. J Pept Sci 2000;6:470–477.

33. Ahmed S, Kaur K. The proteolytic stability and cytotoxicity studies of L-aspartic acid and L-diaminopropionic acid derived β-peptides and a mixed $\alpha$/β-peptide. Chem Biol Drug Des 2009;73(5):545–552.

34. Spatola AF. Peptide backbone modifications: a structure-activity analysis of peptides containing amide bond surrogates, conformational constraints, and related backbone replacements. In: *Chemistry and Biochemistry of Amino Acids, Peptides, and Proteins*. Vol. 7. 1983. p 267–357.

35. Tomita K, Oishi S, Ohno H, Peiper SC, Fujii N. Development of novel G-protein-coupled receptor 54 agonists with resistance to degradation by matrix metalloproteinase. J Med Chem 2008;51(23):7645–7649.

36. Lin W, Theberge CR, Henderson TJ, Zercher CK, Jasinski J, Butcher RJ. Stereoselective formation of a functionalized dipeptide isostere by zinc carbenoid-mediated chain extension. J Org Chem 2009;74(2):645–651.

37. Tamamura H, Hiramatsu K, Ueda S, Wang Z, Kusano S, Terakubo S, Trent JO, Peiper SC, Yamamoto N, Nakashima H, Otaka A, Fujii N. Stereoselective synthesis of [L-Arg-L/D-3-(2-naphthyl)alanine]-type (E)-alkene dipeptide isosteres and its application to the synthesis and biological evaluation of pseudopeptide analogues of the CXCR4 antagonist FC131. J Med Chem 2005;48(2):380–391.

38. Otaka A, Watanabe J, Yukimasa A, Sasaki Y, Watanabe H, Kinoshita T, Oishi S, Tamamura H, Fujii N. SmI2-mediated reduction of β, β-difluoro-α, β-enoates with application to the synthesis of functionalized (Z)-fluoroalkene-type dipeptide isosteres. J Org Chem 2004;69(5):1634–1645.

39. Akaji K, Teruya K, Aimoto S. Solid-phase synthesis of HTLV-1 protease inhibitors containing hydroxyethylamine dipeptide isostere. J Org Chem 2003;68(12): 4755–4763.

40. Yang W, Lu W, Lu Y, Zhong M, Sun J, Thomas AE, Wilkinson JM, Fucini RV, Lam M, Randal M, Shi X, Jacobs JW, McDowell RS, Gordon EM, Ballinger MD. Aminoethylenes: a tetrahedral intermediate isostere yielding potent inhibitors of the aspartyl protease BACE-1. J Med Chem 2006;49(3):839–842.

41. Fu Y, Bieschke J, Kelly JW. E-olefin dipeptide isostere incorporation into a polypeptide backbone enables hydrogen bond perturbation: probing the requirements for Alzheimer's amyloidogenesis. J Am Chem Soc 2005;127(44):15366–15367.

42. Nielsen L, Skrydstrup T. Sequential C-Si bond formations from diphenylsilane: application to silanediol peptide isostere precursors. J Am Chem Soc 2008;130(39):13145–13151.

43. Inokuchi E, Narumi T, Niida A, Kobayashi K, Tomita K, Oishi S, Ohno H, Fujii N. Efficient synthesis of trifluoromethyl and related trisubstituted alkene dipeptide isosteres by palladium-catalyzed carbonylation of amino acid derived allylic carbonates. J Org Chem 2008;73(10):3942–3945.

44. Mittag T, Christensen KL, Lindsay KB, Nielsen NC, Skrydstrup T. Direct entry to peptidyl ketones via SmI2-mediated C-C bond formation with readily accessible N-peptidyl oxazolidinones. J Org Chem 2008;73(3):1088–1092.

45. Elaridi J, Patel J, Jackson WR, Robinson AJ. Controlled synthesis of (S,S)-2,7-diaminosuberic acid: a method for regioselective construction of dicarba analogues of multicystine-containing peptides. J Org Chem 2006;71(20):7538–7545.

46. Engstrom K, Henry R, Hollis LS, Kotecki B, Marsden I, Pu Y, Wagaw S, Wang W. An efficient, stereoselective synthesis of the hydroxyethylene dipeptide isostere core of the HIV protease inhibitor A-792611. J Org Chem 2006;71(14):5369–5372.

47. Benedetti F, Berti F, Norbedo S. Epoxyalcohol route to hydroxyethylene dipeptide isosteres. Stereodivergent synthesis of the diamino alcohol core of Ritonavir and its C-2 epimer. J Org Chem 2002;67(24):8635–8643.

48. Cody WL, He JX, Reily MD, Haleen SJ, Walker DM, Reyner EL, Stewart BH, Doherty AM. Design of a potent combined pseudopeptide endothelin-A/endothelin-B receptor antagonist, Ac-DBhg$^{16}$-Leu-Asp-Ile-[NMe]Ile-Trp$^{21}$ (PD 156252): examination of its pharmacokinetic and spectral properties. J Med Chem 1997;40(14): 2228–2240.

49. Haviv F, Fitzpatrick TD, Swenson RE, Nichols CJ, Mort NA, Bush EN, Diaz G, Bammert G, Nguyen A. Effect of N-methyl substitution of the peptide bonds in luteinizing hormone-releasing hormone agonists. J Med Chem 1993;36(3):363–369.

50. Ali FE, Bennett DB, Calvo RR, Elliott JD, Hwang S, Ku TW, Lago MA, Nichols AJ, Romoff TT. Conformationally constrained peptides and semipeptides derived from RGD as potent inhibitors of the platelet fibrinogen receptor and platelet aggregation. J Med Chem 1994;37(6):769–780.

51. Biron E, Chatterjee J, Ovadia O, Langenegger D, Brueggen J, Hoyer D, Schmid HA, Jelnick R, Gilon C, Hoffman A, Kessler H. Improving oral bioavailability of peptides by multiple N-methylation: somatostatin analogues. Angew Chem Int Edit 2008;47(14):2595–2599.

52. Roemer D, Buescher HH, Hill RC, Pless J, Bauer W, Cardinaux F, Closse A, Hauser D, Huguenin R. A synthetic enkephalin analog with prolonged parenteral and oral analgesic activity. Nature 1977;268:547–549.

53. Bruehlmeier M, Garayoa EG, Blanc A, Holzer B, Gergely S, Tourwe D, Schubiger PA, Blauenstein P. Stabilization of neurotensin analogues: effect on peptide catabolism, biodistribution and tumor binding. Nucl Med Biol 2002;29(3):321–327.

54. Sagan S, Karoyan P, Lequin O, Chassaing G, Lavielle S. N- and C α-methylation in biologically active peptides: synthesis, structural and functional aspects. Curr Med Chem 2004;11(21):2799–2822.

55. Hruby VJ. Design in topographical space of peptide and peptidomimetic ligands that affect behavior. a chemist's glimpse at the mind-body problem. Acc Chem Res 2001;34(5):389–397.

56. Toniolo C. Conformationally restricted peptides through short-range cyclizations. Int J Pept Protein Res 1990;35:287–300.

57. Jianwen A, Feng LAT, Marshall GR. Chimeric protein engineering. Int J Pept Res Ther 2007;13(1-2):151–160.

58. Casanovas J, Zanuy D, Nussinov R, Aleman C. Intrinsic conformational characteristics of *α*, α-diphenylglycine. J Org Chem 2007;72(6):2174–2181.

59. Dangoor D, Biondi B, Gobbo M, Vachutinski Y, Fridkin M, Gozes I, Rocchi R. Novel glycosylated VIP analogs: synthesis, biological activity, and metabolic stability. J Pept Sci 2008;14:321–328.

60. Hruby VJ, Gehrig CA. Recent developments in the design of receptor-specific opioid peptides. Med Res Rev 1989;9:343–401.

61. Polt R, Porreca F, Szabo LZ, Bilsky EJ, Davis P, Abbruscato TJ, Davis TP, Horvath R, Yamamura HI, Hruby VJ. Glycopeptide enkephalin analogs produce analgesia in mice: evidence for penetration of the blood-brain barrier. Proc Natl Acad Sci U S A 1994;91:7114–7118.

62. Negri L, Lattanzi R, Tabacco F, Orru L, Severini C, Scolaro B, Rocchi R. Dermorphin and deltorphin glycosylated analogues: synthesis and antinociceptive activity after systemic administration. J Med Chem 1999;42:400–404.

63. Falciani C, Lozzi L, Pini A, Corti F, Fabbrini M, Bernini A, Lelli B, Niccolai N, Bracci L. Molecular basis of branched peptides resistance to enzyme proteolysis. Chem Biol Drug Des 2007;69(3):216–221.

64. Bracci L, Falciani C, Lelli B, Lozzi L, Runci Y, Pini A, De Montis MG, Tagliamonte A, Neri P. Synthetic peptides in the form of dendrimers become resistant to protease activity. J Biol Chem 2003;278(47):46590–46595.

65. Meng H, Krishnaji ST, Beinborn M, Kumar K. Influence of selective fluorination on the biological activity and proteolytic stability of glucagon-like peptide-1. J Med Chem 2008;51(22):7303–7307.

66. Hirschmann RF, Nicolaou KC, Angeles AR, Chen JS, Smith AB. The β-D-glucose scaffold as a β-turn mimetic. Acc Chem Res 2009;42:1511–1520.

67. Horan PJ, Mattia A, Bilsky EJ, Weber S, Davis TP, Yamamura HI, Malatynska E, Appleyard SM, Slaninova J. Antinociceptive profile of biphalin, a dimeric enkephalin analog. J Pharmacol Exp Ther 1993;265(3):1446–1454.

68. Abbruscato TJ, Williams SA, Misicka A, Lipkowski AW, Hruby VJ, Davis TP. Blood-to-central nervous system entry and stability of biphalin, a unique double-enkephalin analog, and its halogenated derivatives. J Pharmacol Exp Ther 1996;276(3):1049–1057.

69. Haskell-Luevano C, Toth K, Boteju L, Job C, de Castrucci AML, Hadley ME, Hruby VJ. β-Methylation of the $Phe^7$ and $Trp^9$ melanotropin side chain pharmacophores affects ligand-receptor interactions and prolonged biological activity. J Med Chem 1997;40:2740–2749.

70. Benedetti F, Berti F, Budal S, Campaner P, Dinon F, Tossi A, Agrirova R, Genova P, Atanassov V, Hinkov A. Synthesis and biological activity of potent HIV-1 protease inhibitors based on Phe-Pro dihydroxyethylene isosteres. J. Med. Chem. 2012;55:3900–3910.

71. Inokuchi E, Oishi S, Kubo T, Ohno H, Shimura K, Matsuoka M, Fujii N. Potent CXR4 Antagonists Containing Amidine Type Peptide Bond Isosteres. ACS Med. Chem. Lett. 2011;2:477–480.

72. Kobayashi K, Oishi S, Hayashi R, Tomita K, Kubo T, Tanahara N, Ohno H, Yoshikawa Y, Furuya T, Hoshimo M, Fujii N. Structure-Activity Relationship Study of a CXC Chemokine Receptor Type 4 Antagonist, FC131, Using a Series of Alkene Dipeptide Isosteres. J. Med. Chem. 2012;55:2746–2757.

73. Madsen JLH, Anderson TL, Santamaria S, Nagase H, Enghild JJ, Skrydstrup T. Synthesis and Evaluation of Silanediols as Highly Selective Uncompetitive Inhibitors of Human Neutrophil Elastase. J. Med. Chem. 2012;55:7900–7908.

74. Min GK, Hernandez D, Skrydstrup T. Efficient Routes to Carbon-Silicon Bond Formation for the Synthesis of Silicon-Containing Peptides and Azasilaheterocycles. Accounts Chem. Res. 2013;46:457–470.

75. Proteau-Gagne A, Nadon JF, Bernard S, Guérin B, Gendron L, Dory YL. Kinetic deconjugation: gateway to the synthesis of Xxx-Gly (E)-alkene dipeptide isosteres. Tetrahedron Lett. 2011;52:6603–6605.

76. Wangsell F, Gustafsson K, Kvarnstrom I, Borkakoti N, Edlund M, Jansson K, Lindberg J, Hallberg A, Rosenquist A, Samuelsson B. Synthesis of potent BACE-1 inhibitors incorporating a hydroxyethylene isostere as central core. Eur. J. Med. Chem. 2012;45: 870–882.

77. Watanabe D, Koura M, Saito A, Yanai H, Nakamura Y, Okada M, Sato A, Taguchi T. Copper mediated defluorinative allylic alkylation of difluorohomoallyl alchohol derivatives directed to an efficient synthetic method for (Z)-fluoroalkane dipeptide isosteres. J. Flourine Chem. 2011;132:327–338.

78. Yamagishi T, Kinbara A, Okubo N, Sato S, Yokomatsu T. Diastereoselective synthesis of the Leu-Pro type phosphinyl dipeptide isosteres tetrahedron asymmetr. 2011;76:5472–5476.

# 8

# DELIVERY OF PEPTIDE DRUGS

JEFFREY-TRI NGUYEN AND YOSHIAKI KISO

*Department of Medicinal Chemistry, Kyoto Pharmaceutical University, Kyoto, Japan*

## 8.1 INTRODUCTION

In Chapter 5, we have discussed methods used to predict and enhance the pharmacodynamic properties of a lead drug. That is, the chapter emphasized on improving the pharmacological activity, that is, potency of peptide drugs. We urge the unfamiliar readers to read our disclaimers and about peptide nomenclature in Sections 5.1 and 5.1.1 in the aforementioned chapter, before proceeding further. In this chapter, we will concentrate our discussion on enhancing the pharmacokinetic properties of peptide drugs with an emphasis on membrane permeability.

## 8.2 LIPINSKI'S RULE OF FIVE

Without a doubt, the most convenient route of drug administration is the oral route. To enhance the oral bioavailability of an active lead drug, one must realize that oral bioavailability involves several factors, such as gastrointestinal transit and absorption, chemical stability in the gastrointestinal tract, and the first-pass effect of gut wall and liver metabolism.

In 1997, Christopher A. Lipinski formulated a rule of thumb to evaluate if a drug has properties that would make it a likely orally active drug in humans [1]. "Lipinski's Rule of Five" described the physicochemical features for the pharmacokinetics of

*Peptide Chemistry and Drug Design*, First Edition. Edited by Ben M. Dunn.

a drug in the human body, including its absorption, distribution, metabolism, and excretion (ADME). The rule states that, in general, an orally bioavailable drug should have no more than one violation of the following criteria.

- A molecular weight under 500 g/mol
- A calculated log *P* of less than 5
- Low hydrogen bond potential
    - Not more than 10 hydrogen bond acceptors
    - Not more than five hydrogen bond donors

### 8.2.1 Molecular Size

When relying solely on passive diffusion to cross membranes, large molecules are poor candidates for good oral bioavailability. Peptide drugs are generally perceived as large molecules and would have difficulty crossing membranes. Most researchers correlate molecular size with molecular weight, and have set out the general rule of thumb that orally bioavailable drugs should be less than 500 g/mol. This description has been further refined by others to orally bioavailable drugs with a molecular weight between 160 to 480 g/mol [2]. As we have described in Chapter 5 we noticed that most orally bioavailable peptide drugs are comprised of three to five residues that fits into three to five subsites of the active site.

An aspect of our work on β-secretase inhibitors and Alzheimer's disease will be used to illustrate methods of reducing the molecular size of a peptide design. Amyloid precursor protein (APP) is normally cleaved by α-secretase and γ-secretase leading to fragments that undergo a cascade of events that lead to neuroprotective processes (Figure 8.1). In Alzheimer's disease patients, APP is cleaved by β-secretase instead of α-secretase, resulting in amyloid β-peptides being produced. The subsequent aggregation of these peptide amyloid β-peptide fragments leads to the pathology of the disease. Based on the sequence of the four amino acids preceding and following the cleavage site of the pathology-proned Swedish-type mutant APP, we optimized the natural amino acid residues of the octapeptide inhibitor design for inhibitory activity against β-secretase and derived a potent inhibitor, KMI-008 [3]. We used two methods to reduce the size of the octapeptide inhibitor [4]. In the first method, we synthesized compounds in which one amino acid was systematically removed at a time from the N-terminal, then from the C-terminal. As we would expect, the shorter analogs exhibited lower β-secretase inhibition. A nearly complete loss of inhibitory activity on the removal of a residue indicated that the position of the residue was important for active site recognition. In the second method, we synthesized analogs compounds in which only one natural amino acid in the model drug, KMI-008, was substituted to glycine. Because glycine does not have a side-chain, any near loss of β-secretase inhibition suggested that the interactions between the side-chain of the residue and its associated subsite were important at the affected position. Both methods suggested that a smaller analog of inhibitor KMI-008 was a pentapeptide. The resulting pentapeptide was optimized at the two end-terminals

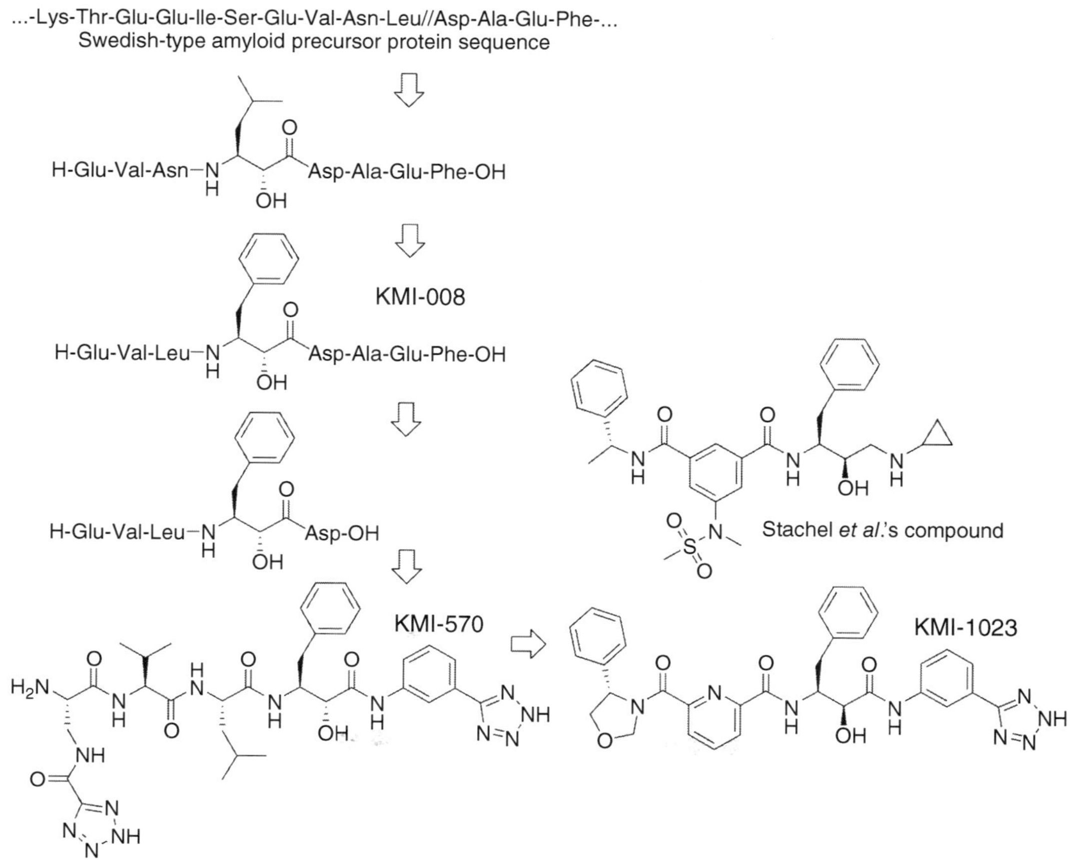


**Figure 8.1** β-Secretase inhibitors used to discuss methods of designing smaller inhibitors.

to derive a potent β-secretase inhibitor, KMI-570 [5]. A wonderful discovery of a potent non-peptide inhibitor of β-secretase by another research group [6] inspired us to shift our focus on non-peptides. From our inhibitor KMI-570, we derived a potent non-peptide β-secretase inhibitor, KMI-1023, which was considerably smaller in size [7].

### 8.2.2 Lipophilicity

Lipophilicity of a compound plays a significant role in its absorption in the gastrointestinal tract and subsequent metabolism in the gut wall and liver [8]. As an overall measure of lipophilicity, the log $P$ value can be experimentally determined or estimated by calculations, where the partition coefficient, $P$, is a ratio of concentrations of an unionized compound between $n$-octanol and water [9]. As it pertains to passive diffusion across membranes, only the unionized form of the compound will traverse the membrane. Lipinski's calculated log $P$ rule, in which a drug would most likely be orally bioavailable, was elaborated to a range of −0.4 to +5.6 by other research groups [2]. This range suggests that lipophilic compounds are expected to exhibit improved membrane permeability when compared to hydrophilic compounds. One should note that there are several methods of estimating log $P$, and Lipinski's rule relies on the *calculated log P* method.

There is a method of estimating lipophilicity that an improvement over log $P$. As a characteristic that is needed for oral bioavailability, the most relevant measure of lipophilicity with regard to oral absorption by passive diffusion is the compound's log $D$ value [10]. The distribution coefficient, $D$, is the ratio of the sum of the concentrations of all forms of the compound between $n$-octanol and water. Thus, while log $P$ only considers the unionized form of the compounds, log $D$ takes into account both ionized and unionized forms of the compound. log $D$ values can be estimated from calculations using the compound's log $P$ and p$K_a$ values [11]. It is noteworthy that, as with log $P$ values [9], p$K_a$ values [12] can also be mathematically predicted. log $D$ is pH dependent, and the organ where absorption mostly takes place is the small intestine. Considering that the pH in the small intestine is about 6.5, log $D$ at pH 6.5 is relevant to passive drug diffusion across the small intestine to the blood circulatory system. Once the drug enters the bloodstream, it encounters a different pH environment of about 7.4. To account for acidic and basic compounds, the difference between the fractions of the neutral form at pH 6.5 (small intestine pH) and pH 7.4 (blood pH) is described as $\Delta\log D = \log D_{6.5} - \log D_{7.4}$ and would thus account for two different biological environments [13]. Compounds with positive $\Delta\log D$ values are acidic, whereas compounds with negative values are basic. Acidic compounds tend to have better bioavailability characteristics, because in the acidic pH 6.5 environment of the small intestine, acidic compounds have a higher fraction in the unionized form that can be absorbed. Moreover, once absorbed, in the more basic pH 7.4 environment of blood, the acidic compounds have a lower fraction in the unionized form that can enter the liver where the compounds are degraded. In other words, acidic compounds have a lower risk than basic compounds of entering the liver and being degraded. The optimum range for $\log D_{6.5}$ is estimated to be values between −0.3 and +0.7, and takes into account both absorption and first-pass effect [10]. As another benefit for slightly acidic drugs, highly ionized drugs, either acidic or basic, may also cause patient discomfort due to direct irritation of the gastrointestinal lining. Taken together what we have discussed, slightly acidic drugs are favored for improved gastrointestinal absorption, less first-pass metabolism, and less mucosal irritation.

In general, hydrophobic compounds are often favored for pharmacological activity over hydrophilic compounds due to desolvation entropy [14]. Simply put, a hydrophobic compound is more entropically favored to release water molecules before binding to the often hydrophobic active site of the target biological substance. Hydrophobic compounds need to spend less energy to part with water because they have fewer interactions with water. Interestingly, compounds with high hydrogen bond potentials can interact with water and would thus exhibit unfavored desolvation entropy. Thus, hydrogen bond potential is related to hydrophilicity. Hence, lipophilicity is preferred in both pharmacodynamics and pharmacokinetics. One of the goals of rational drug design is to optimize lipid solubility for membrane permeation while retaining a significant pharmacological activity. However, simply increasing the lipid solubility of a drug may have undesired effects such as decreasing water solubility and bioavailability, increasing plasma protein binding with a high affinity, and increasing uptake by the liver and spleen macrophages.

***8.2.2.1 Plasma Protein Binding*** Lipophilic drugs have a higher risk of unwanted binding with untargeted biological substances, such as plasma proteins. Such inadvertent binding delays and prevents the drug from reaching its target site of action. Hence, the less bound a drug is, the more efficiently it can traverse cell membranes. Acidic and neutral drugs will primarily bind to albumin, which is basic, or to lipoprotein when albumin becomes saturated. Basic drugs bind to $\alpha_1$-glycoprotein, which is acidic. Only the unbound drug exhibits pharmacologic effects, is metabolized and is excreted. Generally speaking, protein binding should be minimized to reduce unpredictable pharmacokinetic factors.

The activity of a thrombin inhibitor is lower if it has high plasma protein binding [15]. Dabigatran is a univalent direct thrombin inhibitor that was derived from a peptide drug. In the design of dabigatran, a carboxylate function was purposely implemented to increase hydrophilicity, which would decrease plasma protein binding and increase inhibitory activity (Figure 8.2) [16]. The carboxylate function was attached such that it would not greatly affect the interactions between the drug and the target enzyme, thrombin. Dabigatran etexilate is the prodrug form of dabigatran. Indeed, for certain cases, a fine tuning of a drug design could potentially reduce plasma protein binding.

In the case of inhibitors of the human immunodeficiency virus (HIV) protease, hepatic metabolism becomes another factor to consider along with plasma protein binding. Most HIV protease inhibitors are peptide drugs that are often quite lipophilic as a consequence of rational drug design to improve pharmacodynamic effects. Although the lipophilic character is desired for the drug to enter the infected cell and reach the target HIV protease, lipophilicity increases plasma protein binding.

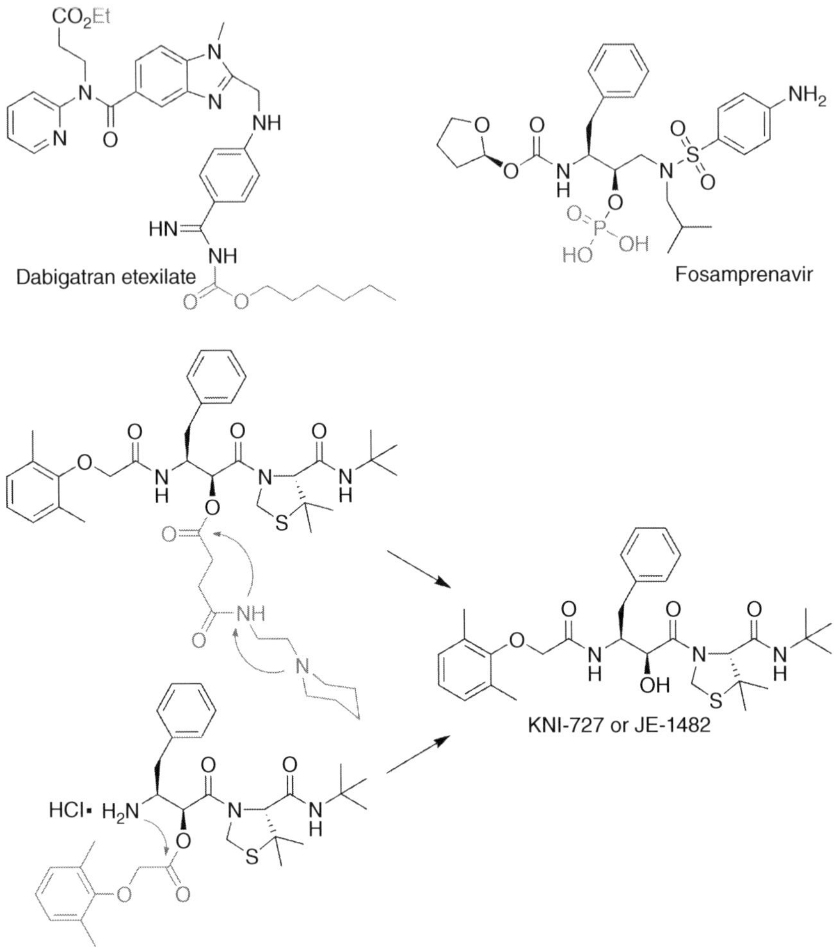


**Figure 8.2** Dabigatran etexilate and HIV protease inhibitors used to discuss lipophilicity. The protective groups are drawn in gray.

All commonly used Food and Drug Administration (FDA)-approved HIV protease inhibitors are over 85% bound to plasma protein. This high protein binding decrease drug efficacy, and a larger quantity of the drug would need to be given to compensate. This increase in *pill burden* subsequently introduces risks of adverse drug reactions, compliance, and cost issues. Unlike most protease inhibitors, indinavir only has 60% plasma protein binding. However, as previously mentioned, unbound drugs are more susceptible to metabolism, and HIV protease inhibitors are readily metabolized by the liver. Hence, despite its lower plasma protein binding profile, hepatic metabolism of indinavir greatly reduces its biological half-life to an impractical 2 h. The fine balance between plasma protein binding and hepatic metabolism has yet to be resolved.

*8.2.2.2* ***Water Solubility*** It is noteworthy that a drug may be poorly absorbed if its water solubility is very low, and this effect is often dose dependent. However, one should recall that a hydrophilic drug also tends to have higher clearance than a lipophilic drug, which has higher membrane permeability (Section 8.2.2). Ionized or polar compounds are deemed more likely to be water soluble. Of course, the choice of salt form for ionized compounds would affect the extent of solubilization. It should be noted that the water solubility factor has already been taken into account by the distribution coefficient, because water solubility correlates well with log $D_{6.5}$, a determinant of lipophilicity (Section 8.2.2). Moreover, one should not forget that from a very simplistic viewpoint, the word "hydrophilic" suggests that the compound would "love to be in water." As we will discuss in Section 8.3.2, permeation enhancers are often implemented in the formulation of a drug to improve its solubility in extra and intracellular fluids.

A way of improving water solubility in a peptide drug is to introduce a water solubilization moiety. Phospholipids are a major component of cell membranes by forming a lipid bilayer within the membrane. Generally speaking, phospholipids have an amphipathic character where the "head" of the molecule is a hydrophilic phosphate group, while the "tail" is lipophilic. Amprenavir is an HIV protease inhibitor with over 90% plasma protein binding and requires a standard twice-a-day dose of 1200 mg or eight capsules. Fosamprenavir is a phosphate ester prodrug of amprenavir (Figure 8.2). Much like a phospholipid, the structure of amprenavir can be considered as the lipophilic "tail" and the phosphate group as the hydrophilic "head." As the prodrug is incorporated inside the gastrointestinal tract's epithelium, the phosphate group prevents the prodrug from fully crossing into the bloodstream, because fosamprenavir is too hydrophilic, until the body's alkaline phosphatase slowly metabolizes the prodrug to the more lipophilic parent drug, amprenavir. Consequently, fosamprenavir is a slow-release version of amprenavir that reduces the "pill burden" of the standard regimen of amprenavir. It is noteworthy that plasma protein binding for fosamprenavir is still theoretically 90% because conversion to the parent drug, amprenavir, is needed before reaching the bloodstream.

As discovered by our research group, other available methods of improving oral bioavailability by increasing water solubility of HIV protease inhibitors revolve around an oxygen-to-nitrogen acyl migration reaction, with or without a water-solubility-promoting cleavable spacer [17]. Figure 8.2 illustrates two examples of the oxygen-to-nitrogen acyl migration strategy for the parent drug KNI-727, a potent HIV protease inhibitor. The depicted spacer demonstrated an improvement in water solubility from less than 0.01 mg/mL in the parent drug to 92 mg/mL in the prodrug with a conversion half-life of 35 min at 37 °C in pH 7.4 phosphate buffered saline. Different spacers would produce different prodrugs with different water solubility and conversion time values. This means that the water solubility and conversion time of the prodrug can be controlled by the structural features of the spacer. In the exemplified cleaner strategy that does not require a spacer, water solubility was 13 mg/mL with a conversion half-life of less than 1 min. Thus, we have shown that prodrugs of drugs with little water solubility could exhibit much improved water solubility profiles and modifiable conversion time.

### 8.2.3 Chemical Stability

Inside the body, the drug can have diverse type of metabolic reactions that deactivate the drug and assist in the removal of the drug from the body. Lipinski's rule attempts to associate the drug's susceptibility to metabolic reactions with its ability to form hydrogen bonds. Hydrogen bond potential is characterized as hydrogen bond acceptors and donors. A hydrogen atom attached to a relatively electronegative atom is a hydrogen bond donor. An electronegative atom is a hydrogen bond acceptor, regardless of whether it is bonded to a hydrogen atom or not. This electronegative atom is usually fluorine, oxygen, or nitrogen. Drugs with a high hydrogen-bond potential have a higher risk of undergoing acid–base reactions or reactions that are catalyzed by enzymes. These reactions often change the chemical structure of the drug, thereby deactivating the drug and increase the hydrophilicity of the drug thus facilitating clearance of the drug from the body.

In contrast to Lipinski's oversimplified rule on hydrogen bond acceptors and donors, several structural characteristics, that is, chemical functional groups have been strongly correlated with oral bioavailability. Certain functional groups are more susceptible to transformations in the gut wall, liver, or conjugated in several ways. For a functional group, the significance in reducing bioavailability is related to the metabolic reactivity of the function. Structural functions that can undergo metabolic reactions have been parameterized into quantitative structure–activity relationship equations to predict oral bioavailability [13]. One should note that functionally reactive groups and hydrogen bond potential contribute to hydrophilicity.

Readily oxidized entities, thiols and dihydropyridines, have the most pronounced effect on oral bioavailability. The reactions are rapid and produce hydrophilic metabolites that are readily cleared from the body. Peptide drugs containing thiol containing amino acid cysteine are often promptly deactivated. Biotransformations that have highly significant effects on reducing oral bioavailability include phenolic hydroxy groups (Ar-OH), sulfonamide groups ($SO_2NH_2$) and groups that are susceptible to hydrolytic cleavage by acid–base or enzymatic conditions. Phenolic hydroxy groups are prone to various transformations and conjugation. The formation of a glucuronide metabolite is an example of a sugar-conjugated phenol group. Natural amino acid tyrosine contains a phenolic hydroxy function. Sulfonamide groups ($SO_2NH_2$) can be *N*-acetylated by *N*-acetyltransferase and conjugated. Functional groups that are known to easily undergo hydrolysis include esters, lactones, β-lactams, and alkyl carbamates.

Metabolic carbon oxidative processes play a significant role in reducing bioavailability, where para-hydroxylation of an activated aromatic ring has the largest effect, followed by aryl methyl and allylic groups contributing to a lesser extent. Reduction of ketones may contribute to the same extent as aromatic para-hydroxylation on decreasing bioavailability. Alcoholic hydroxy groups (R-OH) can undergo conjugation and oxidation, although their effects in reducing oral bioavailability are considered moderate. As examples, amino acids serine and threonine each possesses an alcoholic hydroxy group.

N-dealkylation does not greatly affect bioavailability because the process is usually rate-limiting. However, certain compounds, such as lidocaine and ketamine, may undergo rapid dealkylation. Reduction of aromatic nitro groups slightly decreases oral bioavailability.

The reactivity of aromatic and heterocyclic amines, hydrazines, hydrazones, and amidines to metabolic acetylation and oxidation seems to be proportional to their p$K_a$ values. In other words, their significance in metabolic deactivation is related to their p$K_a$ values.

Issues with metabolic transformations and conjugations can be addressed by further improving on the structure of the peptide drug. As an example of dealing with metabolism, the design of an HIV protease inhibitor from our research group, KNI-764, was inspired by the symmetrical design of HIV-1 protease inhibitor lopinavir and the dual activity of nelfinavir against HIV-1 and HIV-2 proteases (Figure 8.3) [18]. KNI-764 is a highly potent HIV-1 protease inhibitor with excellent antiviral activity in infected cells, moderate oral bioavailability, low cytotoxicity, and effectiveness against HIV-2 protease and several HIV-1 resistant strains. In spite of these promising traits, during early clinical trials, KNI-764 was found to be metabolized by not only the cytochrome P450-3A4 cytochrome P450 (CYP3A4) mixed-function oxidase system but also processed by uridine 5′-diphospho-glucuronosyltransferase (UDP-glucuronosyltransferase or UGT) 2B7 at the $P_2$ phenol function [19]. Although the family of CYP3A enzymes is primarily

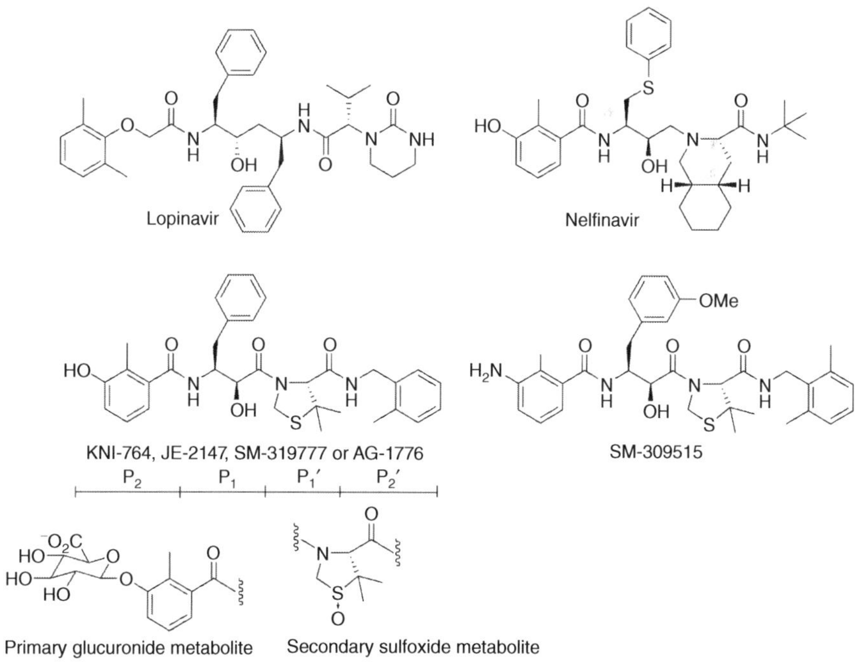


**Figure 8.3** HIV protease inhibitors used to discuss peptide drug metabolism.

found in the liver, it is also located in the epithelial lining of the intestine and contributes significantly to the first-pass metabolism of many drugs. To deal with CYP3A4, it is common practice to coadminister ritonavir, an HIV protease inhibitor that acts as a competitive inhibitor for CYP3A4, in order to maintain elevated plasma levels of the primary HIV protease inhibitor, that is, KNI-764 [20]. However, conjugation by UGT results in a glucuronide metabolite that is more hydrophilic and, therefore, more easily excreted than the original inhibitor [19]. Ritonavir cannot inhibit UGT activity. While conjugation occurs for inhibitor KNI-764, on the other hand, in the case of nelfinavir, which also possessed a $P_2$ phenol function, glucuronidation at the $P_2$ residue was not observed. Speculating that the steric hindrance of the extended $P_1$ aromatic ring system in nelfinavir might be effective to prevent glucuronidation of the $P_2$ phenol function, a slightly bulkier $P_1$ aromatic function was substituted for compound KNI-764. To further greatly reduce the risk of glucuronidation at the $P_2$ position, the $P_2$ phenol moiety was replaced by a 3-amino-2-chlorobenzoate moiety. It was also found that a slight bulk increase by a $P_2'$ 2,6-dimethylbenzyl moiety could improve stability against glucuronidation. It is noteworthy that considering the symmetry of the dimeric HIV protease, the new $P_2'$ moiety is very similar to the $P_2$ moiety of lopinavir. On top of glucuronidation by UGT, human liver microsomes, such as CYP, can oxidize the $P_2$ thiazolidine residue to a thiazolidine sulfoxide moiety. Replacement of the $P_2$ sulfur atom by a methylene function to avoid oxidation by microsomes, unfortunately, led to compounds with considerably lower inhibitory activity against HIV protease, and this change was not adopted. Nonetheless, the resulting structural modifications afforded orally bioavailable inhibitor SM-309515 that is more resistant to glucuronidation, oxidation and overall metabolism than inhibitor KNI-764, especially when coadministered with ritonavir. Moreover, inhibitor SM-309515 is also more potent than inhibitor KNI-764 due to superior permeation across cell membranes, because it is more lipophilic. Hence, several slight structural modifications led to an inhibitor that displays a desirable pharmacokinetic profile and strong antiviral activity, when boosted with ritonavir, an inhibitor of HIV protease that also blocks the activity of drug-degrading CYP34A.

Peptide drugs are more susceptible to degradation by peptidases than other classes of drugs. In order to avoid recognition and premature degradation of the peptide drug by peptidases in the body, our research group often exchange the natural amino acids found in the peptide drug with their isosteres. These isosteres have similar physical or chemical properties that hopefully impart similar or higher biological activity to the parent drug. These replacements are intended to lower the risk of unexpected metabolism by peptidases without lowering the potency of the peptide drug. During the drug-optimization process of our β-secretase inhibitors, isosteres of the carboxylate function were evaluated of which the 1*H*-tetrazol-5-yl isostere is depicted in Figure 8.1 [5, 21]. Other illustrated examples from our peptide drugs include the $P_1$ norstatine inhibitory unit that resembles leucine, as well as $P_1$ isosteres of phenylalanine such as the phenylnorstatine inhibitory unit found in inhibitors KMI-008 and KMI-570, and the allophenylnorstatine unit in inhibitors KMI-1023, KNI-727, KNI-764, KNI-10166, and KNI-10635 (Figures 8.1–8.4). $P_1'$ isosteres of proline,

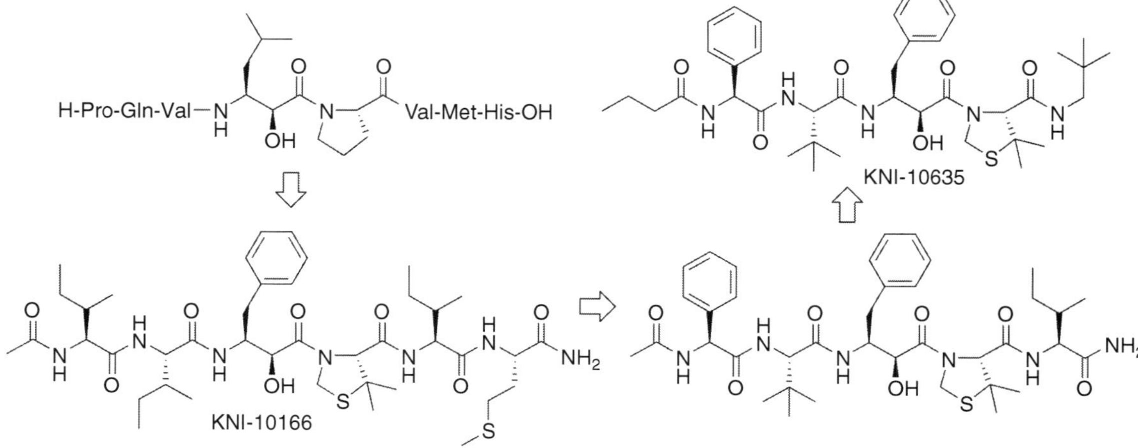


**Figure 8.4** HTLV-I protease inhibitors used to discuss isosteres.

such as (*R*)-5,5-dimethyl-1,3- thiazolidine-4-carboxylate, can also be found in our designs (Figures 8.2–8.4). In our study on the human T-cell lymphotrophic/leukemia virus type 1 (HTLV-I), using similar rational drug design strategies that we have previously described with β-secretase inhibitors (Section 8.2.1), we derived an HTLV-I protease inhibitor, KNI-10166, which was laden with isoleucine residues (Figure 8.4) [22]. To replace these isoleucine residues, L-(+)-α-phenylglycine, *tert*-leucine and 2,2-dimethylpropylamine were exchanged as isosteres to derive a potent HTLV-I protease inhibitor, KNI-10635 [23]. To sum up, natural amino acid residues in a peptide drug can be exchanged for their respective isosteres to reduce the risk of premature digestion by peptidases.

Other than modifying the peptide drug to reduce its risk of enzymatic degradation, enzyme inhibitors can be added to the formulation. These enzyme inhibitors can either directly inhibit the peptidases, or indirectly remove ions that are needed for peptidase hydrolysis. We have already mentioned HIV protease inhibitor, ritonavir, as an enzyme inhibitor of CYP3A4, and will expand the list to other enzyme inhibitors in Section 8.3.1.

### 8.2.4 Routes of Administration

Lipinski's rule of five essentially describes the three requirements that are needed for good bioavailability, regardless if the drug is given by mouth or not. The rules can be generalized in the following statements.

- Lipophilic drugs are more likely to cross membranes.
- Large drugs cannot squeeze themselves across membranes.
- Drugs should reach their target sites of action without being prematurely deactivated by side-reactions.

Keeping these key points in mind, other routes of administration can also be examined. There are many routes of drug administration. For example, the US FDA recognizes 111 distinct routes of administration. For simplicity, we will refer to the routes of administration as parenteral, topical, and enteral. The parenteral route involves piercing the skin or mucous membrane for the drug to reach the bloodstream, meaning injectable drugs. The topical route refers to ear, eye, hair, nail, rectal, skin, and vaginal products. The enteral route would include all medications that would pass by the throat including orally inhaled, nebulized, oral, and sublingal medications.

## 8.3 APPROACHES TO DELIVERING PEPTIDE DRUGS

Solubilization and internalization of drugs into cells is a fairly broad topic. Our discussion on solubilizing agents is applicable to all routes of administration, namely, parenteral, topical, and enteral. Even with the parenteral routes where several physical barriers have been bypassed, the peptide drug may need to be internalized into the target cells. Improvement in solubilization can be as simple as creating a salt form of the peptide drug, or as complicated as using a masking carrier agent to solubilize and transport the peptide drug across membranes.

Several drugs are currently available on the pharmaceutical market as transdermal patches as a noninvasive entry of drugs into the systemic circulation through the skin. Such drugs include estrogen in hormone replacement therapy, nicotine as a smoking cessation aid, and antimuscarinic scopolamine for the management of nausea and motion sickness. However, because normal skin is permeable only to small lipophilic molecules, peptide drugs, being hydrophilic and macromolecular, do not readily penetrate the skin. To facilitate the permeation of peptide drugs into the skin, chemical permeation enhancers and enzyme inhibitors, which are meant to prevent hydrolysis of the peptide drug, are under investigation. Phonophoresis or sonophoresis may assist the transport of peptide drugs under the influence of ultrasound. A promising technique is iontophoresis, in which a constant low level electric current is used to push a charged peptide drug through the skin. A combination of the above techniques may one day be able to reliably deliver peptide drugs across the skin to the systemic blood circulation in human. Until that day comes, we will examine a series of transendothelial routes that has been proven to be effective and approved by the FDA: mucosal peptide drug delivery.

In mucosal drug delivery, it can be generalized that the intrinsic membrane permeability for a hydrophilic peptide drug follows the order of intestinal, nasal, bronchial, tracheal, vaginal, rectal, corneal, buccal, sublingual, and skin, ranked from highest to lowest permeability where absorptions through the intestinal and nasal surface are comparable. Enzymatic degradation seems to be less rapid in nasal than rectal mucosa. The two most favored mucosal surfaces for nonoral peptide drug delivery are the nasal and bronchial mucosa. The pharmacokinetic profiles of the same drug delivered through different routes are different and adjustments to therapeutic levels are definitively needed. As with all route of drug delivery, intersubject variability, that is, physiological differences from one person to the other will play a key factor and dosage regimens will need to be tailored to the patient.

### 8.3.1 Enzyme Inhibitors

There are two major barriers to mucosal administration. The first is an enzymatic barrier that degrades the peptide drug. To lessen the risk of enzymatic degradation, enzyme inhibitors are incorporated in the formulation. These broad-spectrum peptidase inhibitors competitively bind to the active sites of proteolytic enzymes to prevent enzymatic hydrolysis of the peptide drug. Mucosal proteolytic enzymes are composed of exopeptidases, such as mono and diaminopeptidases, and endopeptidases, such as serine, cysteine, and aspartic peptidases. Examples of common peptidase inhibitors include amastatin, aprotinin, bestatin, boroleucine, borovaline, leupeptin, pepstatin, and trypsin inhibitors. As previously mentioned (Section 8.2.3), HIV protease inhibitor ritonavir is often coadministered with a primary HIV protease inhibitor to competitively inhibit the degradation of CYP3A4. Antibacterial agents, such as azelaic acid, fusidic acid, and puromycin, exhibit activity against peptidases. Other enzyme inhibitors, such as *p*-chloromercuribenzoate, phenylmethylsulfonyl fluoride, thiomersal, and chelate metal ions, are essential for proteolytic activity.

### 8.3.2 Permeation Enhancers

The second barrier is the physical mucosal and endothelial layers. The bioavailability of mucosal delivered peptide drugs can be dramatically improved by solubilizing the drug with permeation enhancers that improve solubility, enhance membrane fluidity, or open tight junctions [24]. Tight junctions are the gaps between the margins of adjacent endothelial cells with several transmembrane proteins that project into and seal the gaps. The need for an adjuvant to enhance the penetration of peptide drugs in order to obtain adequate absorption for practical use is especially true for larger molecules and those having relatively high water solubility. However, most absorption enhancers can potentially damage the mucosa, especially when used continuously or chronically, because the increase in membrane permeability may cause unpleasant sensations or local irritation [25].

There are many different types of permeation enhancers. Mucolytic agents, such as amino acid *N*-acetyl-L-cysteine, reduce the viscosity and tenacity of mucus, the first layer, to allow surfactant molecules to diffuse more efficiently onto the endothelial membrane, the second layer, to increase membrane fluidity and mucosal permeability. Interestingly, dornase alfa is a highly purified recombinant human deoxyribonuclease I that acts as a mucolytic agent by hydrolyzing the DNA present in the sputum and mucus of cystic fibrosis patients and thereby reducing the viscosity of the lungs to promote improved clearance of secretions. This mucolytic peptide drug is administered to the lungs of cystic fibrosis patients through a nebulizer. Bile salts are fat emulsifying agents that cause lysis of membranes. Certain bile salts, such as sodium glycocholate, also act as enzyme inhibitors. Fatty acids, such as palmitic acid and oleic acid, as well as lipids are used to moisturize and soften the cell layers, thereby increasing membrane fluidity. Signal transduction substances, such as enzyme protein kinase C, regulates the tight junctions, as junction modulating peptides to permit the entry of the drug into the bloodstream.

The use of a mixed solvent system, that is, cosolvent, could facilitate drug solubilization but has not shown any clear advantage when one considers the higher risk of mucosal irritability [26]. Evaluated solvents include gelatin, glycerol, ethanol, propylene glycol, and polyethylene glycol (PEG).

***8.3.2.1 Peptide Drug Transporters*** The general concept of conjugated dissolution agents has been previously glanced upon, when we discussed methods of improving the solubility of a drug by deriving a more water soluble salt or prodrug (Section 8.2.2.2). In developing a salt form, one should ascertain that the salt form of the peptide drug would not cause intolerable mucosal irritation. In the unfortunate event that mucosal irritation does occur, protecting the ionizable functional group with a readily cleavable protective group could circumvent the problem.

Inert polysaccharides can be intentionally conjugated to the drug to form a more hydrophilic and solubilizable prodrug. It should be noted that if the drug is still conjugated when it is in the bloodstream, it is more readily cleared from the body. Ideally, the prodrug should convert to the parent drug once it reaches the bloodstream. This glucuronidation issue has been touched on when we discussed the metabolism of

HIV protease inhibitor KNI-764 (Section 8.2.3). Polysaccharides have a fair number of hydroxy groups that can conjugate to carboxylate drugs by esterification. More often, oxidation of a hydroxy group is performed to obtain a reactive aldehyde, which then can be more readily conjugated to the parent drug. Polysaccharides such as alginate, cellulose, chitosan, pectin, cyclodextrin, dextrans, and inulin are biodegradable carriers that can release the parent drug [27]. In particular to peptides, chitosan displayed an improved paracellular route of absorption, that is, transport between cells, of peptide drugs [28]. Moreover, analogs of chitosan can be derived and evaluated. Interestingly, cyclodextrins form dynamic molecular inclusion complexes, in which the lipophilic part of the peptide drug can be incorporated into the lipophilic cavity of cyclodextrin. In simpler words, cyclodextrin hides the drug inside of its funnel-like structure.

While polysaccharides are conjugated to improve hydrophilicity, liposaccharides can be used to increase the lipophilicity of peptide drugs and thereby improve their membrane permeability. Lipidic α-amino acids possessing a long alkyl side-chain have been used to provide protection for peptide drugs from enzymatic attack and improve oral absorption [29]. To balance out the high lipophilicity, lipoamino acids can be conjugated to mono and polysaccharides, resulting in amphipathic liposaccharides. Other than being conjugated to the parent drug, liposaccharides can form particulate structures and used as colloidal carriers for oral administered drugs. From a similar viewpoint as lipopolysaccharides, conjugating a fatty acid to a peptide drug could enhance the bioavailability and membrane permeability of the drug [30]. A common way of linking the fatty acyl group to the peptide drug is to use 2-amino-2-hydroxymethyl-1,3-propanediol (Tris) [31] while other methods such as chemoselective ligation [32] have been investigated.

Liposomes are phospholipids vesicles that act as drug carriers. Liposomes usually contain an aqueous core that stores the hydrophilic drug. Lipid spheres that do not contain aqueous materials are called *micelles*. Liposomes are biodegradable and offer possibilities for structural modifications to induce specific cell targeting. Targeting is achieved by placing a marker on the surface of the liposome and the peptide drug is hidden within the phospholipid vesicle. Interestingly, some studies have identified certain peptides as markers.

As an alternative to liposomes, a peptide drug can be encapsulated within a synthetic or semisynthetic polymeric sphere. These PEG conjugated systems are named according to the molecular size of the end product, such as *microspheres* or *nanoparticles*. As with liposomes, marker molecules can also be used.

Instead of dissolving peptide drugs in PEG, peptide drugs can be directly conjugated to PEG. Conjugation of a drug to PEG often increases the circulation half-life of the drug to over 50-fold. As for larger complexes, systemically injected colloidal particles, such as liposomes and nanoparticles are usually rapidly cleared by the liver and spleen macrophages. However, the half-life of the particle in the circulation can be substantially lengthened when PEG is introduced in the colloidal formulation. Considering that nanoparticles are condensed complexes of PEG, the addition of flexible PEG chains would reduce clearance of the drug complex. Penetration can be further enhanced by coating the colloidal particles with a surfactant. However, one should be careful about the potential toxic effect that surface active agents may exert.

As an interesting use of peptides as a drug carrier, our research group has been investigating poly-L-arginine as an oligoarginine-based cargo-transporter system to carry a drug across membranes [33]. HIV-1 transactivator of transcription (Tat) protein derived Tat(48–60), penetratin, and other arginine-rich peptides have cell-penetrating capabilities [34, 35]. Cell-penetrating peptides consist of an amphipathic α-helix and contain alternating and discrete hydrophobic domains and, most peculiarly, positively charged domains [36]. The structure is created by repeating sequences of a charged amino acid such as arginine or lysine followed by a series of hydrophobic residues. High cell membrane permeability of these cell-penetrating peptides is mostly attributed to clusters of consecutive basic amino acids, especially the oligoarginine residues with their highly cationic guanidine clusters that can interact with the negative charges on the surface of cells, and thereby promoting membrane permeation [35, 37]. A sequence of seven consecutive residues of arginine seems to exhibit higher cell penetration. Although the precise internalization mechanism of cell-penetrating peptides has not been clarified, a mechanism by macropinocytosis has been proposed where the peptides are able to "worm" their way directly through the cell membrane [38]. Hence, peptide drugs that are conjugated to a self-cleavable oligoarginine carrier can be transported across membranes, and once crossed, the parent drug is released. However, guanidine function may potentially cause mucosal irritation leading to drug intolerability. It has yet to be demonstrated that drugs conjugated to oligoarginines does not cause significant mucosal irritation due to the cationic clusters.

Once drugs are internalized into cells, the overall drug absorption can be diminished by efflux transporters such as large glycosylated membrane proteins, P-glycoproteins (P-gps), located primarily in the apical membrane of epithelial cells of the small intestine and in various other tissues throughout the body. These efflux transporter systems actively pump drugs and other compounds from cells back out. The P-gp efflux transporter system does not seem to select drugs by size, although most basic or uncharged drugs are more effectively transported. Once again, acidic drugs are favored for bioavailability (Section 8.2.2). P-gp efflux inhibitors include the drugs cyclosporin A, pluronic acid, quinidine, rifampin, and verapamil, and metabolic inhibitors ouabain and 2,4-dinitrophenol. Rifampin is the more popular drug used to modulate efflux by P-gps. Of special interest, pluronic acid is commonly used in a formulation of pluronic lecithin organogel to improve the skin penetration of nonsteroidal anti-inflammatory agents, ibuprofen, and diclofenac acid.

### 8.3.3 Delivery of Peptide Drugs across the Blood–Brain Barrier

The targeting of peptide drugs to the central nervous system (CNS) is a formidable obstacle. The delivery of peptide-based drugs to the brain is limited not only by the general bioavailability issues, which we have thus far discussed, but also by the presence of the blood–brain barrier (BBB). BBB is located at the level of the endothelial layer of the brain microvascular capillaries [39]. Structurally, the BBB has higher transendothelial electrical resistance in its blood vessels than peripheral vessels, has reduced vesicular transport, and is well encased. As with the other membranes, an enzymatic barrier is present in the BBB, along with higher number of certain types of enzymes in brain microvessels than peripheral vessels.

Substances move in and out of the brain through influx (blood-to-brain) or efflux (brain-to-blood) systems. Substances move from blood to brain either by paracellular or transcellular transport. In the paracellular pathway, substances leak between endothelial cells by paracellular diffusion. As an example of cell migration, white blood cells may cross the BBB adjacent to, or by modifying, the tight junctions. Very small amounts of water-soluble compounds may also traverse, on a partial or full opening of the aqueous diffusional paracellular pathway when the tight junctions are modulated. In certain neurodegenerative disorders such as stroke, HIV encephalitis, Alzheimer's disease, multiple sclerosis, and bacterial meningitis, the tight junctions are disrupted leading to additional symptoms and accelerated disease progression [40]. Moreover, opportunistic bacterial and viral pathogens, such as *Streptococcus pneumoniae* and HIV type 1, may take advantage of the inflamed BBB to enter the brain [41]. Hypoxia and the generation of reactive oxygen species, that is, oxygen radicals could open the tight junctions, resulting in an excessive penetration of white blood cells that release proteases, such as metalloproteases, to induce cytotoxic and vasogenic edema [42]. Nevertheless, the entry of substances by unregulated pathways or by leakage is limited, when it concerns the BBB of healthy individuals.

The transcellular pathway entails that substances enter the endothelial cells on the blood side and exit the cells on the brain side. Certain lipid soluble substances may pass through by transcellular passive diffusion. Saturable transport of essential polar solutes such as amines, amino acids, glucose, monocarboxylates, nucleosides, and small peptides cross by the solute carrier-mediated pathway, which may be passive or secondarily active, and unidirectional, bidirectional, or cotransport/exchanger. Viruses and essential proteins such as cytokines, insulin, leptin, and transferrin are transported through specific receptor-mediated transcytosis. Positively charged macromolecules such as avidin, cationized albumin, and histone may nonspecifically induce adsorptive-mediated transcytosis by electrostatic interactions with the negatively charged plasma membrane surface. This adsorptive-mediated transcytosis entry may be a reasonable explanation for the properties of cationic cell-penetrating peptides that we have previously mentioned (Section 8.3.2.1).

Active efflux carriers, also known as adenosine triphosphate (ATP) binding cassette (ABC) transporters, may pump out a wide range of passively penetrating substances from the endothelial cells to blood or brain. Amphiphilic, basic, and lipid soluble drugs seem to have a higher risk of efflux. We have briefly touched on the P-gp carrier-mediated efflux system that pumps from the endothelial cells into blood (Section 8.3.2.1). Most HIV protease inhibitors fall victim to the P-gp efflux transporter, possibly due to their high lipophilicity [43]. Other transporters include multidrug resistance-associated proteins (MRPs) that are known to efflux glucuronide, glutathione, and sulfate drugs along with HIV protease inhibitors saquinavir, ritonavir, and lopinavir. The breast cancer resistance protein (BCRP) is also an efflux transporter.

***8.3.3.1 Strategies for Peptide Drug Transport into the Brain*** One strategy of keeping high therapeutic levels of a neuropeptide drug in the brain is to prevent the metabolism of the drug in the periphery. Levodopa is a precursor of dopamine and is used in the management of Parkinson's disease. Levodopa is an endogeneous substance that crosses the BBB through the large neutral amino acid transporter, also known as the L-system [44]. A common strategy to increase brain levels of levodopa is to inhibit peripheral enzymes that either degrade levodopa or convert levodopa to dopamine in the periphery. Preventing the peripheral conversion and degradation would leave more levodopa for cerebral conversion. The L-system is selective to amino-acid-like compounds with a bulky hydrophobic side-chain [45]. Several amino acid drugs or drugs that are structurally similar to amino acids that are transported by the L-system include α-methyldopa, another derivative from dopamine biosynthesis; analogs of γ-aminobutyric acid (GABA) biosynthesis such as baclofen and gabapentin; as well as anticancer agents such as melphalan and D,L-2-amino-7-bis[(2-chloroethyl)amino]-1,2,3,4-tetrahydro-2-naphthoic acid (D,L-NAM). Various amino acid drugs such as acivicin, phosphonoformate-tyrosine conjugates, and nitrosoarginine derivatives are also carried by the L-system.

Conjugation of polysaccharides was briefly discussed in Section 8.3.2.1. As it pertains to the brain, glycosylation of peptide drugs has been shown to enhance BBB permeability [46]. Indeed, glycosylation seems to be useful in enhancing biodistribution to the brain, through increased stability, reduced clearance, and improved BBB transport. This enhanced transport is believed to be due to adsorptive endocytosis [47]. However, due to the restrictive nature of the BBB, the extent of glycosylation is usually restricted from one to three glucose moieties.

The lipophilicity of drugs correlates strongly with CNS permeability [48]. Structurally modifying a peptide drug to be more lipophilic would increase its passive diffusion across the BBB [49]. Halogenation of peptide drugs can enhance lipophilicity and the subsequent BBB permeability [50]. However, rendering a peptide more lipophilic would mean that the resulting analog tends to be more extensively plasma protein bound and may have an increase affinity for efflux transporters at the BBB. There is also a risk that the more lipophilic analog or prodrug may have a higher uptake into other tissues, that is, low target selectivity. The characteristics needed to improve peptide drug bioavailability that we have previously described in Section 8.2 are definitively relevant to BBB permeability. Additional physicochemical properties that are disfavored for BBB penetration include significantly high polarity, a polar surface in excess of 80 $Å^2$, a high Lewis bond strength, a high potential for hydrogen bond formation, numerous rotatable bonds, and highly branched compounds [51]. As another thermodynamic benefit of compounds with fewer rotatable bonds, there is less flexibility in the drug [14]. Because the drug was designed to make the free conformation of the inhibitor similar to its bound conformation, there is a minimal loss of conformation entropy.

Numerous successful applications of liposome formulations to transport peptide drugs across the BBB have been reported [52]. As liposomes are highly lipophilic, they are favored for traversing the BBB. Although the exact mechanism for liposomes to cross the BBB is yet to be determined, a reasonable assumption is transcellular passive diffusion. Often, a combination of techniques is employed such as surfactant-coated PEGylated immunoliposomes. Similar to liposomes, polymeric nanoparticles have been investigated as a multicomponent BBB delivery system with some success [53]. Of interest, cell-penetrating Tat-peptide may also be attached to the surface of either liposomes or nanoparticles [54]. This attachment appears to greatly facilitate the internalization by cells, despite the relatively complex and large structures of the colloidal particles. With the use of cell-penetrating peptides, the permeation of the lipophilic particles may be attributed to both adsorptive-mediated and passive diffusinal transcytosis.

Nanogels are made from a network of cross-linked hydrophilic nonionic PEG chains and cationic polyethyleneimine chains [55]. Hence, this emulsion system exhibits combined properties of a hydrophilic nonionic network and a swollen polyelectrolyte network. Further refinement of this system could lead to a delivery of drugs and biomacromolecules to the brain.

Another strategy of maintaining high peptide drug levels in the brain is to synthesize novel analogs of BBB-permeable neuropeptides. Met-enkephalin is an endogenous opioid analgesic (Tyr-Gly-Gly-Phe-Met) that is prone to enzyme degradation in both brain and plasma [56]. A derivative was reported as being more enzymatically stable, more bioavailable, and more BBB permeable [57]. PEGylation of the derivative led to an overall higher brain uptake over time due to reduction of plasma clearance, reduction of hepatic metabolism, and reduction of P-gp binding leading to decrease in brain efflux [58]. PEG may induce inhibition of P-gp. Indeed, this improved overall bioavailability outweighs the decrease in blood-to-brain transport due to structural modifications.

An alternative BBB transporter system entails the conjugation of a peptide drug with a marker that has high affinity for characteristics or receptors in the brain. We have briefly mentioned the use of markers with liposomes and nanoparticles in Section 8.3.2.1. Transferrin receptor is involved in the transport of iron into the brain and is constituently expressed at the BBB at higher levels than other capillary beds. A murine monoclonal antibody to rat transferrin was conjugated with peptides, and it was demonstrated that the conjugation increased brain uptake of the peptides [59]. A main concern with this strategy is that the antibody is not inert and has effects on iron-transferrin delivery across the BBB [60]. Of course, there is always the possibility of taking advantage of surface active agents, liposome, PEG, and marker in a single formulation of surfactant-coated PEGylated immunoliposomes. The use of a viral vector (virosome) with inherent capability of penetrating endothelial cells of brain capillaries has been demonstrated [61]. Cell-penetrating peptide Tat has been shown to carry heterologous proteins across the BBB [62]. Another approach is to use cell-mediated delivery [63]. White blood cells are recruited during an inflammatory response to a pathogen. HIV protease inhibitor indinavir nanosuspensions were internalized into bone marrow-derived macrophage lysosomes and the drug was subsequently carried by white blood cells and released in tissue. Although still preliminary, this strategy has a potential application for peptide drug delivery to the brain.

Rather than modifying peptide drugs to allow them to penetrate the BBB, an alternative strategy would be to modify the permeability of the BBB. Modulating efflux transporters, such as P-gp, MRPs, and BCRP, and their signaling mechanism is one method by which drug delivery into the brain could be improved [64]. For example, inhibition of P-gp has been investigated [65] and some P-gp inhibitors were brought up in Section 8.3.2.1. Inhibitors of MRPs include uricosuric drugs such as benzbromarone, probenecid, and sulfinpyrazone. Alternatively, osmotic opening of the BBB can be performed by a hypertonic solution aimed to initiate cell shrinkage and thereby opening the tight junctions [53]. Other methods include biochemical opening of the BBB, and ultrasound and electromagnetic radiation as modulators of the BBB. One should, however, note that modifying the permeability of the BBB also carries the inherent risk that physical damage to the BBB may incur, and that malicious agents and toxins may cross into the brain, such as opportunistic bacteria and viruses.

Of course, one obvious method for circumventing the BBB is invasive peptide drug delivery using neurosurgery-based infusion of drugs using catheters, pumps, and reservoirs [66]. The drug is directly injected into the brain parenchyma or cerebral spinal fluid (CSF). Needless to say, this method would most likely damage brain tissues and raises the risk for infections.

Intranasal delivery of peptide drugs has been proposed as a potential strategy. The olfactory epithelium is located in the upper posterior part of the human nasal cavity with its nerve cells directly projected into the olfactory bulb of the brain. These nerve cells provide a direct connection between the external environment and the brain without the hindrance from the BBB. Thus, if a peptide drug could be transferred along the olfactory nerve cells, it can bypass the BBB and directly enter the brain. Although studies have suggested that this feat is achievable [67], actual delivery of peptide drugs or any drug by this mean has yet to be fully demonstrated. Three neuropeptides, melanocortin (4–10), vasopressin, and insulin, have been administered intranasally to 36 healthy volunteers, resulting in the accumulation of each peptide in the CSF within 30 min [68]. Other than vasopressin, there was no increase in peripheral blood concentrations of the peptides over the observed 30-min period. The delivery of peptide-derived antibiotic cephalexin, and permeability enhancer enzyme hyaluronidase have also been achieved by intranasal application [69]. However, it is likely that transnasal delivery would not be able to deliver the peptides deep into the brain parenchyma, because peptides that are delivered to the CSF would be cleared out into the peripheral blood before there is time for the peptides to be carried to brain tissues.

## 8.4 PARENTERAL PEPTIDE DRUGS

Many peptide and protein drugs cannot be delivered using the topical or enteral routes of administration. Most peptide and protein drugs are restricted to the parenteral route because they are susceptible to enzymatic degradation during absorption or cannot be efficiently absorbed into the systemic circulation, due to their large molecular sizes and charge issues. These charge issues encompass cases in which these drugs are either readily ionized, too polar or hydrophilic to cross membranes to reach the circulatory system. Immunization exemplifies the delivery of protein drugs that must be administered by injection. Although the parenteral route is the most direct route into the main circulatory system and the onset of action is rapid, it is also the most dangerous because it bypasses most of the body's natural barriers and defenses, and exposes the user to health problems such as hepatitis, abscesses, infections, and insoluble particles. As a painful and inconvenient route that requires extreme care during injection, the parenteral route is laden with compliance and adherence problems.

Some problems associated with the needle-and-syringe form of injection have been partially relieved by jet injectors. These gas- or air-powered medical injector devices use a high pressure narrow jet of the injection liquid instead of a hypodermic needle to penetrate the upper layers of the skin. Because the pressure is provided by either a portable air cartridge or a gas tank, the lack of compactness in the device and potential risk of explosion make jet injectors unlikely to be used for patient self-administration. Moreover, patients may still feel pain in the form of burning and stinging sensations because of the drug formulation. These devices are mainly used in vaccination.

Most FDA-approved injectable peptide drugs often share a common characteristic: the peptide is too large to be efficiently absorbed. Large peptide drugs also tend to have a fair number of functional groups that can be metabolically processed, resulting in deactivation or increase clearance of the drugs. Exenatide is an analog of glucagon-like peptide-1, a gastrointestinal hormone incretin that enhances insulin secretion [70]. The 39-residue peptide drug (4187 g/mol) is approved for the treatment of diabetes mellitus type 2 as a twice-daily subcutaneous injection. Another example of an injectable peptide drug is $\alpha_1$-antitrypsin (44,325 g/mol), a human plasma-derived glycoprotein, which is injected once-a-week to manage emphysema [71]. Plasminogen activators (30,000–60,000 g/mol) are serine proteases with blood anticoagulating effects [72]. The class of plasminogen activators includes alteplase, monteplase, reteplase, tenecteplase, urokinase, streptokinase, and anistreplase. Because plasminogen activators are injected, their rapid onset of action fits their role in emergency medicine to treat such acute cases as pulmonary embolism, myocardial infarction, and stroke. Similarly, activated drotrecogin alfa (55,000 g/mol), a recombinant of activated serine protease protein C, is injected in intensive care medicine to treat sepsis by exerting anticoagulating effects [73]. Desirudin and lepirudin (~7000 g/mol) are recombinant hirudins, which is a 65-residue protein [74]. The recombinants are bivalent direct thrombin inhibitors injected as blood anticoagulants in cases where heparin is contraindicated. In common with desirudin and lepirudin, bivaluridin is a synthetic 20-residue peptide analog (2180 g/mol) of hirudin that must be injected [75].

Large peptides used as antibiotic and antiviral agents must also be administered by the parenteral route. Enfuvitide is the only FDA-approved HIV entry or fusion inhibitor in 2003, which disrupts the fusion of virus and the target cell to keep the virus out of the cell [76]. The peptide drug is used as a reserve for "salvage" therapy in patients with multidrug resistant HIV. Being a 36-residue peptide (4492 g/mol), enfurvitide must be injected to be effective. Actinomycin D is an antibiotic that is not exploited for its antibiotic properties due to its high toxicity toward genetic material [77]. Consequently, it is commonly used to treat a variety of cancers. As a large bicyclic polypeptide, actinomycin D (1256 g/mol) is administered intravenously. Bleomycin (1416 g/mol) is a glycopeptides antiobiotic that, similar to actinomycin D, is exploited for its toxic effects against DNA, and is thus not used as an antibiotic but as an injectable anticancer agent [78]. Large size antibiotics that are approved by the FDA and used for their antibiotic effects include glycopeptide vancomycin (1449 g/mol), lipopeptide daptomycin (1620 g/mol) and lipoglycopeptide telavancin (1756 g/mol) [79, 80]. Of interest, although vancomycin is a large hydrophilic molecule that poorly crosses the gastrointestinal mucosa, it is given orally for the treatment of pseudomembranous colitis caused by *Clostridium difficile* to reach the site of infection in the colon [79]. In this case, the low oral bioavailability, namely, the lack of intestinal absorption and lack of systemic effect of oral vancomycin, is actually beneficial in the therapy against the Gram-positive bacteria in the colon. From the same line of thought, inhaled vancomycin, administered through nebulizer, has been used off-label to target infections in the upper and lower respiratory tract.

As discussed in Section 8.2.2, although acidic drugs are favored for intestinal absorption and against first-pass metabolism, when the drug is too ionized or polar, gastrointestinal absorption of the drug is most likely low. Argatroban (509 g/mol) is a moderately small univalent direct thrombin inhibitor consisting of three residues (Figure 8.5) [81]. The peptide drug is indicated as a blood anticoagulant when heparin cannot be used. Despite its size, argatroban must be administered intravenously because of charge issues. As previously explained (Section 8.2.2), ionized basic drugs have very low oral bioavailability. Indeed, the highly basic side-chain of the key anchoring arginine residue in argatroban greatly interferes with gastrointestinal absorption and contributes to drug intolerability, in spite of the presence of the carboxylate function as a counter-ion [82]. Interestingly, dabigatran (472 g/mol) is also a small univalent direct thrombin inhibitor with a carboxylate function and a highly basic benzamidine isostere of the guanidine side-chain of arginine, and would presumably have similar gastrointestinal absorption issues as argatroban (Figure 8.2) [16]. However, dabigatran is administered as an etexilate prodrug where the acidic function is ethyl esterified and the basic function is protected by a hexyloxycarbonyl moiety. Consequently, prodrug dabigatran etexilate is an orally bioavailable prodrug that is metabolized in the circulatory system to the active anticoagulant univalent direct thrombin inhibitor, dabigatran. Although dabigatran is non-peptide, it was derived from peptide studies. Similar to dabigatran etexilate, peptide drug melagatran (429 g/mol) is a univalent direct thrombin peptide that becomes orally bioavailable after its benzamidine and carboxylate functions are protected, resulting in prodrug ximelagatran (Figure 8.5) [83]. Unfortunately, ximelagatran has been removed from the pharmaceutical market due to hepatotoxicity in a subpopulation of patients. The concept of using ethyl ester as protective groups of carboxylate functions is also applicable for the antihypertensive agents, angiotensin converting enzyme (ACE) inhibitors. As a dicarboxylate ACE inhibitor, enalaprilat must be

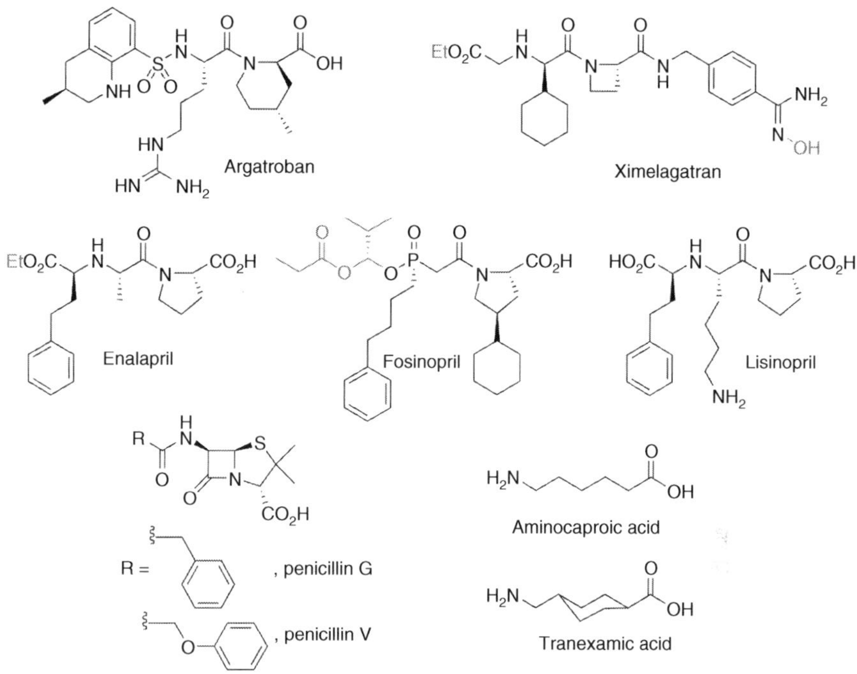


**Figure 8.5** Univalent direct thrombin inhibitors, angiotensin converting enzyme inhibitors, penicillin antibiotics, and blood coagulating agents used to discuss injectable drugs. The protective groups are drawn in gray.

administered intravenously to be therapeutically effective, whereas its ethyl ester prodrug, enalapril (376 g/mol), is orally bioavailable. It is noteworthy that only one of the two carboxylate functions is protected, so that the prodrug is slightly acidic and therefore exhibits improved intestinal absorption. Several FDA-approved ACE inhibitors (368–499 g/mol), namely, benazepril, moexipril, perindopril, quinapril, ramipril, and trandolapril, are marketed as single ethyl ester prodrugs. Similar to dicarboxylate ACE inhibitors, phosphonate-containing fosinoprilat also exhibits low oral bioavailability unless a hydrophobic side-chain was added to modulate its ionization characteristics. Fosinopril, its oral prodrug (564 g/mol), is converted *in vivo* to the active form, fosinoprilat. Of interest, lisinopril (405 g/mol) is an ACE inhibitor with two carboxylate functions and one lysine residue that does not require protective groups to improve its oral bioavailability. Similar to the aforementioned argatroban, the lysine moiety in lisinopril acts as an internal counter-ion to one of the carboxylate function. Unlike argatroban, lysine residue of lisinopril is less basic than arginine residue of argatroban, and consequently, would not greatly interfere with gastrointestinal absorption or discomfort. The classical ACE inhibitor, captopril (217 g/mol), only has one carboxylate function. All in all, ionization problems may be resolved with protective groups.

From three amino acids, L-α-aminoadipic acid, cysteine, and valine, the antibiotic penicillin G (334 g/mol) can be biosynthesized [84]. Penicillin G, also known as benzylpenicillin, is typically given by a parenteral route of administration because it is decayed by hydrochloric acid in the stomach (Figure 8.5) [85]. A slight modification of the benzyl moiety in penicillin G by a phenoxymethyl moiety afforded a less potent yet orally active and orally bioavailable penicillin V (350 g/mol), also referred to as phenoxymethylpenicillin. Thus, stability issues in acid labile drugs are sometimes manageable.

When a drug is only available in an injectable form does not necessarily mean that it is not orally bioavailable. Aminocaproic acid and tranexamic acid are blood coagulating agents by inhibiting fibrinolysis in the treatment of excessive bleeding [86]. Both lysine-derived amino acid drugs were initially available in oral and injectable dosage forms. However, due to economic and patient compliance reasons, the respective drug's manufacturer decided to only one market one dosage form, namely oral form for aminocaproic acid and injectable form for tranexamic acid. The small size (131 and 157 g/mol) and counter-ion characteristics of the drugs would favor their oral bioavailability.

## 8.5 TOPICAL PEPTIDE DRUGS FOR LOCAL EFFECTS

A peptide drug that is meant to act at the site of application has much fewer factors that can reduce its potency and residence time at the target site. Considering that localized effects are desired, topical application of peptide drugs to the skin or eye often only have pharmaceutical issues to address patient's comfort such as local discoloration, irritation, odor, oiliness, and pain. While ophthalmic products are administered directly at the target site, skin products often contain fatty acids, white petroleum jelly or an alcohol to assist in the permeation of the peptide drug across the epidermal layer of the skin.

Many cyclic polypeptide antibiotics such as gramicidin S (gramicidin Soviet), bacitracin, and polymyxin B can be found in topical antibiotics preparations [87]. These large peptides of 1141–1423 g/mol act by disrupting the cellular membranes of bacteria. Being fairly large, their oral bioavailability would most likely be quite low.

Of course, there are products that are more therapeutically effective in the topical formulation. For example, collagenase from *Clostridium histolyticum* bacteria is approved by the FDA as a sterile enzymatic debridment ointment to remove collagen from dead tissue and aerate the healing wound. Eflornithine (182 g/mol) is an ornithine decarboxylase inhibitor that was originally developed to treat trypanosomiasis, commonly known as *sleeping sickness* [88]. However, it was discovered that the amino acid drug is effective in retarding hair-growth, and the drug was subsequently marketed as a dermatological cream to reduce unwanted facial hair in women.

### 8.5.1 Cosmeceutical Peptides

In 1984, Albert Kligman invented the term "cosmeceutical" to describe products that have the characteristics of "cosmetics" for cleansing, beautifying, promoting attractiveness, or altering the appearance, and "pharmaceutics" for mitigating or preventing diseases [89]. In general, cosmeceuticals are topical creams and lotions designed to fight the effects of aging skin and rejuvenate its appearance. In the United States, cosmetics are not regulated by the FDA, and cosmeceutical manufacturers have no incentive to make drug claims that would result in scrutiny by the FDA. Although cosmeceuticals are not officially listed as "drugs" due to marketing reasons, they exhibit drug effects. Cosmeceutical peptides are classified as signal peptides, neurotransmitter-affecting peptides and carrier peptides.

Signal peptides increase dermal remodeling by directly stimulating human dermal skin fibroblast production of collagen, inhibiting collagenase, and increasing ground substance production. Most cosmeceutical signal peptides such as biopeptide-CL (Gly-His-Lys), biopeptide-EL (Val-Gly-Val-Val-Ala-Pro-Gly), palmitoyl oligopeptide (Val-Gly-Val-Ala-Pro-Gly), palmitoyl pentapeptide-3 (Lys-Thr-Thr-Lys-Ser), and palmitoyl tripeptides-3/5 (Lys-Val-Lys) are linked with palmitic acid, a fatty acid, to enhance delivery through the epidermal layer of the skin to reach the target dermal layer [90]. Signal peptide lipospondin (Lys-Phe-Lys) is linked with elaidic acid, the *trans* isomer of oleic acid, a fatty acid [91]. In general, signal peptides are moderately small in size of less than eight residues, and are often coupled with a fatty acid to facilitate permeation through the skin's epidermis.

Neurotransmitter-affecting peptides decrease muscle contraction by inhibiting acetylcholine release at the muscular function. These peptides are mimics of the botulinum neurotoxins. Only botulinum neurotoxin type A is approved by the FDA for subcutaneous, intradermal, and intramuscular injection for facial wrinkles [92]. Botulinum neurotoxin type B share a similar therapeutic goal through a different mechanism of action [93]. Permeation of the single-chain neurotoxic botulinum polypeptide depends on cleavage by proteases to a heavy and light chain. The heavy chain binds to a high affinity receptor on the presynaptic nerve terminal to enable internalization of the bound toxin into the cell, where the activated light chain functions as a zinc-dependent endopeptidase. In other words, the heavy chain is a carrier while the light chain is the active agent. Although there are claims of effectiveness from several topical neurotransmitter-affecting peptides, whether these superficially applied products can penetrate deep enough to reach the target site is highly questionable, especially when one considers that manufacturers' reports are usually not peer-reviewed and may lack important scientific information.

Carrier peptides stabilize and deliver important trace elements, such as copper, that are required for wound healing, angiogenesis, and various other enzyme processes that are necessary for maintaining the dermis. The aforementioned signal peptide, biopeptide-CL (Gly-His-Lys), acts mainly as a copper carrier peptide to improve the appearance of fine lines as well as increase skin density and thickness [94].

## 8.6 INTRANASAL PEPTIDE DRUG DELIVERY

The nasal cavity is covered by a well-vascularized thin mucosa that would permit a peptide drug to quickly be transferred across the single epithelial cell layer directly to the systemic blood circulation without first-pass hepatic and intestinal metabolism. In consideration that peptides are extensively degraded in the gastrointestinal tract and liver, intranasal administration of peptide drugs offer an attractive route of delivery. However, this is a relative improvement over the oral route, because peptide problems associated with high polarity and susceptibility to enzyme degradation would still need to be addressed [95].

There are limitations to the nasal route. There is a limited volume of drug that can be sprayed into the nasal cavity, and thus only potent drugs are good candidates for this route. Continuous or frequent administration could cause harmful long-term effects on the nasal epithelium. In the past, there were concerns that the amount of drug absorbed could vary greatly from one person to another because of upper airway infections, sensory irritation of the nasal mucosa, nasal inflammation, amount drug that gets swallowed instead of being retained in the nasal cavity, and the method of spraying [96]. However, it came to a general understanding that the variability in the amount absorbed after nasal administration should be similar to that after oral administration [97].

Vaccine delivery has been achieved via the nasal route. FluMist is the trade name of the first and only live attenuated vaccine for influenza that has been approved by the FDA. This cold-adapted temperature-sensitive influenza virus product is given once or twice over the influenza season through a syringe sprayer. The attenuated vaccine viruses replicate in the nasopharynx to induce protective immunity.

Most peptide drugs that are delivered through the nasal route are peptide hormones. Buserelin (1300 g/mol) is a gondadotropin-releasing hormone agonist used for the treatment of hormone-responsive cancers such as prostate or breast cancer, estrogen-dependent conditions, and in assisted reproduction. Although a nasal spray formulation of buserelin is available in countries such as Canada, nasal buserelin has not been approved by the FDA for its use in the United States. Approved by the FDA, nafarelin (1322 g/mol) is a gonadotropin-releasing hormone agonist much like the aforementioned buserelin. Nasal nafarelin is indicated for the treatment of estrogen-dependent conditions and central precocious puberty, and in assisted reproduction.

Calcitonin is a 32-residue linear polypeptide hormone that reduces blood calcium and thereby opposes the effects of the parathyroid hormone (PTH). The nasal spray of salmon-derived calcitonin (3432 g/mol) is indicated in postmenopausal osteoporosis.

Desmopressin is a moderate size (1069 g/mol) nonapeptide synthetic D-arginine analog of the natural pituitary hormone 8-arginine vasopressin , an antidiuretic hormone (ADH) affecting renal water conservation (Figure 8.6). Desmopressin may be taken through the nasal route for treating coagulation disorders and replacement of ADH in central diabetes insipidus patients. The former use of the nasal desmopressin in bedwetting has been banned by the FDA following reports of hyponatrial death. Although a similar substance, lypressin (lysine vasopressin), was available for nasal administration, it was discontinued by the manufacturer.

Oxytocin is a nonapeptide hormone (1007 g/mol) that acts as a neurotransmitter in the brain (Figure 8.6). As a drug, oxytocin has a half-life of 3 min in blood, and can be administered nasally to stimulate breastfeeding, although its efficacy is dubious

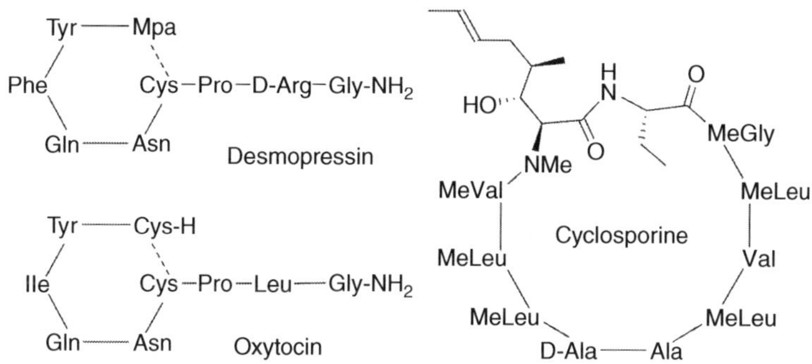


**Figure 8.6** Cyclic peptide drugs.

[98]. Consequently, the production of nasal oxytocin has been discontinued in the US pharmaceutical market.

Glucagon is a 29-residue peptide hormone (3483 g/mol), whose action is opposite to that of insulin. Glucagon is administered by injection to treat hypoglycemia, such as insulin-induced hypoglycemia, in emergency cases when the victim cannot take glucose orally. As an unapproved method, systemic administration of glucagon through the nasal cavity has been proposed. Several human studies suggest that nasal glucagon, especially when it is solubilized with an absorptive enhancer such as bile salt glycocholate, is effective to treat insulin-induced hypoglycemia as a safer and faster method than oral glucose in an unconscious patient [99]. Obviously, no one would claim that intranasal glucagon is more efficacious than its injectable form.

## 8.7 ENTERAL PEPTIDE DRUGS

The pharmaceutical market is consumer driven. Patients demand affordable and hassle-free dosage forms. Their dosage form of choice is the oral dose. They prefer to avoid fiddling with cumbersome devices. In fact, these finicky patients prefer solid dosage forms over liquids, and among the solid oral dosage forms, smaller pills are favored. The liquid dosage form is best suited for children. The lay person is sometimes more interested on whether a drug is easier to carry around than its effectiveness. To meet the needs of its consumers and promote drug utilization compliance, pharmaceutical companies drive to reduce the size of peptide drugs while maintaining acceptable therapeutic effects and drug tolerability. Indeed, the pursuit of a small solid dosage formulation is the ultimate goal of most endeavors in pharmaceutics.

As a result of trying to meet the needs of the consumers, most orally bioavailable peptide drugs are fairly small and are mostly comprised of three to five residues to anchor in their respective pockets at the active site of the enzyme or receptor. Most amino acid drugs have a higher chance for oral bioavailability, mainly because of their very small sizes. As we have previously described (Sections 8.4 and 8.5), aminocaproic acid, eflornithine, and tranexamic acid are orally bioavailable, despite the fact that manufacturers may not market them in oral form. Other FDA-approved amino acid drugs that can be administered by mouth include synthetic thyroid hormones, and drugs derived from the biosynthesis of epinephrine such as levodopa, carbidopa, and α-methyldopa, and those that are derived from the biosynthesis of GABA such as baclofen, gabapentin, γ-hydroxybutyric acid (GHB or sodium oxybate) and pregabalin. orally bioavailable tripeptide drugs include those that we have already discussed, namely, ACE inhibitors and HIV protease inhibitors (Sections 8.2.2.1, 8.2.2.2, 8.2.3, and 8.4). Most derivatives of penicillin and cephalosporin antibiotics, in other words β-lactam analogs (Section 8.4), are semipeptide drugs that have been rationally modified to be orally bioavailable.

Interestingly, desmopressin acetate (Section 8.6), a synthetic nonapeptide of ADH, is available as an injection, nasal spray and oral tablet (Figure 8.6). This peptide drug is therapeutically effective in the oral formulation due to its exceptional potency, small cyclic structure (a sulfur bridge is formed between the two cysteine residues), nonnatural D-arginine residue and/or acetate salt form. Comparisons between the different dosage forms affirm the drug's relative low oral bioavailability, in that the bioavailability of desmopressin oral tablets is about 5% compared to that of the intranasal spray, and approximately 0.16% compared to that of its intravenous formulation. Intranasal desmopressin has about 10% of the potency of the injectable form. These low relative bioavailability values suggest that the effectiveness of oral desmopressin, despite its low oral bioavailability, is attributed to its inherent potency. Desmopressin is natural ADH with a change in the arginine configuration. Indeed the D-configuration has resulted in less vasopressor activity and decreased action on visceral smooth muscle relative to enhanced antidiuretic effect. Consequently, clinically effective antidiuretic doses are usually below the threshold effects on vascular or visceral smooth muscle. In simpler words, desmopressin has a lower undesired risk of increasing blood pressure when compared with ADH, and exhibits a relatively more potent antidiuretic effect.

Another exceptional peptide drug that is orally bioavailable is cyclosporine. Cyclosporine is a cyclic undecapeptide (1203 g/mol) with immunosuppressive properties. The drug is available in injectable, topical, and oral forms (Figure 8.6). The topical route is for local effects while the injectable and oral forms are for systemic effects. Because the peptide drug has very low water solubility, suspension, and emulsion forms of the drug were developed. Unfortunately, blood–drug level monitoring is recommended because absorption by the gastrointestinal tract is variable from one person to another. A key property of the drug that may contribute to its feasible oral bioavailability is that it is mainly distributed outside of blood. Moreover, even when the drug is in blood, 90% of the drug is protein bound, and bound mainly to lipoproteins. An explanation that ties the extravascular distribution and high protein-binding properties of cyclosporine is its high lipophilicity. Interestingly, because of the cyclic structure of cyclosporine and the fact that 7 out of 11nitrogens of cyclosporine's peptide amide bond are methylated, the peptide drug is more resistant to peptidase metabolism.

## 8.8 DIFFERENT ROUTES OF ADMINISTRATION FOR INSULIN

Diabetes mellitus is one of the more common contemporary diseases in these modern ages. This chronic disease is yet not curable, and type 1 diabetic patients are required to supplement their bodies with analogs of a 51-residue peptide hormone, insulin (5808 g/mol), in order to lower their blood–glucose levels.

Most type 1 diabetes mellitus patients have to subcutaneously inject insulin multiple times each day while changing their sites of administration, that is, site rotation, to keep the skin healthy. As an alternative, insulin pumps are "electrical injectors" that are attached to a temporarily implanted catheter or cannula. This injectable method of insulin delivery requires care and effort to use correctly. Another choice for diabetic patients is the use of jet injectors (Section 8.4) that would offer the possibility of pulsatile insulin where insulin is injected in pulses instead continuous infusion.

The lungs offer a large surface for the absorption of therapeutic peptides or proteins. Being able to move sufficient amount of drug from the mouth to the lower respiratory tract has been a major setback for most research. In 2006, the FDA approved an inhalable powdered form of insulin, Exubera [100]. The product is most likely formulated as a PEG complexed insulin, so that it can effectively reach beyond the upper respiratory system to the lung lining and be absorbed into the capillaries. PEGylation would provide sufficient kinetics to allow the droplets to go beyond the upper respiratory system because PEG is thick and flexible. PEGylation of insulin as a physical barrier allows a slow clearance of insulin from blood, resulting in a longer acting medicinal effect and potentially reduces toxicity while permitting longer dosing intervals. However, in reality, the inhalable insulin is short-acting and an injection of long-acting insulin is still required at night. The product was not cost-effective when compared with injected short-acting insulin [101]. Accurate dosing with inhalable insulin is a concern [102]. Following its commercial launch in the United Kingdom, the National Institute of Health and Clinical Excellence does not recommend the product for routine use except when there is "proven injection phobia diagnosed by a psychiatrist or psychologist." Due to bad publicity resulting in low sale figures, Exebura was withdrawn from the US pharmaceutical market in year following its North American release. This end-turn suggests that although delivery of peptide drugs through the lungs is applicable, it is currently not an economically viable approach.

Intranasally or rectally administered insulin is under investigation [103].

Although the stratified epithelium of the oral cavity is much less permeable than that of the nasal mucosa, the buccal and sublingual mucosae are more accessible and robust. Moreover, the oral cavity would most likely be favored over the nasal cavity for peptide drugs that need to be delivered continuously or chronically. A method of administration of insulin by the buccal route is in Phase III clinical trials. Oral-lyn is a device that sprays a high velocity fine-particle aerosol of insulin into the mouth to widely deposit particles of insulin over the oral mucosa. Since the particles are very fine and move very fast, the insulin molecules delivered through this system cross the top-most layers of the epithelial membrane, passing though the other layers and are absorbed into the bloodstream with the assistance of permeation enhancers. Current trial results suggest that oral spray insulin produces rapid absorption and metabolic control that are comparable to that of subcutaneously injected insulin [104]. Another company is developing a sublingually administered rapid-acting oral formulation of insulin, VIAtab, is in Phase II clinical trials. The charged surface of insulin is temporarily masked by an undisclosed excipient to improve sublingual absorption. The same company is using the same technology to develop sublingual salmon calcitonin and sublingual PTH 1–34 to treat osteoporosis.

The strategy of masking insulin is also used to develop orally swallowed insulin. Oradel nanoparticle insulin incorporates insulin in protective polysaccharides coated with vitamin $B_{12}$ molecules. Delivery across the walls of the small intestine is based on the body's natural transport system for vitamin $B_{12}$, and the nanoparticles are broken down in the bloodstream to release insulin. Oradel nanoparticle insulin is in Phase I clinical trials. Another company, Oramed Pharmaceuticals, is developing an enteric coated formulation of insulin that would prevent breakdown of insulin until the pill reaches the intestines. The enteric coated insulin is in Phase II clinical trials.

## 8.9 DISCUSSION

As pharmacists and pharmaceutical scientists, we, the authors, have tried to cover the different aspects of peptide drug delivery from its structural and formulation designs to patient compliance. Three key factors determine drug bioavailability: molecular size, lipophilicity, and side-reactivity.

Large peptide drugs have a lower chance for passive transport from one side of the membrane to the other. Despite this lower risk, the scientific community has demonstrated that transporters, such as liposomes, can successfully carry large drugs into membrane cells and out on the other side. Of course, there is a limit as to how large a drug can be, so that it can efficiently be loaded onto a transporter.

Lipophilicity is a major determinant for membrane permeability. Hydrophilic and charged drugs can only squeeze in the tight junctions between cells during paracellular transport. Considering that the paracellular pathway does not occur as often as the transcellular pathway, much research has focused on improving the lipophilicity of a drug. Charged functions on a drug can be neutralized by a salt or protected with a readily cleavable moiety. Often, acidic drugs are favored over basic drugs in terms of membrane permeability and mucosal irritation. Although, generally speaking, high lipophilicity favors transcellular transport, too much lipophilic character could lead to complications such as low water solubility in body fluids, high plasma protein binding, high uptake by the liver and spleen macrophages, and high efflux from the brain. Lipophilic drugs can be glycosylated, or polar groups can be added in nonessential regions of the drug to increase its hydrophilicity. When all are taken into considerations, finding an appropriate balance between hydrophilicity and lipophilicity would greatly improve drug bioavailability.

Peptide drugs with high side-reactivity are readily metabolized by peptidases and other enzymes. The difficulty in reducing the premature metabolic processing by masking or eliminating highly reactive functions is that these functions are often needed for high pharmacological activity. Hence, as with hydrophilic and lipophilic balancing, increasing pharmacodynamic effects could be offset by an undesired pharmacokinetics profile.

Other than enhancing drug design to optimize bioavailability, additives can be incorporated in the dosage formulation. Hydrophilic drugs can be hidden in lipophilic carrier complexes. The transient time of a drug in the body can be improved by PEGylation of the carrier. One strategy of decreasing premature degradation of the drug by peptidases is to add peptidase inhibitors. The surface of the body's membrane can be rendered more receptive for permeation with fat emulsifiers, mucolytic agents, membrane moisturizers, membrane softeners, and surface active agents. An appealing method for membrane penetration is to implement self-cleavable cell-penetrating peptides that would carry an attached drug across the cell. Alternatively, the drug can be conjugated to an endogeneous substance that is naturally transported across the membrane. Somewhat risky methods for membrane penetration across the BBBs are to either inhibit the body's efflux transporters, or to rely on neurosurgical administration.

In the current US pharmaceutical market, several peptide drugs are available for different routes of administration. When one looks beyond topical applications to the skin or eyes, peptide drugs are available as injectables for large drugs, intranasal formulations for several classes of peptide hormones, and oral forms for smaller drugs. Some large peptide drugs with peculiar biophysicochemical characteristics, such as desmopressin and cyclosporine, have therapeutically effect through the oral route, despite the fact that they may not have high or reliable oral bioavailability. The most promising route of administration for peptide drugs seems to be the intranasal route.

The delivery of insulin exemplifies the relentless effort to deliver the peptide hormone via different routes. Although insulin is commonly injected subcutaneously using needle and syringe, alternative injection devices are available. Inhalable insulin was available at one time. Other routes of administration currently under evaluation include the buccal, intranasal, oral, rectal, and sublingual routes for insulin.

## ACKNOWLEDGMENTS

The writing chapter was supported in part by the Frontier Research Program, the 21st COE (Century Centers of Excellence) Program from the Ministry of Education, Culture, Sports, Science and Technology (MEXT) – Japan. We are also grateful for the relentless efforts from researchers, pharmaceutical firms, health-care providers, patients, and supporters in their endeavors to seek applicable and viable solutions to many medical conditions and diseases.

## REFERENCES

1. Lipinsky CA, Lombardo F, Dominy BW, Feeney PJ. Experimental and computational approaches to estimate solubility and permeability in drug discovery and development settings. Adv Drug Deliv Rev 1997;23:3–25.

2. Oprea TI, Davis AM, Teague SJ, Leeson PD. Is there a difference between leads and drugs? A historical perspective. J Chem Inf Comput Sci 2001;41:1308–1315.

3. Shuto D, Kasai S, Kimura T, Liu P, Hidaka K, Hamada T, Shibakawa S, Hayashi Y, Hattori C, Szabo B, Ishiura S, Kiso Y. KMI-008, a novel β-secretase inhibitor containing a hydroxymethylcarbonyl isostere as a transition-state mimic: design and synthesis of substrate-based octapeptides. Bioorg Med Chem Lett 2003;13:4273–4276.

4. Kimura T, Shuto D, Kasai S, Liu P, Hidaka K, Hamada T, Hayashi Y, Hattori C, Asai M, Kitazume S, Saido TC, Ishiura S, Kiso Y. KMI-358 and KMI-370, highly potent and small-sized BACE1 inhibitors containing phenylnorstatine. Bioorg Med Chem Lett 2004;14:1527–1531.

5. Kimura T, Hamada Y, Stochaj M, Ikari H, Nagamine A, Abdel-Rahman H, Igawa N, Hidaka K, Nguyen J-T, Saito K, Hayashi Y, Kiso Y. Design and synthesis of potent β-secretase (BACE1) inhibitors with $P_1'$ carboxylic acid bioisosteres. Bioorg Med Chem Lett 2006;16:2380–2386.

6. Stachel SJ, Coburn CA, Steele TG, Jones KG, Loutzenhiser EF, Gregro AR, Rajapakse HA, Lai M-T, Crouthamel M-C, Xu M, Tugusheva K, Lineberger JE, Pietrak BL, Espeseth AS, Shi X-P, Chem-Dodson E, Holloway MK, Munshi S, Simon AJ, Lawrence K, Vacca JP. Structure-based design of potent and selective cell-permeable inhibitors of human β-secretase (BACE-1). J Med Chem 2004;47:6447–6450.

7. Hamada Y, Ohta H, Miyamoto N, Yamaguchi R, Yamani A, Hidaka K, Kimura T, Saito K, Hayashi Y, Ishiura S, Kiso Y. Novel non-peptidic and small-sized BACE1 inhibitors. Bioorg Med Chem Lett 2008;18:1654–1658.

8. (a) Austel V, Kutter E. In: Topliss JG, editor. *Quantitative Structure-Activity Relationships of Drugs*. New York: Academic Press; 1983. p 437–496; (b) Lien EJ. In: Ariens EJ, editor. *Drug Design*. Vol. V. New York: Academic Press; 1975. Chapter 3.

9. (a) Hansch C, Leo A, editors. *Substituent Constants for Correlation Analysis in Chemistry and Biology*. New York, NY: John Wiley & Sons, Ltd; 1979; (b) Hansch C, Leo A, Hoekman D, editors. *Exploring QSARs Fundamentals and Applications in Chemistry and Biology*. Washington, DC: American Chemical Society; 1995. Chapter 8.

10. Dressman JB, Amidon GL, Fleisher DJ. Absorption potential: estimating the fraction absorbed for orally administered compounds. J Pharm Sci 1985;74:588–589.

11. Scherrer RA, Howard SM. Use of distribution coefficients in quantitative structure-activity relationships. J Med Chem 1977;20:53–58.

12. (a) Perrin DD, Dempsey B, Serjeant EP, editors. *pKa Prediction for Organic Acids and Bases*. Chapman & Hall: London; 1981; (b) Fraczkiewicz R. In silico prediction of ionization. In: Testa B, van de Waterbeemd H, editors. *Comprehensive Medicinal Chemistry II*. Vol. 5. Amsterdam, The Netherlands: Elsevier; 2007.

13. Yoshida F, Topliss JG. QSAR models for drug human oral bioavailability. J Med Chem 2000;43:2575–2585.

14. Freire E. Do enthalpy and entropy distinguish first in class from best in class? Drug Discov Today 2008;13:869–874.

15. Tucker TJ, Lumma WC, Lewis SD, Gardell SJ, Lukas BJ, Baskin EP, Woltmann R, Lynch JJ, Lyle EA, Appleby SD, Chen I-W, Dancheck KB, Vacca JP. Potent noncovalent thrombin inhibitors that utilize the unique amino acid D-dicyclohexylalanine in the $P_3$ position. Implications on oral bioavailability and antithrombotic efficacy. J Med Chem 1997;40:1565–1569.

16. Hauel NH, Nar H, Priepke H, Ries U, Stassen J-M, Wienen W. Structure-based design of novel potent nonpeptide thrombin inhibitors. J Med Chem 2002;45:1757–1766.
17. (a) Matsumoto H, Soham Y, Kimura T, Hayashi Y. Controlled drug release: new water-soluble prodrugs of an HIV protease inhibitor. Bioorg Med Chem Lett 2001;11:605–609; (b) Hamada Y, Ohtake J, Sohma Y, Kimura T, Hayashi Y, Kiso Y. New water-soluble prodrugs of HIV protease inhibitors based on O-N intramolecular acyl migration. Bioorg Med Chem 2002;10:4155–4167; (c) Hamada Y, Matsumoto H, Kimura T, Hayashi Y, Kiso Y. Effect of the acyl-groups on O → N acyl migration in the water-soluble prodrugs of HIV-1 protease inhibitor. Bioorg Med Chem Lett 2003;13:2727–2730; (d) Sohma Y, Hayashi Y, Ito T, Matsumoto H, Kimura T, Kiso Y. Development of water-soluble prodrugs of the HIV-1 protease inhibitor KNI-727: importance of the conversion time for higher gastrointestinal absorption of prodrugs based on spontaneous chemical cleavage. J Med Chem 2003;46:4124–4135; (e) Hamada Y, Matsumoto H, Yamaguchi S, Kimura T, Hayashi Y, Kiso Y. Water-soluble prodrugs of dipeptide HIV protease inhibitors based on O to N intramolecular acyl migration: design, synthesis and kinetic study. Biorg Med Chem 2004;12:159–170.
18. Yoshimura K, Kato R, Yusa K, Kavlick MF, Maroun V, Nguyen A, Mimoto T, Ueno T, Shintani M, Falloon J, Masur H, Hayashi H, Erickson J, Mitsuya H. JE-2147: a dipeptide protease inhibitor (PI) that potently inhibits multi-PI-resistant HIV-1. Proc Natl Acad Sci U S A 1999;96:8675–8680.
19. (a) Mimoto T, Terashima K, Nojima S, Takaku H, Nakayama M, Shintani M, Yamaoka T, Hayashi H. Structure-activity and structure-metabolism relationships of HIV protease inhibitors containing the 3-hydroxy-2-methylbenzoylallophenylnorstatine structure. Bioorg Med Chem 2004;12:281–293; (b) Mimoto T, Nojima S, Terashima K, Takaku H, Shintani M, Hayashi H. Structure-activity relationships of novel HIV-1 protease inhibitors containing the 3-amino-2-chlorobenzoyl-allophenylnorstatine structure. Biorg Med Chem 2008;16:1299–1308.
20. Kempf DJ, Marsh KC, Kumar G, Rodrigues AD, Denissen JF, McDonald E, Kukulka MJ, Hsu A, Granneman GR, Baroldi PA, Sun E, Pizzuti D, Plattner JJ, Norbeck DW, Leonard JM. Pharmacokinetic enhancement of inhibitors of the human immunodeficiency virus protease by coadministration with ritonavir. Antimicrob Agents Chemother 1997;41:654–660.
21. Hamada Y, Igawa N, Ikari H, Ziora Z, Nguyen J-T, Yamani A, Hidaka K, Kimura T, Saito K, Hayashi Y, Ebina M, Ishiura S, Kiso Y. β-Secretase inhibitors: Modification at the $P_4$ position and improvement of inhibitory activity in cultured cells. Bioorg Med Chem Lett 2006;16:4354–4359.
22. (a) Maegawa H, Kimura T, Arii Y, Matsui Y, Kasai S, Hayashi Y, Kiso Y. Identification of peptidomimetic HTLV-1 protease inhibitors containing hydroxymethylcarbonyl (HMC) isostere as the transition state mimic. Bioorg Med Chem Lett 2004;14:5925–5929; (b) Kimura T, Nguyen J-T, Maegawa H, Nishiyama K, Arii Y, Matsui Y, Hayashi Y, Kiso Y. Chipping at large, potent human T-cell leukemia virus type 1 protease inhibitors to uncover smaller, equipotent inhibitors. Bioorg Med Chem Lett 2007;17:3276–3280.
23. (a) Nguyen J-T, Zhang M, Kumada H-O, Itami A, Nishiyama K, Kimura T, Cheng M, Hayashi Y, Kiso Y. Truncation and non-natural amino acid substitution studies on HTLV-I protease hexapeptidic inhibitors. Bioorg Med Chem Lett 2008;18:366–370; (b) Zhang M, Nguyen J-T, Kumada H-O, Kimura T, Cheng M, Hayashi Y, Kiso Y. Synthesis and activity of tetrapeptidic HTLV-I protease inhibitors possessing different $P_3$-cap

moieties. Bioorg Med Chem 2008;16:5795–5802; (c) Zhang M, Nguyen J-T, Kumada H-O, Kimura T, Cheng M, Hayashi Y, Kiso Y. Locking the two ends of tetrapeptidic HTLV-I protease inhibitors inside the enzyme. Bioorg Med Chem 2008;16:6880–6890.

24. (a) Merkus FW, Verhoef JC, Marttin E, Romeijn SG, van derKuy PH, Hermen WA, Schipper NG. Cyclodextrins in nasal drug delivery. Adv Drug Deliv Rev 1999;36:41–57; (b) Bagger MA, Nielsen HW, Bechgaard E. Nasal bioavailability of peptide T in rabbits: absorption enhancement by sodium glycocholate and glycofuro. Eur J Pharm Sci 2001;14:69–74; (c) Davis SS, Illum L. Absorption enhancers for nasal dug delivery. Clin Pharmacokinet 2003;42:1107–1128.

25. (a) Lee VH, Yamamoto A, Kompella UB. Mucosal penetration enhancers for facilitation of peptide and protein drug absorption. Crit Rev Ther Drug Carrier Syst 1991;8:91–192; (b) Kompella UB, Lee VH. Delivery systems for penetration enhancement of peptide and protein drugs: design considerations. Crit Rev Ther Drug Carrier Syst 1991;46:211–245.

26. Costantino HR, Illum L, Brandt G, Johnson PH, Quay SC. Intranasal delivery. Physicochemical and therapeutic aspects. Int J Pharm 2007;337:1–24.

27. Sinha VR, Kumria R. Internat. Polysaccharides in colon-specific drug delivery. J Pharm 2001;224:19–38.

28. Bernkop-Schnürch A. Chitosan and its derivatives: potential excipients for peroral peptide delivery systems. Int J Pharm 2000;194:1–13.

29. Drouillat B, Hillery AM, Dekany G, Falconer R, Wright K, Toth I. Novel liposaccharide conjugates for drug and peptide delivery. J Pharm Sci 1998;87:25–30.

30. Hashizume M, Douen T, Murakami M, Yamamoto A, Takada K, Muranishi S. Improvement of large intestinal absorption of insulin by chemical modification with palmitic acid in rats. J Pharm Pharmacol 1991;44:555–559.

31. Whittaker RG, Hayes PJ, Bender VJ. A gentle method for linking Tris to amino acids and peptides. Pept Res 1993;6:125–128.

32. Bonnet D, Thiam K, Loing E, Melnyk O, Gras-Masse H. Synthesis by chemoselective ligation and biological evaluation of novel cell-permeable PKC-zeta pseudosubstrate lipopeptides. J Med Chem 2001;44:468–471.

33. (a) Hayashi Y, Takayama K, Suehisa Y, Fujita T, Nguyen J-T, Futaki S, Yamamoto A, Kiso Y. Development of oligoarginine-drug conjugates linked to new peptidic self-cleavable spacers toward effective intestinal absorption. Bioorg Med Chem Lett 2007;17:5129–5132; (b) Takayama K, Suehisa Y, Fujita T, Nguyen J-T, Futaki S, Yamamoto A, Kiso Y, Hayashi Y. Oligoarginine-based prodrugs with self-cleavable spacers for intestinal absorption. Chem Pharm Bull 2008;56:1515–1520.

34. (a) Derossi DA, Joliot H, Chassaing G, Prochiantz A. The third helix of the Antennapedia homeodomain translocates through biological membranes. J Biol Chem 1994;269:10444–10450; (b) Vivès E, Brodin P, Lebleu B. A truncated HIV-1 Tat protein basic domain rapidly translocates through the plasma membrane and accumulates in the cell nucleus. J Biol Chem 1997;272:16010–16017; (c) Snyder EL, Dowdy SF. Recent advances in the use of protein transduction domains for the delivery of peptides, proteins and nucleic acids *in vivo*. Expert Opin Drug Delivery 2005;2:43–51; (d) Rothbbard JB, Jessop TC, Wender PA. Adaptive translocation: the role of hydrogen bonding and membrane potential in the uptake of guanidinium-rich transporters into cells. Adv Drug Deliv Rev 2005;57:495–504; (e) Zorko M, Langel Ü. Cell-penetrating peptides: mechanism and kinetics of cargo delivery. Adv Drug Deliv Rev 2005;57:529–545; (f) Futaki S, Suzuki T, Ohashi W, Yagami T, Tanaka S, Ueda K, Sugiura Y. Arginine-rich

peptides. An abundant source of membrane-permeable peptides having potential as carriers for intracellular protein delivery. J Biol Chem 2001;276:5836–5840; (g) Futaki S. Oligoarginine vectors for intracellular delivery: design and cellular-uptake mechanisms. Biopolymers 2006;84:241–249.

35. (a) Wender PA, Mitchell DJ, Pattabiraman K, Pelkey ET, Steinman L, Rothbard JB. The design, synthesis, and evaluation of molecules that enable or enhance cellular uptake: peptoid molecular transporters. Proc Natl Acad Sci U S A 2000;97:13003–13008; (b) Wright LR, Rothbard JB, Wender PA. Guanidinium rich peptide transporters and drug delivery. Curr Protein Pept Sci 2003;4:105–124.

36. Bodor N, Buchwald P. Brain-targeted drug delivery: experiences to date. Am J Drug Targ 2003;1:13–26.

37. (a) Console S, Marty C, Garcia-Echeverria C, Schwendener R, Ballmer-Hofer K. Antennapedia and HIV transactivator of transcription (TAT) "protein transduction domains" promote endocytosis of high molecular weight cargo upon binding to cell surface glycosaminoglycans. J Biol Chem 2003;278:35109–35114; (b) Goncalves E, Kitas E, Seelig J. Binding of oligoarginine to membrane lipids and heparan sulfate: structural and thermodynamic characterization of a cell-penetrating peptide. Biochemistry 2005;44:2692–2702.

38. a Suzuki T, Futaki S, Niwa M, Tanaka S, Ueda K, Sugiura Y. Possible existence of common internalization mechanisms among arginine-rich peptides. J Biol Chem 2002;277:2437–2443; (b) Wadia JS, Stan RV, Dowdy SF. Transducible TAT-HA fusogenic peptide enhances escape of TAT-fusion proteins after lipid raft macropinocytosis. Nat Med 2004;10:310–315; (c) Nakase I, Niwa M, Takeuchi T, Sonomura K, Kawabata N, Koike Y, Takehashi M, Tanaka S, Ueda K, Simpson JC, Jones AT, Sugiura Y, Futaki S. Cellular uptake of arginine-rich peptides: roles for macropinocytosis and actin rearrangement. Mol Ther 2004;10:1011–1022.

39. Egleton RD, Davis TP. Development of Neuropeptide drugs that cross the blood–brain barrier. NeuroRx 2005;2:44–53.

40. Calabria AR, Shusta EV. Blood–brain barrier genomics and proteomics: elucidating phenotype, identifying disease targets and enabling brain drug delivery. Drug Discov Today 2006;11:792–799.

41. (a) Fiala M, Looney DJ, Stins M, Way DD, Zhang L, Gan X, Chiappelli F, Schweitzer ES, Shapshak P, Weinand M, Graves MC, Witte M, Kim KS. TNF-alpha opens a paracellular route for HIV-1 invasion across the blood–brain barrier. Mol Med 1997;3:553–564; (b) Tsao N, Chang WW, Liu CC, Lei HY. Development of hematogenous pneumococcal meningitis in adult mice: the role of TNF-alpha. FEMS Immunol Med Microbiol 2002;32:133–140.

42. Lagrange P, Romero IA, Minn A, Revest PA. Transendothelial permeability changes induced by free radicals in an in vitro model of the blood–brain barrier. Free Radic Biol Med 1999;27:667–672.

43. Varatharaja L, Thomas SA. The transport of anti-HIV drugs across blood-CNS interfaces: summary of current knowledge and recommendations for further research. Antivir Res 2009. DOI: 10.1016/j.antiviral.2008.12.013.

44. Wade LA, Katzman R. 3-*O*-Methyldopa uptake and inhibition of L-dopa at the blood–brain barrier. Life Sci 1975;17:131–136.

45. Begley DJ. Delivery of therapeutics agents to the central nervous system: the problems and the possibilities. Pharmacol Ther 2004;104:29–45.

46. Elmagbari NO, Egleton RD, Palian MM, Lowery JJ, Schmid WR, Davis P, Navratilova E, Dhanasekaran M, Keyari CM, Yamamura HI, Porreca F, Hruby VJ, Polt RL, Bilsky EJ. Antinociceptive structure-activity studies with enkephalin-based opioid glycopeptides. J Pharmacol Exp Ther 2004;311:290–297.

47. Palian MM, Boguslavsky VI, O'Brien DF, Polt R. Glycopeptidemembrane interactions: glycosyl enkephalin analogues adopt turn conformations by NMR and CD in amphipathic media. J Am Chem Soc 2003;125:5823–5831.

48. Levin VA. Relationship of octanol/water partition coefficient and molecular weight to rat brain capillary permeability. J Med Chem 1980;23:682–684.

49. Hansen DW Jr, Stapelfeld A, Savage MA, Reichman M, Hammond DL, Haaseth RC, Mosberg HI. Systemic analgesic activity and δ-opioid selectivity in [2,6-dimethyl-Tyr1,DPen2,D-Pen5]enkephalin. J Med Chem 1992;35:684–687.

50. Weber SJ, Greene DL, Sharma SD, Yamamura HI, Kramer TH, Burks TF, Hruby VJ, Hersh LB, Davis TP. Distribution and analgesia of [3*H*][D-Pen2, D-Pen5]enkephalin and two halogenated analogs after intravenous administration. J Pharmacol Exp Ther 1991;259:1109–1117.

51. Doan KMM, Humphreys JE, Webster LD, Wring SA, Shampine LJ, Searbit-Singh CJ, Adkinson KK, Polli J. Passive permeability and P-glycoprotein-mediated efflux differentiate central nervous system (CNS) and non-CNS marketed drugs. J Pharm Exp Ther 2002;303:1029–1037.

52. Tiwari SB, Amiji MM. A review of nanocarrier-based CNS delivery systems. Curr Drug Deliv 2006;3:219–232.

53. Brasnjevic I, Steinbusch HWM, Schmitz C, Martinez-Martinez P. Delivery of peptide and protein drugs over the blood–brain barrier. Prog Neurobiol 2009. DOI: 10.1016/j.pneurobio.2008.12.002.

54. (a) Torchilin VP, Rammohan R, Weissig V, Leuchenko TS. TAT peptide on the surface of liposomes affords their efficient intracellular delivery even at low temperature and in the presence of metabolic inhibitors. Proc Natl Acad Sci U S A 2000;98:8786–8791; (b) Lewin M, Carleso N, Tung C-H, Tang X-W, Cory D, Scadden DT, Weissleder R. Tat peptide-derivatized magnetic nanoparticles allow in vivo tracking and recovery of progenitor cells. Nat Biotechnol 2000;18:410–414.

55. Vinogradov SV, Batrakova EV, Kabanov AV. Nanogels for oligonucleotide delivery to the brain. Bioconjug Chem 2004;15:50–60.

56. Dupont A, Cusan L, Garon M, Alvarado-Urbina G, Labrie F. Extremely rapid degradation of [3*H*] methionine-enkephalin by various rat tissues *in vivo* and *in vitro*. Life Sci 1977;21:907–914.

57. Williams SA, Abbruscato TJ, Hruby VJ, Davis TP. Passage of a δ-opioid receptor selective enkephalin, [D-penicillamine2,5] enkephalin, across the blood–brain and the blood-cerebrospinal fluid barriers. J Neurochem 1996;66:1289–1299.

58. Witt KA, Huber JD, Egleton RD, Roberts MJ, Bentley MD, Guo L, Wei H, Yamamura HI, Davis TP. Pharmacodynamic and pharmacokinetic characterization of poly(ethylene glycol) conjugation to met-enkephalin analog [D-Pen2, D-Pen5]-enkephalin (DPDPE). J Pharmacol Exp Ther 2001;298:848–856.

59. Friden PM, Walus LR, Musso GF, Taylor MA, Malfroy B, Starzyk RM. Anti-transferrin receptor antibody and antibody-drug conjugates cross the blood–brain barrier. Proc Natl Acad Sci U S A 1991;88:4771–4775.

60. Ueda F, Raja KB, Simpson RJ, Trowbridge IS, Bradbury MW. Rate of 59Fe uptake into brain and cerebrospinal fluid and the influence thereon of antibodies against the transferrin receptor. J Neurochem 1993;60:106–113.

61. Jiang C, Koyabu N, Yonemitsu Y, Shimazoe T, Watanabe S, Naito M, Tsuruo T, Ohtani H, Sawada Y. *In vivo* delivery of glial cell-derived neurotrophic factor across the blood–brain barrier by gene transfer into brain capillary endothelial cells. Hum Gene Ther 2003;14:1181–1191.

62. Schwarze SR, Ho A, Vocero-Akbani A, Doway S. *In vivo* protein transduction: delivery of a biologically active protein into the mouse. Science 1999;285:1569–1572.

63. (a) Dou H, Morehead J, Destache CJ, Kingsley JD, Shlyakhtenko L, Zhou Y, Chaubal M, Werling J, Kipp J, Rabinow BE, Gendelman HE. Laboratory invest tigations for the morphologic, pharmacokinetic, and anti-retroviral properties of indinavir nanoparticles in human monocyte-derived macrophages. Virology 2007;358:148–158; (b) Kabanov AV, Gendelman HE. Nanomedicine in the diagnosis and therapy of neurodegenerative disorders. Prog Polym Sci 2007;32:1054–1082.

64. (a) Bauer B, Hartz AM, Fricker G, Miller DS. PregnaneXreceptor up-regulation of P-glycoprotein expression and transport function at the blood–brain barrier. Mol Pharmacol 2004;66:413–419; (b) Imai Y, Ishikawa E, Asada S, Sugimoto Y. Estrogen-mediated post transcriptional down-regulation of breast cancer resistance protein/ABCG2. Cancer Res 2005;65:596–604; (c) Wang H, Zhou L, Gupta A, Vethanayagam RR, Zhang Y, Unadkat JD, Mao Q. Regulation of BCRP/ABCG2 expression by progesterone and 17beta-estradiol in human placental BeWo cells. Am J Physiol Endocrinol Metab 2006;290:E798–E807.

65. Miller DS, Bauer B, Hartz AM. Modulation of P-glycoprotein at the blood–brain barrier: opportunities to improve central nervous system pharmacotherapy. Pharmacol Rev 2008;60:196–209.

66. (a) Bobo RH, Laske DW, Akbasak A, Morrison PF, Dedrick RL, Oldfield EH. Convection-enhanced delivery of macromolecules in the brain. Proc Natl Acad Sci U S A 1994;91:2076–2080; (b) Kroll RA, Pagel MA, Muldoon LL, Roman-Goldstein S, Neuwelt EA. Increasing volume of distribution to the brain with interstitial infusion: dose, rather than convection, might be the most important factor. Neurosurgery 1996;38:746–752.

67. (a) Mathison S, Nagilla R, Kompella UB. Nasal route for direct delivery of solutes to the central nervous system: fact or fiction? J Drug Target 1998;1998(5):415–441; (b) Illum L. Is nose-to-brain transport of drugs in man a reality? J Pharm Pharmacol 2004;56:3–17; (c) Westin UE, Bostrom E, Grasjo J, Hammarlund-Udenaes M, Bjork E. Direct nose-to-brain transfer of morphine after nasal administration to rats. Pharm Res 2006;23:565–572.

68. Born J, Lange T, Kern W, McGregor GP, Bickel U, Fehm HL. Sniffing neuropeptides: a transnasal approach to the human brain. Nat Neurosci 2002;5:514–516.

69. (a) Sakane T, Akizuki M, Yamashita S, Nadai T, Hashida M, Sezaki H. Transport of cephalexin to the cerebrospinal fluid directly from the nasal cavity. J Pharm Pharmacol 1991;43:449–451; (b) Okuyama S. The first attempt at radioisotopic evaluation of the integrity of the nose-brain barrier. Life Sci 1997;60:1881–1884.

70. Knudsen LB. Glucagon-like peptide-1: the basis of a new class of treatment for type 2 diabetes. J Med Chem 2004;47:4128–4134.

71. Gettins PG. Serpin structure, mechanism, and function. Chem Rev 2002;102: 4751–47804.
72. Sheehan JJ, Tsirka SE. Fibrin-modifying serine proteases thrombin, tPA, and plasmin in ischemic stroke: A review. Glia 2005;50:340–350.
73. PROWESS Study Group. Efficacy and safety of recombinant human activated protein C for severe sepsis. N Engl J Med 2001;344:699–709.
74. Rydell TJ, Tulinsky A, Bode W, Huber R. Refined structure of the hirudin-thrombin complex. J Mol Biol 1991;221:583–601.
75. REPLACE-2 Study Group. Bivalirudin and provisional glycoprotein IIb/IIIa blockade compared with heparin and planned glycoprotein IIb/IIIa blockade during percutaneous coronary intervention: REPLACE-2 randomized trial. JAMA 2003;289:853–863.
76. Greenberg ML, Cammack N. Resistance to enfuvirtide, the first HIV fusion inhibitor. J Antimicrob Chemother 2004;54:333–340.
77. Sobell H. Actinomycin and DNA transcription. Proc Natl Acad Sci U S A 1985;82:5328–5331.
78. Wmezawa H, Maeda K, Takeuchi T, Okami Y. New antibiotics, Bleomycin A and B. J Antibiot (Tokyo) Ser A 1966;19:200–209.
79. Moellering RC Jr. Vancomycin: a 50-year reassessment. Clin Infect Dis 2006;42:S3–S4.
80. (a) Tally FP, DeBruin MF. Development of daptomycin for gram-positive infections. J Antimicrob Chemother 2000;46:523–526; (b) Higgins DL, Chang R, Debabov DV, Leung J, Wu T, Krause KM, Sandvik E, Hubbard JM, Kaniga K, Schmidth DE Jr, Gao Q, Cass RT, Karr DE, Benton BM, Humphrey PP. Telavancin, a multifunctional lipoglycopeptide, disrupts both cell wall synthesis and cell membrane integrity in methicillin-resistant *Staphylococcus aureus*. Antimicrob Agents Chemother 2005;49:1127–1134.
81. (a) Okamoto S, Hijikata A. Potent inhibition of thrombin by the newly synthesized arginine derivative no. 805. The importance of stereo-structure of its hydrophobic carboxamide portion. Biochem Biophys Res Commun 1981;101:440–446; (b) Okamoto S, Kinjo K, Hijikata A. Thrombin inhibitors. 1. Ester derivatives of $N^{\alpha}$-(arylsulfonyl)-L-arginine. J Med Chem 1980;23:827–830.
82. Rewinkel JBM, Adang AEP. Strategies and progress towards the ideal orally active thrombin inhibitor. Curr Pharm Design 1999;5:1043–1075.
83. Ho SJ, Brighton TA. Ximelagatran: direct thrombin inhibitor. Vasc Health Risk Manag 2006;2:49–58.
84. Drews J. Drug discovery: a historical perspective. Science 2000;287:1960–1964.
85. Arrowood S, Hoyt AM Jr, Sepaniak MJ. Monitoring of benzylpenicillin decomposition in gastric contents by capillary zone electrophoresis. J ChromatogrB: Biomed Sci Appl 1992;583:105–110.
86. (a) Royston D. Blood-sparing drugs: aprotinin, tranexamic acid, and epsilon-aminocaproic acid. Int Anesthesiol Clin 1995;33:155–179; (b) Mannucci P. Hemostatic drugs. N Engl J Med 1998;339:245–253.
87. (a) Konashev M. The dicovery of Gramicidin S: the intellectual transformation of G.F. Gause from biologist to researcher of antibiotics and on its meaning for the fate of Russian genetics. Hist Phil Life Sci 2001;23:137–150; (b) Johnson B, Anker H. Meleney F Bacitracin: a new antibiotic produced by a member of the B subtilis group. Science 1945;102:376–377; (c) Cardoso LS, Araujo MI, Góes AM, Pacífico LG, Oliveira RR,

Oliveira SC. Polymyxin B as inhibitor of LPS contamination of *Schistosoma mansoni* recombinant proteins in human cytokine analysis. Microb Cell Fact 2007;6:1.

88. Van Nieuwenhove S, Schechter PJ, Declercq J, Bone G, Burke J, Sjoerdsma A. Treatment of gambiense sleeping sickness in the Sudan with oral DFMO (DL-α-difluoromethylornithine), an inhibitor of ornithine decarboxylase; first field trial. Trans R Soc Trop Med Hyg 1985;79:692–698.
89. Kligman A. The future of cosmeceuticals: an interview with Albert Kligman, MD, PhD. Interview by Zoe Diana Draelos. Dermatol Surg 2005;31:890–891.
90. (a) Maquart FX, Pickart L, Laurant M, Gillery P, Monboisse JC, Borel JP. Stimulation of collagen synthesis in fibroblast cultures by the tripeptidecopper complex glycyl-L-histadyl-L-lysine-$Cu^{2+}$. FEBS Lett 1988;238:343–346; (b) Senior RM, Griffen GL, Mecham RP, Wrenn DS, Prasad KU, Urry DW. Val-Gly-Val-Ala-Pro-Gly, a repeating peptide in elastin, is chemotactic for fibroblasts and monocytes. J Cell Biol 1984;99:870–874; (c) Tajima S, Wachi H, Uemura Y, Okamoto K. Modulation by elastin peptide VGVAPG of cell proliferation and elastin expression in human skin fibroblasts. Arch Dermatol Res 1997;289:489–492; (d) Lintner K. Promoting production in the extracellular matrix without compromising barrier. Cutis 2002;70(6 Suppl):13–16.
91. Cauchard JH, Berton A, Godeau G, Hornebeck W, Bellon G. Activation of latent transforming growth factor beta 1 and inhibition of matrix metalloproteinase activity by thrombospondin-like tripeptides linked to elaidic acid. Biochem Parmacol 2004;67:2013–2022.
92. Vartanian AJ, Dayan SH. Facial rejuvenation using botulinum toxin A: a review and updates. Facial Plast Surg 2004;20:11–19.
93. Vamauchi P, Lowe N. Botulinum toxin types A and B: comparison of efficacy, duration, and dose-ranging studies for the treatment of facial rhytides and hyperhidrosis. Clin Dermatol 2004;22:34–39.
94. (a) Wegrowski Y, Maquart FX, Borel JP. Stimulation of sulfated glycosaminoglycan synthesis by the tripeptide-copper complex glycyl-L-histidyl-L-lysine-$Cu^{2+}$. Life Sci 1992;51:1049–1056; (b) Buffoni F, Pino R, Dal Pozzo A. Effect of tripeptide-copper complexes on the process of skin wound healing and on cultured fibroblasts. Arch Int Pharacodyn Ther 1995;330:345–360.
95. (a) Hussain A, Fara JJ, Aramaki Y, Troulove JE. Hydrolysis of leucine enkephalin in the nasal cavity of the rat—a possible factor in the low bioavailability of nasally administered peptides. Biochem Biophys Res Commun 1985;133:923–928; (b) Lee VH. Enzymatic barriers to peptide and protein absorption. Crit Rev Ther Drug Carrier Syst 1988;5:69–97; (c) Sarkar MN. Drug metabolism in the nasal mucosa. Pharm Res 1992;9:1–9.
96. Kublik H, Vidgren MT. Nasal delivery systems and their effect on deposition and absorption. Adv Drug Deliv Rev 1998;29:157–177.
97. (a) Aerodiol Study Group. Efficacy and acceptability of intranasal 17 beta-oestradiol for menopausal symptoms: Randomised dose–response study. Lancet 1999;353:1574–1578; (b) Coda BA, Rudy AC, Archer SM, Wermeling DP. Pharmacokinetics and bioavailability of single-dose intranasal hydromorphone hydrochloride in healthy volunteers. Anesth Analg 2003;97:117–123.
98. Fewtrell MS, Loh KL, Blake A, Ridout DA, Hawdon J. Randomised, double blind trial of oxytocin nasal spray in mothers expressing breast milk for preterm infants. Arch Dis Child Fetal Neonatal Ed 2006;91:F169–F174.

99. (a) Freychet L, Rizkalla SW, Desplanque N, Basdevant A, Zirinis P, Tchobroutsky G, Slama G. Effect of intranasal glucagon on blood glucose levels in healthy subjects and hypoglycaemic patients with insulin-dependent diabetes. Lancet 1988;18:1364–1366; (b) Rosenfalck AM, Bendtson I, Jorgensen S, Binder C. Nasal glucagon in the treatment of hypoglycaemia in type 1 (insulin-dependent) diabetic patients. Diab Res Clin Pract 1992;17:43–50; (c) Pontiroli AE, Calderara A, Perfetti MG, Bareggi SR. Pharmacokinetics of intranasal, intramuscular and intravenous glucagon in healthy subjects and diabetic patients. Eur J Clin Pharmacol 1993;45:555–558; (d) Stenninger E, Aman J. Intranasal glucagon treatment relieves hypoglycaemia in children with type 1 (insulin-dependent) diabetes mellitus. Diabetologia 1993;36:931–935; (e) Hvidberg A, Djurup R, Hilsted J. Glucose recovery after intranasal glucagon during hypoglycaemia in man. Eur J Clin Pharmacol 1994;46:15–17; (f) Pontiroli AE. Peptide hormones: review of current and emerging uses by nasal delivery. Adv Drug Deliv Rev 1998;29:81–87.
100. McMahon GT, Arky RA. Inhaled insulin for diabetes mellitus. N Engl J Med 2007;356:497–502.
101. Black C, Cummins E, Royle P, Philip S, Waugh N. The clinical effectiveness and cost-effectiveness of inhaled insulin in diabetes mellitus: a systematic review and economic evaluation. Health Technol Assess 2007;11:1–126.
102. Cefalu W, Skyler J, Kourides I, Landschulz W, Balagtas C, Cheng S, Gelfand R. Inhaled human insulin treatment in patients with type 2 diabetes mellitus. Ann Intern Med 2001;134:203–207.
103. (a) Aungst BJ, Rogers NJ, Shefter E. Comparison of nasal, rectal, buccal, sublingual and intramuscular insulin efficacy and the effects of a bile salt absorption promoter. Pharmacol Exp Ther 1988;244:23–27; (b) Lalej-Bennis D, Boillot J, Bardin C, Zirinis P, Coste A, Escudier E, Chast F, Peynegre R, Selam JL, Slama G. Efficacy and tolerance of intranasal insulin administered during 4 months in severely hyperglycemic Type 2 diabetic patients with oral drug failure: a cross-over study. Diabet Med 2001;18:614–618.
104. Guevara-Aquirre J, Guevara M, Saavedra J, Mihic M, Modi P. Oral spray insulin in treatment of type 2 diabetes: a comparison of efficacy of the oral spray insulin (Oralin) with subcutaneous (SC) insulin injection, a proof of concept study. Diabetes Metab Res Rev 2004;20:472–478.

# INDEX



**Note:** Page numbers followed by 'f' and 't' indicates figure and table respectively.

*Peptide Chemistry and Drug Design*, First Edition. Edited by Ben M. Dunn.

## Q